Compendium of Plant Genomes

Series Editor

Chittaranjan Kole, President, International Climate Resilient Crop Genomics Consortium (ICRCGC), President, International Phytomedomics & Nutriomics Consortium (IPNC) and President, Genome India International (GII), Kolkata, India

Whole-genome sequencing is at the cutting edge of life sciences in the new millennium. Since the first genome sequencing of the model plant *Arabidopsis thaliana* in 2000, whole genomes of about 100 plant species have been sequenced and genome sequences of several other plants are in the pipeline. Research publications on these genome initiatives are scattered on dedicated web sites and in journals with all too brief descriptions. The individual volumes elucidate the background history of the national and international genome initiatives; public and private partners involved; strategies and genomic resources and tools utilized; enumeration on the sequences and their assembly; repetitive sequences; gene annotation and genome duplication. In addition, synteny with other sequences, comparison of gene families and most importantly potential of the genome sequence information for gene pool characterization and genetic improvement of crop plants are described.

Sudip Kr. Dutta • Padma Nimmakayala •
Umesh K. Reddy
Editors

The Watermelon Genome

Springer

Editors
Sudip Kr. Dutta
ICAR Research Complex for NEH Region
Indian Council of Agricultural Research
Gangtok, Sikkim, India

Padma Nimmakayala
Gus R. Douglass Institute
West Virginia State University
Institute, WV, USA

Umesh K. Reddy
Department of Biology
West Virginia State University
Institute, WV, USA

ISSN 2199-4781 ISSN 2199-479X (electronic)
Compendium of Plant Genomes
ISBN 978-3-031-34715-3 ISBN 978-3-031-34716-0 (eBook)
https://doi.org/10.1007/978-3-031-34716-0

This Springer imprint is published by the registered company Springer Nature Switzerland AG
The registered company address is: Gewerbestrasse 11, 6330 Cham, Switzerland

Contents

1 **Origin of the Dessert Watermelon** 1
Harry S. Paris

2 **Caveats for Watermelon Whole Genome Sequencing** 17
Purushothaman Natarajan, Padma Nimmakayala, Sudip Kumar Dutta, and Umesh K. Reddy

3 **Watermelon Genetic Resources and Diversity** 23
Nebahat Sari and İlknur Solmaz

4 **The NLR Family of Disease Resistance Genes in Cultivated Watermelon and Other Cucurbits: Opportunities and Challenges** 37
Andres Salcedo, Camilo H. Parada-Rojas, Rafael Guerrero, Madison Stahr, Kimberly N. D'Arcangelo, Cecilia McGregor, Chandrasekar Kousik, Todd Wehner, and Lina M. Quesada-Ocampo

5 **Genetics and Genomics of Fruit Quality Traits of Watermelon** 69
Cecilia McGregor, Samikshya Rijal, Samuel Josiah, and Lincoln Adams

6 **Challenges of Traditional Breeding in Watermelon** 85
Harshawardhan Choudhary, K. Padmanabha, Gograj Singh Jat, and Tusar Kanti Behera

7 **Recent Advances in Genomics, Genetic Resources of Watermelon** 131
Akanksha Jaiswar, Nivedita Rai, Devender Arora, Manisha Malhotra, Sarika Jaiswal, and Mir Asif Iquebal

8 **Health Properties and Breeding for Phytonutrients in Watermelon (*Citrullus lanatus* L.)** 143
Gograj Singh Jat and Umesh K. Reddy

9 **Genomic Resources for Disease Resistance in Watermelon** 155
Brahma Induri, Padma Nimmakayala, and Umesh K. Reddy

1 Origin of the Dessert Watermelon

Harry S. Paris

1.1 Introduction

The dessert watermelon, *Citrullus lanatus* (Thunb.) Matsum. & Nakai (*Cucurbitaceae*), is one of the most widely grown and consumed fruit vegetables. Over 100,000,000 tons are produced annually on more than 3,000,000 ha, in well over 100 countries with 81% of the production in Asia (Levi et al. 2017a; McCreight 2017; FAO 2020; Wehner et al. 2020). Dessert watermelons are popular because of their refreshing watery sweetness, texture, and flavor (Maynard 2001; Aimpoint Research 2020; Ramirez et al. 2020).Watermelons are also increasingly recognized as a good source of nutrition, providing considerable amounts of ascorbic acid (vitamin C), potassium, the antioxidant pigments lycopene and beta-carotene, as well as citrulline, which is a non-protein amino acid having several health benefits (Perkins-Veazie et al. 2012). Yet, the geographic origin, history, and evolution under domestication of the dessert watermelon have been controversial for quite a long time. Using a multidisciplinary approach that encompassed botany, horticulture, cookery, philology, archaeology, iconography, and interpretation of ancient literature, I suggested that the dessert watermelon originated in northeastern Africa (Paris 2015a). Herein, I will summarize and update the understanding of the origin of the dessert watermelon, including the latest findings resulting from genomic investigations of the genus *Citrullus*.

H. S. Paris (✉)
Agricultural Research Organization, Cucurbits Section, Newe Ya'ar Research Center, Ramat Yishay 3009500, Israel
e-mail: hsparis@outlook.co.il

Present Address:
H. S. Paris
P. O. Box 6114, Yoqne'am 2065626, Israel

1.2 Phenotypic Description of *Citrullus*

Citrullus Schrad. ex Eckl. & Zeyh. (*Cucurbitaceae*) is a genus of xerophytic, heat- and sun-loving plants that are readily distinguished from other cucurbits by the pinnatifid shape of their leaf laminae (Paris et al. 2013). The plants are herbaceous, tendril-bearing vines spreading one to three meters in all directions. The flowers are solitary, 2–3 cm in diameter, with five light yellow petals, and each is functional during the morning hours of a single day. Most cultivars are monoecious, but many old or indigenous cultivars are andromonoecious (Wehner et al. 2020). By far, most of the flowers are staminate, with a pistillate or hermaphroditic flower appearing at every seventh or eighth leaf axil (Wehner et al. 2001). The plants usually begin to flower 40–60 days after sowing and are naturally self- and cross-pollinated by bees (Wehner 2008; Njoroge

S. Kr. Dutta et al. (eds.), *The Watermelon Genome*, Compendium of Plant Genomes,
https://doi.org/10.1007/978-3-031-34716-0_1

et al. 2010; Campbell et al. 2019). Ovaries and primordial fruits are lanate, becoming glabrous, smooth, and glossy as they grow. Usually 25–40 days ensue from anthesis to fruit maturity. The very earliest cultivars, which require 65 days from sowing to first harvest, bear small fruits (<5 kg). More time is required for larger fruits of higher quality to ripen. Fruit shape ranges from globular to spherical to oval blocky, to oblong (Maynard 2001; Wehner et al. 2001).

The thin, green, glossy outer layer of the rind, the exocarp, of the watermelon fruit is typically boldly striped or otherwise patterned in two shades of green (Maynard 2001; Wehner et al. 2001). The green colors can range from light to dark. The darker stripes range in breadth from very broad to very narrow, but unlike the stripes of other cucurbits, such as melons (*Cucumis melo* L.), pumpkins, and squash (*Cucurbita pepo* L.), the stripes of watermelons are jaggedly edged (Paris 2015a). Moreover, melons, pumpkins, and squash typically have ten stripes, while watermelons typically exhibit 16–20 stripes. The thick inner layer of the rind, the mesocarp, is wet, white, and hard.

Underneath the rind is the placenta, the endocarp, which is the watery fruit flesh enjoyed by so many people the world over. Early in development, the fruit flesh is hard, white or otherwise pale-colored, and insipid. In most citron watermelons, *Citrullus amarus* Schrad., the fruit flesh remains hard, usually nearly colorless and tasteless to fruit maturity. In dessert watermelons, *C. lanatus*, the flesh of the ripening fruit becomes tender and the range in texture of the ripe fruit flesh has been variously described as crisp, soft or liquefied, and coarsely or finely grained, and accumulates carotenoid pigments and sucrose (Elmstrom and Davis 1981; Brown and Summers 1985; Xu et al. 2012; Soteriou et al. 2014). By modern standards, a minimum of 8% soluble solids is required in dessert watermelons, and the best cultivars can regularly achieve 11% or more soluble solids (Mohr 1986; Maynard 2001; Wehner 2008). Color begins to accumulate in the fruit flesh of dessert watermelons between two and three weeks after anthesis, first around the developing seeds and thereafter gradually spreading throughout the endocarp (Perkins-Veazie et al. 2012; Wang et al. 2016; Zhang et al. 2017). Depending on the genotype, the flesh of ripe watermelon fruits can range in color from red to pink, orange, yellow, a mixture of these colors, green, and white (Gusmini and Wehner 2006). In contrast to the bold changes in fruit color visible internally during ripening, the external indications of fruit ripening are not readily apparent to the consumer. Perhaps the best indication of fruit ripeness is a slight yellowing in some of the lighter stripes. Growers pick the fruits in the field when the adjacent tendrils have shriveled. If stored in a cool, shady place, watermelons can keep for weeks or even months without serious deterioration of their quality, but are subjected to breakage if not handled carefully (Keith-Roach 1924; Rushing 2001).

Each watermelon fruit can contain 200 or more seeds that, to the casual observer, are seemingly scattered throughout the flesh and, to the consumer, are of great annoyance. The seeds of the dessert and citron watermelons are hard, oval, and flat and, depending on the cultivar, generally range in length from 8 to 16 mm, can be black, brown, tan, white, yellow, green, red, or purple, and can be patterned with a second color (Rubatsky 2001; Levi et al. 2017a). The seeds of the colocynth, *Citrullus colocynthis* (L.) Schrad., are similar but smaller, 7–8 mm in length, and those of the egusi watermelon, *C. mucosospermus* (Fursa) Fursa, are characterized by soft, mucilaginous coats that disappear upon desiccation, leaving the seeds devoid of the thick, hard seed coats characteristic of the seeds of the other *Citrullus* species.

1.3 Etymology, Nomenclature, and Taxonomy of *Citrullus*

Citrullus is a xerophytic genus native to the arid regions of Africa. Its species are diploids with 11 pairs of chromosomes, $2n = 2x = 22$. The plants require much heat and direct sunlight in order to thrive and set fruits.

Among cucurbit genera, *Citrullus* has the most confused etymology, nomenclature, and taxonomy (Paris 2015a). As a result, the origin, history, and evolution under cultivation of the dessert watermelon, *C. lanatus*, have been widely misunderstood.

The word *citrullus* is the medieval Latin diminutive of the classical Latin word *citrum*, for the citron, *Citrus medica* L. (*Rutaceae*). The *citrium*, a citron-like fruit (Andrews 1863; Lewis and Short 1951), was the major element in a recipe from the fourth-century cookbook attributed to Apicius. This recipe called for this rather large-fruited, hard cucurbit to be double-cooked and cleared of the hard seeds scattered about within it (Flower and Rosenbaum 1958). This recipe is the earliest-known account of the citron watermelon in Europe (Paris 2015a). During the early medieval period, a smaller green-skinned cucurbit arrived in Europe. This was the cucumber, *Cucumis sativus* L., and it looked like a diminutive citron. In southern Europe, it was appropriately named with the diminutive of *citrum* (citron), *citrullus* (Paris et al. 2011). However, plants well-adapted to the warm climate of southern Europe were much less familiar to botanists of cool northern Europe. The snake melon (*Cucumis melo* L.), which has culinary qualities similar to cucumber, was known in the south as *cucumis*, but it did not succeed in the north, so *cucumis* became applied in the north to the cucumber. The watermelon, which was rare in the north too, became *citrullus*. Perhaps inevitably, then, the binomial *Cucurbita citrullus* was designated by the Swedish botanist Linné for the dessert watermelon and, later, the German botanist H.A. Schrader (1767—1836) adopted *Citrullus* as the scientific designation for the genus of watermelons.

The nomenclature and taxonomy of the genus *Citrullus* have had an at least as confused a pathway as its etymology. Renner et al. (2017) critically traced one major source of this confusion to a paper by Bailey (1930), who synonymized Thunberg's *Momordica lanata*, the citron watermelon of southern Africa, with Linné's *Cucurbita citrullus*, the dessert watermelon in cultivation. Another large contributor to the confusion in the taxonomy of *Citrullus* was experience in attempting controlled hand-pollinations. Self-pollinations, sib-pollinations (pollinations between plants of the same accession), and cross-pollinations between watermelon accessions are never 100% successful. Fruit set and development from hand-pollinations are dependent on a complex set of factors, such that even 50% success can be considered highly satisfactory. So, if pollinations among closely related plants have such a dubious success rate, when wider crosses are made it can be nearly impossible to judge, based on success rate, whether such crosses are indeed within the same species or between different species. And this is the crux of the problem. Crosses between dessert and citron watermelons, and between dessert watermelons and colocynths, have a reasonable chance for successful fruit set and development, under field and greenhouse conditions. The difficulties are experienced in subsequent generations, be they with quasi-linkage (Levi et al. 2003) or sterility (Sain et al. 2002), but these difficulties are not insurmountable and, for over a century, improved dessert watermelon germplasm has been developed through introgression of desirable traits from citron watermelons and colocynths (Wehner et al. 2001; Levi et al. 2006, 2011). Crosses between dessert and egusi watermelons may have fewer difficulties in subsequent generations and, though there are two reports of impaired crossing success in one direction (Fursa 1983; Gusmini et al. 2004), this needs to be more widely confirmed. No reductions of pollen viability were observed in reciprocal hybrids between dessert and egusi watermelons (McGregor and Waters 2013).

Given the confused taxonomy, the dessert and citron watermelons, and later the egusi watermelons, were considered to be botanical varieties of the same species, though the colocynth was considered to be a separate species. Fursa (1972) promulgated changing their status from botanical varieties to subspecies, which further confused the nomenclature with, for example, the dessert watermelon *C. lanatus* var. *lanatus* becoming *C. lanatus* subsp. *vulgaris* and *C. lanatus* var. *citroides* becoming *C. lanatus* subsp. *lanatus*. However, observation of quasi-linkage, sterility, unidirectional success, or lowered success rate in crossing can all be acceptable indicators of the existence of true biological species.

One key element in obtaining an easier grasp of relationships among watermelons as well as a better assessment of dessert watermelon ancestry has been the recognition of seven species within the genus (Chomicki and Renner 2015). This proposal solves the problem of the cumbersome previous nomenclature and would, with the possible exception of the relationship between *C. lanatus* and *C. mucosospermus*, be well in accordance with the biological species concept. Renner et al. (2017) made further order out of the nomenclatural and taxonomic chaos by assigning to species the over 50 subspecific and varietal epithets used for *Citrullus*.

It is difficult to overstate the importance of correct taxonomic identification of germplasm and its particular relevance to accessions of the genus *Citrullus* (Guzzon and Ardenghi 2018). Germplasm collections and databases of *Citrullus* are rife with taxonomic misnaming. Not all *C. lanatus* have sweet fruit flesh, not all *Citrullus* having hard fruit flesh are citrons, and not all *Citrullus* having bitter fruit flesh are colocynths. Moreover, there is quite a bit of overlap among *Citrullus* species for quantitative traits, such as fruit and seed size, and qualitative traits, such as fruit flesh and seed color. For scientific investigations, these accessions need to be correctly identified, otherwise the interpretations of the results will inevitably be flawed. Growing plants from seed samples of the accessions in these collections is necessary in order to determine or confirm the taxonomic identity of the various accessions. Correct naming and taxonomic identification of the accessions in germplasm collections are also fundamental for standardization of the plant genetic-resource databases of various countries. Guzzon and Ardenghi (2018) suggested inauguration of a process of taxonomic and nomenclatural peer review for all new accessions, in order to guarantee that each single accession is reliably identified and usable by all stakeholders.

Complicating the difficulties in achieving correct taxonomic identification of *Citrullus* accessions are observations, using molecular and genomic technologies, that some of these accessions are admixtures, containing elements of more than one species (Angui et al. 2017; Levi et al. 2017b; Mujaju et al. 2018; Guo et al. 2019). This situation is not at all surprising, given the weak crossing barriers among *Citrullus* species. At least since Roman times, fruit or seeds of *Citrullus* have been carried out of native areas and introduced to new areas (Paris 2015a). Over time, this process has accelerated, mostly for sweet dessert watermelons, which are commonly grown in the native areas of citron watermelons, egusi watermelons, colocynths, and of the non-sweet, presumed dessert-watermelon ancestral populations. Dessert watermelons are often grown in close proximity with citron and/or egusi watermelons in some African countries, including Mali (Jensen et al. 2011), Mozambique (Munisse et al. 2011), Namibia (Maggs-Kölling and Christiansen 2003), and Zimbabwe (Mujaju et al. 2018). Quite likely, sweet dessert watermelons have for years interacted genetically with these native watermelon populations. Purely native gene pools may now be scarce or non-existent. Older collections in gene banks are more likely to represent the original native watermelon populations than are more recently collected samples.

1.4 Process of Adoption of Dessert Watermelons by People

To ancient peoples, cucurbits were a source of sustenance and water but, initially, their exploitation must have been difficult. Wild cucurbit fruits are small and, usually, extremely bitter. Quite likely, people nurtured the occasional cucurbit plants that were found to produce non-bitter fruits, non-bitterness in cucurbit fruits being conferred by a single recessive gene (Navot et al. 1990; Paris and Brown 2005; Dogimont 2010–2011; Weng and Wehner 2016–2017). If the non-bitter-fruited plants in the wild population were infrequent, this initial step would have been the first bottleneck in cucurbit evolution under domestication. An additional step was the favoring of plants that produced larger fruits. Fruit size is a quantitatively inherited trait (Sinnott 1937), and stepwise selection for larger fruits would have further narrowed the genetic base of the domesticated cucurbit population. In the case of dessert watermelons, selection for incremental improvements in fruit-flesh palatability, which encompasses sweetness, pleasing flavor, tenderness, and lack of fiber (Maynard 2001; Kyriacou et al. 2018), further and considerably narrowed the genetic base.

Subsequent genetic erosion of dessert watermelons likely occurred when they were diffused by people to regions having rainfall during the growing season, such as some of the warmer parts of Europe and, later, most of eastern Asia and the Americas. Dessert watermelons originate and were first cultivated in arid areas. Their seeds were sown in summer after the brief period of rains and along the banks of rivers and their tributaries, in moist soil, which progressively dried out as the brief rainy season ended or after the rivers crested. This required the watermelon plants to develop an extensive, deep root system, for maintenance of growth through to fruit maturity. Indigenous plants would have been well-adapted to such 'dryland' cultivation. However, centuries ago, when taken to regions having a warm climate but with considerable rainfall during the growing season, the effect of continually wet soil on these dryland-adapted plants was to encourage their vegetative growth and suppress fruit production, with the few fruits produced having impaired quality and reduced sweetness (Karchi 2000). Over many years, people in these rainy areas selected the few plants that did produce good quality fruits, thereby inadvertently selecting those plants that had shallower, less extensive root systems. Hence, modern dessert watermelons are susceptible to drought stress, while local landraces from dryland farming areas are resistant (Said and Fatiha 2018). Although a few accessions of dessert watermelons show some resistance to drought at the seedling stage (Zhang et al. 2011), it is ironic that the present scarcity of drought-resistant dessert watermelon germplasm has led to the suggestion of using citron-watermelon accessions, which produce much larger root systems, as an infrageneric source of tolerance to drought stress in breeding dessert watermelons (Katuuramu et al. 2020). Nonetheless, when subjected to moderate drought stress induced by deficit irrigation during the weeks immediately prior to harvest, modern dessert watermelons respond with lower yields but improved quality and increased vitamin C and citrulline contents of the fruit flesh (Davis et al. 2011; Kuşçu et al. 2015; Kyriacou et al. 2018), as well as improved water-use efficiency (Qin and Leskovar 2020).

The immense popularity of dessert watermelons has led to their being continually selected for improved fruit quality, subjecting their gene pool to a series of bottlenecks over many generations. Though the fruits maintain much outward variation in size, shape, and exterior and interior coloration, the outcome of this long-term selection for quality is that the modern sweet dessert watermelons possess little molecular genetic variation (Levi et al. 2001a, b; Nimmakayala et al. 2014).

1.5 Food History of *Citrullus lanatus*

Food history is a cross-disciplinary subject full of pitfalls, which by and large can be avoided by the judicious use of comparative evidence as context, seasoned with some common sense (Dalby 2003). Traditional sources of evidence often encompass but are not exclusive to artifacts, ancient literature, iconography, and current distribution of wild and primitive relatives. To these can now be added genomics.

I have reviewed in detail the history of the dessert watermelon with regard to artifacts, ancient literature, iconography, and current distribution of its relatives, with a full list of literary support (Paris 2015a). Here I will gloss over this review, citing only the most outstanding literary sources, with updates of results relevant to the origin of the dessert watermelon, including those recently obtained using genomics research.

1.5.1 Prior to the Roman Empire

1.5.1.1 Artifacts

Watermelon artifacts are almost always of seeds, as the hard, thick seed coats are the most likely parts of the entire plant to withstand the rigors of nature and time. Germer (1988) listed findings of watermelon seeds over 4500 years old in Egypt. De Vartavan and Asensi Amoros (1997) listed ten sites in Egypt, dating from the pre-dynastic and dynastic eras, that contain remains of watermelon. Wasylikowa and van der Veen (2004) recovered watermelon seeds, 5000 years old, in an assemblage of wild seeds in southwest Libya. De Vartavan (1999) presented photographs of five individual seeds taken from the tomb of Tutankhamen (1323 BCE); significantly, the seeds have distinctly bulging margins flanking the hilum end, and therefore, these seeds appear to be of watermelon, not colocynth. A watermelon fruit with seeds from approximately 500 BCE was reported from foundation deposits of a temple in Sudan (van Zeist 1983).

Watermelon seeds along with remains of other food plants have been reported from three sites across the Mediterranean, at Krania in central Greece (Margaritis 2006–2007), and on the eastern island of Samos (Kučan 2000), dating to between 800 and 650 BCE, respectively. Also, two watermelon seeds, dated, at the latest, from the second-century BCE were recovered from Sardinia (Sabato et al. 2019).

1.5.1.2 Literature

In the Hebrew Bible, Numbers 11:5, the Children of Israel, in their journey through the Sinai Desert, longed for five vegetables that they ate in the Land of Egypt, the first two mentioned being cucurbits. The first is the *qishu'im*, which are identified as elongate melons, the properties of which are similar to cucumbers (Janick et al. 2007). The second is the *avattihim*, which are identified as watermelons. The remaining vegetables were leeks (kurrats, *Allium kurrat* Schweinf. ex K. Krause, Amaryllidaceae), onions (*A. cepa* L.), and garlics (*A. sativum* L.), respectively. From the context, it is easy to conclude that the watermelons were considered and used in the same way as these other vegetables, fresh, cooked, or pickled, and they were regarded as secondary to the cucumber-like melons. Hence, sweetness does not appear to be among the properties of the watermelons at the time of the writing of the Book of Numbers. Later, in the Land of Israel during a famine, there is an account of the ill effect from the mistaken use of *paqqu'ot*, colocynths, in a pottage (2 Kings 4:39–40).

In Greek writings, *pepon* was the word most used for watermelon (Liddell and Scott 1948; Andrews 1958). The Greek physician, Hippocrates, in his *Regimen* (400 BCE) wrote that the *pepones* are easily digested (Jones 1967). Theophrastus, the botanist, did not use the word *pepon* in his *Enquiry into Plants* (300 BCE) (Hort 1976) but did use *kolokynta*, probably to indicate watermelon, stating that there were no specific cultivars, although some were better and others were worse.

1.5.1.3 Iconography

Depictions of numerous plants and fruits have been preserved in ancient Egyptian tombs (Andrews 1958), though there are relatively few published images of watermelons. An illustration of a round fruit with narrow darker stripes, attached by a thin stem to a short section of vine including two highly dissected leaves, was reproduced by Keimer (1924). Given the shape of the leaves and striping of the fruit, this depiction undoubtedly represents a specimen of *Citrullus*. A second image shows nine round fruits with narrow darker stripes resembling small watermelons neatly placed in a basket adjacent to two large snake melons and two other baskets containing smaller fruits (Feliks 2005, p. 303, provenance not indicated).

Another image, more than 4000 years old, of an oblong, boldly striped fruit laid on a tray, has been identified as representing watermelon, *C. lanatus* (Manniche 1989, p. 92; Janick et al. 2007; Paris 2015a). Renner et al. (2021) presented another image of an oblong striped fruit, also over 4000 years old, identified as of watermelon. However, there is doubt that watermelon is represented in these latter two images, as some melons, *Cucumis melo*, have a similar shape and, when immature and green, have a similar striping pattern (Paris 2021). Moreover, only round fruits are depicted in watermelon images of Roman-era and Byzantine-era iconography (see below).

1.5.2 During the Roman and Early Byzantine Empires

1.5.2.1 Artifacts

Citrullus seeds dating to the Roman period were uncovered at Qusayr al-Qadim on the shore of the Red Sea in Egypt (Cox and van der Veen 2008; van der Veen 2011). These seeds were described as having a rough surface and a photograph of them clearly shows bulges and creases adjacent to the hilum, and thus, these do not appear to be of colocynth. The seeds are relatively small, 10–11 mm in length, too long to be within the range of typical colocynths. Seeds 10–11 mm in length and of rough appearance were recovered from the Roman outpost of Mons Claudians in Egypt, an inland site approximately 300 km to the south, in the eastern desert (van der Veen 1999). Seeds of *Citrullus* dating to the second-century CE have been found in a cave above the Dead Sea near ‘En Gedi in eastern Israel (Kislev and Simhoni 2009). These seeds, too, are 10–11 mm in length and have bulges and creases near the hilum end. Apparently, watermelons were a familiar item in Egypt and neighboring Israel during Roman times.

1.5.2.2 Literature

Hebrew codices of Jewish Law are a rich depository of information concerning the cucurbits of the Land of Israel during the Roman era (Janick et al. 2007; Paris 2015b). The first of these codices is the *Mishna*, which dates to approximately 188 CE. Although the *qishu'im*, the snake melons, are the cucurbit most often discussed in these codices, both the *avattihim*, dessert watermelons, and the *paqqu'ot*, the colocynths, are also considered.

Various passages in these Hebrew writings imply that the *avattihim* were large watery fruits (Paris 2015a, b). These passages also indicate that dessert watermelon fruits were picked when ripe, not immature. Dessert watermelons were to be tithed after they were slicked clean of dirt and dust (Paris 2021) and laid out singly, not on top of one another, implying that these fruits were rather fragile. In the tractate on tithing from the *Mishna*, the dessert watermelons are discussed together with figs, table grapes, and pomegranates, indicating that they belonged to the category of sweet fruits. The seeds of watermelons were not to be tithed, and therefore, they were not a food item, being used only for sowing. Although other cucurbits, which included snake melons, young bottle gourds, and dessert melons, had a postharvest time limit that they could be acceptable as contributions, no time limit was set for watermelons, indicating that they had a longer shelf life.

The *paqqu'ot* fruits, colocynths, were not used as food and therefore not subjected to tithing (Paris 2015a). The oil extracted from the seeds was used for illumination, and the dry shells of the colocynth fruits are mentioned in the tractate on tools.

Writings in Greek by Roman physicians include offhand descriptions of the *pepon* with regard to medicine (Paris 2015a). From these writings, it is clear that the *pepon* is the dessert watermelon. Dioscorides, in the second volume of *On Medical Matters* (ca. 70 CE), wrote that the flesh of the *pepon* is easily digestible and diuretic and that the rind of the *pepon* is to be applied on top of the head of a child suffering from heat stroke. Others, including Galen (ca. 180 CE) and Oribasius (ca. 355 CE), wrote that the *pepon* was abundantly wet, cold, and diuretic and that the best ones were round, as large as a person's head, and sweet. Athenaeus, in *The Learned Banqueters* (ca. 200 CE), quoted verse written by more than 1000 authors from various times and localities across the Greek-speaking world, some of whom used the word *kolokynte* in apparent reference to the watermelon, with one stating that the *kolokynte* adds moisture to the body, another that the best *kolokyntas* are round, very large, sweet (*glukeian*) and easy to digest, and another stated "What a big head he has! As big as a *kolokunte*!".

Dioscorides and other Roman physicians also described various medicinal uses of other cucurbits (Janick et al. 2007). Prominent among these was the flesh of the colocynth fruit, which was most effective when the green color of the rind had faded.

Latin writings of this period discuss the fruits of three species of *Citrullus*, foremost among them the dessert watermelon, *C. lanatus* (Paris 2015a). Pliny, in his *Historia Naturalis* (ca. 77 CE), described the *pepo* as *refrigerant maxime*, a most refreshing and cooling food. Quintus Gargilius Martialis, in his *Medicinae ex Oleribus et Pomis* (ca. 260 CE), wrote that the *pepone* were good to eat after removal of the rind and pits, and he repeated Dioscorides' prescription to use the rind of the *pepon* as an alleviant of sunstroke. Approximately 250 years later, after the fall of the Roman Empire, there are two clear indications of the existence of sweet watermelons in Europe. *De Observantia Ciborum* (ca. 510 CE), of northern Italian provenance, lists 101 foods, including three cucurbits. The *pepone* are listed among the sweet juicy fruits that are usually eaten raw when ripe, the pomegranates, grapes, and figs. According to *De Observatione Ciborum Epistula* (ca. 516 CE), a book intended for the ruler of northeastern France, the *melones* were to be eaten well-ripened, the flesh with the seeds still mixed in being preferable, presumably with the intention of indicating freshly cut fruit.

The fruits of two *Citrullus* species are also mentioned in a cookbook, *De Re Coquinaria*, that is attributed to a cook named Apicius who lived in the first century. The original work has been lost, but a supplemented version (ca. 400 CE) has several recipes for cucurbits, one of which calls for dressed *pepones* et *melones*, dessert watermelons. Another recipe, for a fricassee, calls for a *citrium*, a citron-like fruit, which first has to be 'cleaned inside and outside, diced and boiled'. The citron, *Citrus medica*, is found elsewhere in this work, but Flower and Rosenbaum (1958) indicated that this recipe would make no sense in the case of *Citrus medica*. The word *citrium*, for a citron-like fruit (Andrews 1863; Lewis and Short 1951), is the forerunner of the English word 'citron', *citre* in French, for the citron watermelon, *Citrullus amarus*, which is not eaten raw but is cooked or otherwise processed. Indeed, the citron watermelon, to be included in a fricassee, would require cleaning inside (removal of the seeds).

Pliny also discussed the colocynth, *Citrullus colocynthis*. Like his contemporary, Dioscorides, he described the colocynth fruit flesh as best used for medical preparations after the fruit rind had faded from green to light yellow.

1.5.2.3 Iconography

In Mediterranean lands during the time of the Roman and Byzantine Empires, there are fewer depictions of watermelons than there are of snake melons (Janick et al. 2007), apparently reflecting a lesser appreciation for these fruits. This situation is consistent with the Roman and Jewish writings of this time. Most of the images of cucurbits from this period of time are floor mosaics, and they vary in their quality. Mosaics, though impressive works of art, are by nature of limited resolution, each stone representing a pixel of information, so that the number of stones used to depict an object is a determinant of the resolution of the depiction. Sometimes, not enough stones were used to allow definite identification of plant species.

Three mosaic images of watermelons were presented by Janick et al. (2007). These date from the fourth to sixth centuries and are found in Tunisia and Greece. The watermelons in all three are globular and striped. The slightly flattened fruit shape suggests that the fruits developed from andromonoecious plants. One of the images shows the dark stripes as being broad and, in the other two, the dark stripes are narrow. Two of the images show the fruits as being small, in diameter about the length of a person's hand, approximately 17 cm.

The frequency of occurrence and the taxonomical and morphological diversity of cucurbits in the mosaics of Israel are stunning and greater than those reported for mosaics and all other art forms from any other provincial region of the Roman and Byzantine Empires (Avital and Paris 2014). Of these, the most frequently encountered is melon. Watermelons are shown in 11 images.

All of them are nearly round, some with broad stripes and some with narrow stripes, and appear to be rather small. In the two images having the finest resolution, the stripes are shown as jaggedly edged, as is typical of watermelons.

Stunning Byzantine-era mosaics have been found in neighboring Jordan (Piccirillo 1993). Two of them contain images of globular watermelons, including the depiction of jagged stripes.

1.6 Current Distribution of Wild and Primitive Relatives

Four species of *Citrullus* are cultivated widely to sparingly and three others, not known to be cultivated, grow wild in southern Africa. Wild *Citrullus* invariably has small, spherical fruits and usually bitter, pale-colored, and seedy fruit flesh (Paris 2015a). Nonetheless, even the uncultivated *C. naudinianus*, the most outlying species, is collected and used by local people as a source of water (Maroyi 2020). Of the cultivated species of *Citrullus*, the least cultivated is the colocynth, *C. colocynthis*. The most widespread in cultivation, of course, is the dessert watermelon, *C. lanatus*, but the colocynth, the citron watermelon, and the egusi watermelon have, each in its turn, been considered to be its ancestor.

Citrullus colocynthis, the colocynth, grows wild in the deserts spanning northern Africa to southwestern Asia to India (Levi et al. 2017b) and has been cultivated in semi-arid and arid regions for quite some time, from Spain across northern Africa and through the Middle East to southwestern Asia (Clément-Mullet 1866; Amar and Hazot 2003). Colocynths are small, spherical, and extremely bitter fruits valued for medicinal use of extracts from their dry, spongy pulp and for extraction of the oil from their seeds. Colocynth fruits do not exceed 10 cm in diameter, usually they measure 8 cm in diameter, and as they mature, their exterior fades from brightly striped green to dull, light yellow and their flesh desiccates. The seeds are small, 8 mm long. Dessert watermelons and colocynths have been observed to cross spontaneously in the field (Singh 1978; Fulks et al. 1979).The colocynth was for some time considered to be the progenitor of the dessert watermelon (Whitaker 1933; Singh 1978; Sain et al. 2002), but a number of investigations based on molecular genetics (for example, Jarret and Newman 2000; Levi et al. 2001a, 2013; Dane and Lang 2004) indicate that colocynths are more distant from dessert watermelons than are citron watermelons.

Citrullus amarus, the citron watermelon, grows wild and is widely cultivated in southern Africa (Meeuse 1962). The fruits are usually spherical and weigh several kilograms, but can be oblong and quite large, and vary widely in fruit rind striping and color pattern. Citron watermelons typically have hard, bland, sometimes bitter, watery flesh that is white, light green, yellow, or orange (Rubatsky 2001; Jensen 2012; McGregor 2012; Ngwepe et al. 2019). They are an important source of water for people and wild animals, a main article of cooked food, and provide animal fodder in resource-poor, remote parts of southern Africa (Welman 2011; Mujaju et al. 2011; Jensen 2012). Sightings of citrons growing wild over extensive areas of desert in southern Africa by David Livingstone (1858) created a sensation, and they were considered by many to be ancestral to dessert watermelons (Robinson and Decker-Walters 1997). However, the findings of 3000-year-old artifacts of watermelon in Egypt are contradictory, as farming was not yet underway in southern Africa at that time (Zohary et al. 2012). Massive preferential segregation (Levi et al. 2003) and reduced pollen fertility (McGregor and Waters 2013) occur in progenies of crosses between citron and dessert watermelons.

Citrullus mucosospermus, the egusi watermelon, is indigenous to western Africa. The fruits are usually spherical with bitter, white flesh. They are cultivated for the consumption of their seeds, which vary widely in size and shape (Oyolu 1977; Fursa 1983). Typically, the seeds of egusi watermelons are distinguished from those of other *Citrullus* by lacking thick, hard seed coats. When fresh, the seeds have peculiar soft, moist, mucilaginous seed coats (Fursa 1972, 1983; Jensen et al. 2011; Achigan-Dako et al.

2015). The wild ancestor may be the watermelons known as *birkinda* in the Tombouctou area of Mali (annual rainfall < 200 mm), and these wild plants occur in close proximity to and genetically interact with landraces grown for consumption of the seeds and others grown for consumption of the sweet flesh (Jensen et al. 2011). The egusi watermelon is the closest relative of the dessert watermelon (Guo et al. 2013; Chomicki and Renner 2015).

Citrullus lanatus, the dessert watermelon, has been reported as growing wild in northeastern Africa, where it is known locally as *gurma* or *gurum* (Schweinfurth 1883; Muschler 1912; Andrews 1950; Abdel-Magid 1989; Robinson 2005; Gouda 2007; Mariod et al. 2009; Karrar et al. 2021). Therefore, wild colocynth and wild or primitive dessert watermelons have overlapping natural ranges. The fruits of wild or 'semi-cultivated' dessert watermelon plants are small, spherical, and striped with pale-colored, bland or sweetish flesh and are used as a source of clean water during the long dry season as well as for food and animal fodder. Cultigens indigenous to Kordofan and Darfur in Sudan vary widely in fruit size, color, striping pattern, flesh color and taste, and seed size and color (Keith-Roach 1924; Hassan et al. 1984). The Kordofan watermelons have been designated as a separate subspecies of the dessert watermelon, *C. lanatus* subsp. *cordophanus* (Ter-Avanesyn 1966). Renner et al. (2021), who assembled and annotated the genome of an indigenous primitive, small-fruited, non-bitter, white-fleshed Kordofan watermelon at the chromosome level, suggested that the Kordofan watermelon represents the closest relative of modern sweet watermelons and represents their likely progenitor that early farmers first brought into cultivation.

1.7 Genomics

The watermelon genome was first sequenced for 21 accessions, including 11 dessert watermelons, six egusi watermelons, and four citron watermelons, by Guo et al. (2013). These authors observed that modern dessert watermelons are most closely related to the modern egusi watermelons of western Africa. An interesting observation was that there is a high nucleotide divergence of reproductive barrier genes between dessert and egusi watermelons, which makes a good case for the dessert and egusi watermelons being considered to belong to separate species. Indeed, results of DNA sequencing using 12 *Citrullus* accessions that included two each of dessert and egusi watermelons led to the proposal by Chomicki and Renner (2015) that *Citrullus* can be considered to consist of seven species and that the egusi and dessert watermelons are closely related but separate species, *C. mucosospermus* and *C. lanatus*, respectively. These authors also clocked the divergence time between these sister species to approximately 3.1 million years ago, long before the onset of plant domestication.

Unfortunately, no primitive dessert watermelons from Egypt or Sudan were included in this first compilation of the watermelon genome. Since then, sequencing the genomes of crop-plant species has quickly become less expensive and tedious, so that now hundreds of watermelon accessions have been sequenced.

The *Citrullus* genome was resequenced by Guo et al. (2019) using 414 accessions representing all seven species with high-quality long reads. The results showed two primitive landrace *C. lanatus* accessions originating from Sudan to be in the deepest branch of the *C. lanatus* clade. Furthermore, they revealed that an allele for non-bitterness must have arisen in ancestral *C. lanatus* and that allele today is largely fixed in this species. Subsequent selection for fruit-flesh coloration and sweetness appears to have co-evolved through common genetic components, including an allele for sugar transport, and several genes affecting fruit size have undergone selection during the centuries of cultivation. These authors also observed that *C. colocynthis* and *C. amarus* were more genetically diverse than *C. mucosospermus* and *C. lanatus*, and gene flow was more apt to occur between some species pairs and less so between others.

Genotyping-by-sequencing was used to genotype 1365 *Citrullus* accessions maintained at the US National Plant Germplasm System (Wu et al. 2019). Six accessions derived from Egypt were located in one of the deepest branches of the dessert watermelon phylogenetic tree, again suggesting that modern dessert watermelons originate from northeastern Africa. Four from Sudan, though locating here as well, also showed some affinity to *C. mucosospermus*, suggesting that they could be derived from the same ancient ancestral population.

Recently, a biogeographical and chloroplast haplotype genealogy was inferred from a collection of 135 accessions representing *Citrullus lanatus*, *C. mucosospermus*, *C. amarus*, and *C. colocynthis* (Achigan-Dako et al. 2021). *C. mucosospermus* had the fewest haplotypes, some of which were nearly exclusive to western Africa. *C. lanatus* and *C. mucosospermus* shared haplotypes that were distinct from those of both *C. amarus* and *C. colocynthis*, again confirming that *C. mucosospermus* is the closest relative of *C. lanatus*.

Caution is advised in interpreting genomic data on *Citrullus* obtained from modern plant specimens. Valuable plants or their seeds have been carried by people long distances since ancient times (Muthukumaran 2016). Among the cucurbits, for example, the sponge gourd (*Luffa aegyptiaca* Mill.), a native of Southeast Asia, was mentioned in rabbinic literature from Israel dating to the third-century CE (Paris 2021). Watermelons have also been long-distance travelers for quite some time. The citron watermelon, *Citrullus amarus*, reached the Roman Empire from southern Africa by the fourth-century CE (Paris 2015a). Certainly, the sweet dessert watermelons of northeastern Africa and the Mediterranean would have been noticed and esteemed by visitors, who would have taken some seeds back to where they lived, for planting there. As crossing barriers between *Citrullus* species are weak, *C. lanatus* is likely to have spontaneously crossed and genetically interacted with other watermelon species, especially *C. mucosospermus*, when it was introduced to their native African lands. African watermelon accessions that appear to be admixtures of more than one species have been identified in some molecular and genomic investigations (Angui et al. 2017; Levi et al. 2017b; Mujaju et al. 2018; Guo et al. 2019).

The confounding issue of possible long-term species contamination of the watermelons growing today could be circumvented by genomic analysis of archaeological specimens. Since the 1980s, archaeological plant specimens have been analyzed with molecular techniques to probe the evolution of plants with people. Desiccated plant remains are among the most amenable of archaeological finds for ancient DNA sequencing (Brown et al. 2015) and those of watermelons from Egypt and Sudan and, indeed, most of the native distribution of *Citrullus* fits into this category (Paris 2015a). Archaeological specimens would need to be sequenced and these sequences then compared with sequences from modern germplasm. A major determinant of the outcome of such an investigation would be the array of modern germplasm selected to be used for comparison with the archaeological specimens (Paris 2016). Ancient DNA sequencing and analysis may already be able to infer some phenotypic features of archaeological specimens and track the order and occurrence of selection for desired horticultural traits in ever-finer nuances (Allaby et al. 2015; Przelomska et al. 2020). The resequencing of the watermelon genome (Guo et al. 2019) and its application to ancient DNA analysis offer a golden opportunity for obtaining much improved assessments of dessert watermelon origin, including defining the geographic area and genetic constitution of dessert watermelon populations over time.

1.8 Conclusion

The weight of the multidisciplinary evidence, which encompasses archaeology, literature, linguistics, iconography and, lately, genomics, suggests that the dessert watermelon was first domesticated and cultivated in northeastern Africa. Selection for sweetness succeeded by no later than the second-century CE and selection for colored fruit flesh by the sixth-century CE.

Dessert watermelons have been cultivated in the Nile Valley for several thousand years and have undergone major genetic bottlenecks in their evolution under domestication, presumably with stepwise selection for non-bitterness, increased fruit size, and improved fruit-flesh palatability including increased sweetness and coloration, resulting in a narrow genetic base. Subsequently, with their diffusion to areas of high rainfall, they were further bottlenecked to adaptation to mesic environments. As all Roman and early Byzantine images of watermelon fruits show them as round and of modest size, usually striped, the two much earlier images from ancient Egypt of oblong striped fruits perhaps are not of watermelons but might be depictions of vegetable melons. Genomics has and will continue to make important contributions to furthering the understanding of watermelon history in context with other disciplines related to archaeobotany and food history. Archaeogenomics, the genomic analysis of archaeobotanical artifacts, potentially can make an enormous contribution to improving the understanding of the evolutionary interaction of watermelons and humans.

References

Abdel-Magid A (1989) Plant domestication in the Middle Nile Basin, vol 523. BAR Int Ser, Oxford, pp 1–10, 21, 41–48

Achigan-Dako EG, Avohou ES, Linsoussi C, Ahanchede A, Vodouhe RS, Blattner FR (2015) Phenetic characterization of *Citrullus* sp. (Cucurbitaceae) and differentiation of egusi-type (*C. mucosospermus*). Genet Resour Crop Evol 62:1159–1179

Achigan-Dako EG, Degbey H, Hale I, Blattner FR (2021) Georeferenced phylogenetic analysis of a global collection of wild and cultivated *Citrullus* species. Ecol Evol 11:1918–1936

Aimpoint Research (2020) Annual consumer research findings [Fact Sheet]. National Watermelon Promotion Board. https://www.watermelon.org/wp-content/uploads/2020/04/NWPBWinter2020CONSUMERFINALFORWEB.pdf

Allaby RG, Gutaker R, Clarke AC, Pearson N, Ware R, Palmer SA, Kitchen JL, Smith O (2015) Using archaeogenomic and computational approaches to unravel the history of local adaptation in crops. Phil Trans R Soc B 370:20130377

Amar Z, Hazot I (2003) Avattiah hapaqqu'a bekhalkalat ha'ir 'Azza ba'et hahadasha [Colocynth in the economy of the city of Gaza in modern times, in Hebrew]. Qorot 16:103–119

Andrews EA (1863) A copious and critical Latin-English lexicon. Sampson Low, Son & Co, London

Andrews AC (1958) Melons and watermelons in the classical era. Osiris 12:368–375

Andrews FW (1950) The flowering plants of the Anglo-Egyptian Sudan. T Buncle, Arbroath, pp 163–168

Angui CMV, Koffi KK, Koffi KG, Flamand MC, Bertin P, Baudoin JP, Djè Y, Zoro BIA (2017) Microsatellite-based genetic diversity among accessions of *Citrullus* spp. collected from 36 countries. Aust J Crop Sci 11:1033–1041

Avital A, Paris HS (2014) Cucurbits depicted in Byzantine mosaics from Israel 350–600 CE. Ann Bot 114:203–222

Bailey LH (1930) Three discussions in Cucurbitacae. Gent Herbar 2:175–186

Brown AC, Summers WL (1985) Carbohydrate accumulation and color development in watermelon. J Am Soc Hortic Sci 110:683–686

Brown TA, Cappellini E, Kistler L, Lister DL, Oliveira HR, Wales N, Schlumbaum A (2015) Recent advances in ancient DNA research and their implications for archaeobotany. Veget Hist Archaeobot 24:207–214

Campbell JW, Stanley-Stahr C, Bammer M, Daniels JC, Ellis JD (2019) Contribution of bees and other pollinators to watermelon (*Citrullus lanatus* Thunb.) pollination. J Apicult Res 58:597–603

Chomicki G, Renner SS (2015) Watermelon origin solved with molecular phylogenetics including Linnean material: another example of museomics. New Phytol 205:526–532

Clément-Mullet JJ (1866) Le livre de l'agriculture d'Ibn-al-Awam, vol 2. A. Franck, Paris, p 226

Cox A, van der Veen M (2008) Changing foodways: watermelon (*Citrullus lanatus*) consumption in Roman and Islamic Quseir al-Qadim, Egypt. Veget Hist Archaeobot 17:S181–S189

Dalby A (2003) Food in the ancient world from A to Z. Routledge, London, pp x–xii

Dane F, Lang P, Bakhtiyarova R (2004) Comparative analysis of chloroplast DNA variability in wild and cultivated *Citrullus* species. Theor Appl Genet 108:958–966

Davis A, Webber CL III, Fish WW, Wehner TC, King S, Perkins-Veazie P (2011) L-citrulline levels in watermelon cultigens tested in two environments. HortScience 46:1572–1575

De Vartavan C (1999) Hidden fields of Tutankhamun. Triade Exploration, London, p 22

De Vartavan C, Asensi Amoros V (1997) Codex of ancient Egyptian plant remains. Triade Exploration, London, pp 77–79

Dogimont C (2011) 2011 Gene list for melon. Cucurbit Genet Coop Rep 33–34:104–133

Elmstrom GW, Davis PL (1981) Sugars in developing and mature fruits of several watermelon cultivars. J Am Soc Hortic Sci 106:330–333

FAO (2020) fao.org/faostat/en/#data/QC/visualize. Last accessed 23 Dec 2020

Feliks Y (2005) Talmud Yerushalmi: Massekhet Ma'asrot [The Jerusalem Talmud: Tractate Tithing, in Hebrew]. Bar-Ilan University Press, Ramat Gan, pp 39–40, 55–67, 287, 303

Flower B, Rosenbaum E (1958) The Roman cookery book: a critical translation of the art of cooking by Apicius. Harrap, London, pp 74–79,112–115

Fulks BK, Scheerens JC, Bemis WP (1979) Natural hybridization of two *Citrullus* species. J Hered 70:214–215

Fursa TB (1972) On the taxonomy of genus *Citrullus* Schrad. Botan Zhurn 57:31–41

Fursa TB (1983) A new species of watermelon *Citrullus mucosospermus* (Fursa) Fursa. Trudy Prikl Bot Genet Selek 81:108–112

Germer R (1988) Katalog der Altägyptischen Pflanzenreste der Berliner Museeun. Ägyptologische Abhandlungen, vol. 47. Otto Harrassowitz, Wiesbaden, pp 43–44

Gouda M (2007) Diversity of local genetic resources of watermelon, *Citrullus lanatus* (Thunb.) Matsum. & Nakai, in Sudan. MSc Thesis, University of Uppsala

Guo S, Zhang J, Sun H, Salse J, Lucas WJ, Zhang H, Zheng Y, Mao L, Ren Y, Wang Z, Min J, Guo X, Murat F, Ham BK, Zhang Z, Gao S, Huang M, Xu Y, Zhong S, Bombarely A, Mueller LA, Zhao H, He H, Zhang Y, Zhang Z, Huang S, Tan T, Pang E, Lin K, Hu Q, Kuang H, Ni P, Wang B, Liu J, Kou Q, Hou W, Zou X, Jiang J, Gong G, Klee K, Schoof H, Huang Y, Hu X, Dong S, Liang D, Wang J, Wu K, Xia Y, Zhao X, Zheng Z, Xing M, Liang X, Huang B, Lv T, Wang J, Yin Y, Yi H, Li R, Wu M, Levi A, Zhang X, Giovannoni JJ, Wang J, Li Y, Fei Z, Xu Y (2013) The draft genome of watermelon, *Citrullus lanatus*, and resequencing of 20 diverse accessions. Nat Genet 45:51–58

Guo S, Zhao S, Sun H, Wang X, Wu S, Lin T, Ren Y, Gao L, Deng Y, Zhang J, Lu X, Zhang H, Shang J, Gong G, Wen C, He N, Tian S, Li M, Liu J, Wang Y, Zhu Y, Jarret R, Levi A, Zhang X, Huang S, Fei Z, Liu W, Xu Y (2019) Resequencing of 414 cultivated and wild watermelon accessions identifies selection for fruit quality traits. Nat Genet 51:1616–1623

Gusmini G, Wehner TC (2006) Qualitative inheritance of rind pattern and flesh color in watermelon. J Hered 97:177–185

Gusmini G, Wehner TC, Jarret RL (2004) Inheritance of egusi seed type in watermelon. J Hered 95:268–270

Guzzon F, Ardenghi NMG (2018) Could taxonomic misnaming threaten the ex situ conservation and the usage of plant genetic resources? Biodivers Conserv 27:1157–1172

Hassan MS, Dinar HM, El Hussein SA, Ahmed AK, Geneif AA (1984) Indigenous horticultural germplasm of western Sudan. Plant Genet Resour Newsl 59:4–11

Hort A (1976) Theophrastus: enquiry into plants. William Heinemann, London

Janick J, Paris HS, Parrish DC (2007) The cucurbits of Mediterranean antiquity: identification of taxa from ancient images and descriptions. Ann Bot 100:1441–1457

Jarret RL, Newman M (2000) Phylogenetic relationships among species of *Citrullus* and the placement of *C. rehmii* De Winter as determined by internal transcribed spacer (ITS) sequence heterogeneity. Genet Resour Crop Evol 47:215–222

Jensen BD, Touré FM, Hamattal MA, Touré FA, Nantoumé AD (2011) Watermelons in the sand of Sahara: cultivation and use of indigenous landraces in the Tombouctou region of Mali. Ethnobotanic Res Applic 9:151–162

Jensen BD (2012) African watermelons and their uses, In: Sari N, Solmaz I, Aras V (eds) Cucurbitaceae 2012. Proceedings of the 10th Eucarpia meeting on genetics and breeding of Cucurbitaceae. Cukurova University, Adana, pp 264–274

Jones WHS (1967) Hippocrates, vol 4. William Heinemann, London, pp 332–337

Karchi Z (2000) Development of melon culture and breeding in Israel. In: Katzir N, Paris HS (eds) Proceedings of Cucurbitaceae 2000, the 7th Eucarpia Meeting on Cucurbit Genetics & Breeding. Acta Hortic 510:13–17

Karrar E, Ahmed IAM, Manzoor MF, Sarpong F, Wei W, Wang X (2021) Gurum seeds: a potential source of edible oil. Eur J Lipid Sci Technol 123:2000104

Katuuramu DN, Wechter WP, Washington ML, Horry M, Cutulle MA, Jarret RL, Levi A (2020) Phenotypic diversity for root traits and identification of superior germplasm for root breeding in watermelon. HortScience 55:1272–1279

Keimer L (1924) Die Gartenpflanzen im Alten Ägypten. Hoffmann & Campe, Hamburg, p 170

Keith-Roach E (1924) Adventures among the lost tribes of Islam in eastern Darfur. Natl Geograph Mag 45: 41–73

Kislev M, Simhoni O (2009) Mezonam shel shokhne me'arat Har Yishay–me'arat miflat bizman mered Bar-Kokhva. [Food of the inhabitants of the Har Yishay cave, a cave of refuge at the time of the Bar-Kokhva rebellion, in Hebrew.] Mehqare Yehuda we-Shomeron [Judea and Samaria Research Studies] 18:201–214

Kučan D (2000) Rapport synthétique sur les recherches archéobotaniques dans le sanctuaire d'Héra de l'île de Samos. Pallas 52:99–108

Kuşçu H, Turhan A, Özmen N, Aydınol P, Büyükcangaz H, Demir AO (2015) Deficit irrigation effects on watermelon (*Citrullus vulgaris*) in a sub humid environment. J Anim Plant Sci 25:1652–1659

Kyriacou MC, Leskovar DI, Colla G, Rouphael Y (2018) Watermelon and melon fruit quality: the genotypic and agro-environmental factors implicated. Sci Hortic 234:393–408

Levi A, Thomas CE, Keinath AP, Wehner TC (2001a) Genetic diversity among watermelon (*Citrullus*

lanatus and *Citrullus colocynthis*) accessions. Genet Resour Crop Evol 48:559–566
Levi A, Thomas CE, Wehner TC, Zhang X (2001b) Low genetic diversity indicates the need to broaden the genetic base of cultivated watermelon. HortScience 36:1096–1101
Levi A, Thomas CE, Newman M, Zhang X, Xu Y, Wehner TC (2003) Massive preferential segregation and non random assortment of linkage groups produce quasi-linkage in an F2 mapping population of watermelon [abstract]. HortScience 38:782
Levi A, Thomas CE, Thies JA, Simmons AM, Ling K, Harrison HF Jr (2006) Novel watermelon breeding lines containing chloroplast and mitochondrial genomes derived from the desert species *Citrullus colocynthis*. HortScience 41:463–464
Levi A, Wechter WP, Massey LM, Carter L, Hopkins D (2011) Genetic linkage map of *Citrullus lanatus* var. *citroides* chromosomal segments introgressed into the watermelon cultivar Crimson Sweet (*Citrullus lanatus* var. *lanatus*) genome. Am J Plant Sci 2:93–110
Levi A, Thies JA, Wechter WP, Harrison HF, Simmons AM, Reddy UK, Nimmakayala P, Fei Z (2013) High-frequency oligonucleotides: targeting active gene (HFO–TAG) markers revealed wide genetic diversity among *Citrullus* spp. accessions used for enhancing disease or pest resistance in watermelon cultivars. Genet Resour Crop Evol 60:427–440
Levi A, Jarret R, Kousik S, Wechter WP, Nimmakayala P, Reddy UK (2017a) Genetic resources of watermelon. In: Grumet R, Katzir N, Garcia-Mas J (eds) Genetics and genomics of Cucurbitaceae. Springer International, Cham, Switzerland, pp 87–110
Levi A, Simmons AM, Massey L, Coffey J, Wechter WP, Jarret RL, Tadmor Y, Nimmakayala P, Reddy UK (2017b) Genetic diversity in the desert watermelon *Citrullus colocynthis* and its relationship with *Citrullus* species as determined by high-frequency oligonucleotides-targeting active gene markers. J Am Soc Hortic Sci 142:47–56
Lewis CT, Short C (1951) A Latin dictionary. Clarendon Press, Oxford, p 346
Liddell HG, Scott R (1948) A Greek–English Lexicon. Clarendon Press, Oxford, p 1364
Livingstone D (1858) Missionary travels and researches in South Africa. Harper, New York, pp 54–55
Maggs-Kölling GL, Christiansen JL (2003) Variability in Namibian landraces of watermelon (*Citrullus lanatus*). Euphytica 132:251–258
Manniche L (1989) An ancient Egyptian herbal. British Museum, London, pp 91–93
Margaritis E (2006–2007) Archaeobotanical investigations at the geometric site of Krania, southern Pieria in Macedonia, Greece. Talanta (Amsterdam) 38–39: 123–132
Mariod AA, Ahmed YM, Matthäus B, Khaleel G, Siddig A, Gabra AM, Abdelwahab SI (2009) A comparative study of the properties of six Sudanese cucurbit seeds and seed oils. J Am Oil Chem Soc 86:1181–1188
Maroyi A (2020) Chemical properties, biological activities and ethnobotanical uses of *Citrullus naudinianus* (Sond.) Hook. f.: an indigenous melon species of South Africa. Med Plants 12:182–191
Maynard DN (2001) An introduction to the watermelon. In: Maynard DN (ed) Watermelons: characteristics, production and marketing. ASHS Press, Alexandria, VA, pp 9–20
McCreight JD (2017) Cultivation and uses of cucurbits. In: Grumet R, Katzir N, Garcia-Mas J (eds) Genetics and genomics of Cucurbitaceae. Cham, Switzerland, Springer International, pp 1–12
McGregor CE (2012) *Citrullus lanatus* germplasm of southern Africa. Israel J Plant Sci 60:403–413
McGregor CE, Waters V (2013) Pollen viability of F_1 hybrids between watermelon cultivars and disease-resistant, infraspecific crop wild relatives. HortScience 48:1428–1432
Meeuse ADJ (1962) The Cucurbitaceae of southern Africa. Bothalia 8(1):57–58
Mohr HC (1986) Watermelon breeding. In: Bassett MJ (ed) Breeding vegetable crops. Avi Publishing, Westport, CT, pp 37–66
Mujaju C, Zborowska A, Werlemark G, Garkava-Gustavsson L, Andersen SB, Nybom H (2011) Genetic diversity among and within watermelon (*Citrullus lanatus*) landraces in southern Africa. J Hortic Sci Biotechnol 86:353–358
Mujaju C, Musango R, Garkava-Gustavsson L (2018) Molecular characterisation reveals the existence of a hybridogenous intermediary form between sweet watermelon and cow melon forms of watermelon. Am J Plant Sci 9:1245–1259
Munisse P, Andersen SB, Jensen BD, Christiansen JL (2011) Diversity of landraces, agricultural practises and traditional uses of watermelon (*Citrullus lanatus*) in Mozambique. Afr J Plant Sci 5:75–86
Muschler R (1912) A manual flora of Egypt. R. Friedlaender, Berlin, pp 937–939
Muthukumaran S (2016) An ecology of trade: Tropical Asian cultivars in the ancient Middle East and the Eastern Mediterranean. Doctoral thesis, University College London
Navot N, Sarfatti M, Zamir D (1990) Linkage relationships of genes affecting bitterness and flesh color in watermelon. J Hered 81:162–165
Ngwepe RM, Mashilo J, Shimelis H (2019) Progress in genetic improvement of citron watermelon (*Citrullus lanatus* var. *citroides*): a review. Genet Resour Crop Evol 66:735–758
Nimmakayala P, Levi A, Abburi L, Abburi VL, Tomason YR, Saminathan T, Vajja VG, Malkaram S, Reddy R, Wehner TC, Mitchell SE, Reddy UK (2014) Single nucleotide polymorphisms generated by genotyping by sequencing to characterize genome-wide diversity, linkage disequilibrium, and selective sweeps in cultivated watermelon. BMC Genom 15:767
Njoroge GN, Gemmill B, Bussmann R, Newton LE, Ngumi VM (2010) Diversity and efficiency of wild

pollinators of watermelon (*Citrullus lanatus* (Thunb.) Mansf.) at Yatta (Kenya). J Appl Hortic 12:35–41
Oyolu C (1977) A quantitative and qualitative study of seed types in egusi (*Colocynthis citrullus* L.). Trop Sci 19:55–62
Paris HS (2015a) Origin and emergence of the sweet dessert watermelon, *Citrullus lanatus*. Ann Bot 116:133–148
Paris HS (2015b) The identity, characteristics and vending of cucurbit crops in Israel of Roman times. In: McWilliams M (ed) Food & markets, proceedings of the Oxford symposium on food and cookery 2014. Prospect Books, London, pp 299–306
Paris HS (2016) Overview of the origins and history of the five major cucurbit crops: issues for ancient DNA analysis of archaeological specimens. Veget Hist Archaeobot 25:405–414
Paris HS (2021) Delu'e ha-ma'akhal shel Ḫazal: Zihuyam ha-taqsonomi. [Taxonomical identity of the edible cucurbits of the Mishna, Tosefta and Talmud]. Jewish Stud Internet J 20: https://jewish-faculty.biu.ac.il/files/jewish-faculty/shared/JSIJ20/paris.pdf (in Hebrew)
Paris HS, Brown RN (2005) The genes of pumpkin and squash. HortScience 40:1620–1630
Paris HS, Janick J, Daunay MC (2011) Medieval herbal iconography and lexicography of *Cucumis* (cucumber and melon, Cucurbitaceae) in the Occident, 1300–1458. Ann Bot 108:471–484
Paris HS, Daunay MC, Janick J (2013) Medieval iconography of watermelons in Mediterranean Europe. Ann Bot 112:867–879
Perkins-Veazie P, Davis A, Collins JK (2012) Watermelon: from dessert to functional food. Israel J Plant Sci 60:395–402
Piccirillo M (1993) The mosaics of Jordan. American Center for Oriental Research, Amman
Przelomska NAS, Armstrong CG, Kistler L (2020) Ancient plant DNA as a window into the cultural heritage and biodiversity of our food system. Front Ecol Evol 8:74
Qin K, Leskovar DI (2020) Assessments of humic substances application and deficit irrigation in triploid watermelon. HortScience 55:716–721
Ramirez JL, Du X, Wallace RW (2020) Investigating sensory properties of seven watermelon varieties and factors impacting refreshing perception using quantitative descriptive analysis. Food Res Intl 138:109681
Renner SS, Sousa A, Chomicki G (2017) Chromosome numbers, Sudanese wild forms and classification of the watermelon genus *Citrullus*, with 50 names allocated to seven biological species. Taxon 66:1393–1405
Renner SS, Wu S, Perez-Escobar OA, Silber MV, Fei Z, Chomicki G (2021) A chromosome level genome of a Kordofan melon illuminates the origin of domesticated watermelons. Proc Natl Acad Sci USA 118: e2101486118
Robinson J (2005) Desertification and disarray: the threats to plant genetic resources of southern Darfur, western Sudan. Plant Genet Resour 3:3–11
Robinson RW, Decker-Walters DS (1997) Cucurbits. CAB International, Wallingford, pp 65–97, 152–163
Rubatsky VE (2001) Origin, distribution and uses. In: Maynard DN (ed) Watermelons: characteristics, production and marketing. ASHS Press, Alexandria, VA, pp 21–26
Rushing JW, Fonseca JM, Keinath AP (2001) Harvesting and postharvest handling. In: Maynard DN (ed) Watermelons: characteristics, production and marketing. ASHS Press, Alexandria, VA, pp 156–164
Sabato D, Peña-Chocarro L, Ucchesu M, Sarigu M, Del Vais C, Sanna I, Bacchetta G (2019) New insights about economic plants during the 6th–2nd centuries BC in Sardinia, Italy. Veg Hist Archaeobot 28:9–16
Said EM, Fatiha H (2018) Genotypic variability in fruits characters of Moroccan watermelon cultivars (*Citrullus lanatus*) cultivars under well and limited watered conditions. Hortic Int J 2:378–381
Sain RS, Joshi P, Divakara Sastry EV (2002) Cytogenetic analysis of interspecific hybrids in genus *Citrullus* (Cucurbitaceae). Euphytica 128:205–210
Schweinfurth G (1883) The flora of ancient Egypt. Nature 28(709):109–114
Singh AK (1978) Cytogenetics of semi-arid plants, III. A natural interspecific hybrid of Cucurbitaceae (*Citrullus colocynthis* Schrad. × *C. vulgaris* Schrad). Cytologia 43:569–574
Sinnott EW (1937) The relation of gene to character in quantitative inheritance. Proc Natl Acad Sci USA 23:224–227
Soteriou GA, Kyriacou MC, Siomos AS, Gerasopoulos D (2014) Evolution of watermelon fruit physicochemical and phytochemical composition during ripening as affected by grafting. Food Chem 165:282–289
Ter-Avanesyn DV (1966) Arbuz Kordofanskyi *Citrullus lanatus* Mansf. ssp. cordophanus Ter-Avan. Botan Zhurn 51:423–426
Van der Veen M (1999) The food and fodder supply to Roman quarry settlements in the eastern desert of Egypt. In: van der Veen M (ed) The exploitation of plant resources in ancient Africa. Kluwer Academic/Plenum, New York, pp 171–183
Van der Veen M (2011) Consumption, trade and innovation, exploring the botanical remains from the Roman and Islamic ports at Quseir al-Qadim, Egypt. Africa Magna, Frankfurt-am-Main, pp 106–107, 179–185
Van Zeist W (1983) Fruits in foundation deposits of two temples. J Archaeol Sci 10:351–354
Wang N, Liu S, Gao P, Luan F, Davis AR (2016) Developmental changes in gene expression drive accumulation of lycopene and β-carotene in watermelon. J Am Soc Hortic Sci 141:434–443
Wasylikowa K, van der Veen M (2004) An archaeological contribution to the history of watermelon, *Citrullus lanatus* (Thunb.) Matsum. & Nakai (syn. *C. vulgaris* Schrad). Veget Hist Archaeobot 13:213–217
Wehner TC (2008) Watermelon. In: Prohens J, Nuez F (eds) Handbook of plant breeding, vegetables I. Springer, New York, pp 381–418

Wehner TC, Shetty NV, Elmstrom GW (2001) Breeding and seed production. In: Maynard DN (ed) Watermelons: characteristics, production and marketing. ASHS Press, Alexandria, VA, pp 27–73

Wehner TC, Naegele RC, Myers JR, Dhillon NPS, Crosby K (2020) Cucurbits, 2nd edn. CABI, Wallingford, Oxfordshire

Welman M (2011) *Citrullus lanatus* (Thunb.) Matsum & Nakai http://www.plantzafrica.com/plantcd/citrullanat.html

Weng Y, Wehner TC (2017) Cucumber gene catalog 2017. Cucurbit Genet Coop Rep 39–40:17–54

Whitaker TW (1933) Cytological and phylogenetic studies in the Cucurbitaceae. Bot Gaz 94:780–790

Wu S, Wang X, Reddy U, Sun H, Bao K, Gao L, Mao L, Patel T, Ortiz C, Abburi VL, Nimmakayala P, Branham S, Wechter P, Massey L, Ling KS, Kousik C, Hammar SA, Tadmor Y, Portnoy V, Gur A, Katzir N, Guner N, Davis A, Hernandez AG, Wright CL, McGregor C, Jarret R, Zhang X, Xu Y, Wehner TC, Grumet R, Levi A, Fei Z (2019) Genome of 'Charleston Gray', the principal American watermelon cultivar, and genetic characterization of 1,365 accessions in the U.S. National Plant Germplasm System watermelon collection. Plant Biotechnol J 17:2246–2258

Xu Y, Guo S, Liu J, Fei Z, He H, Zhang H, Ren Y, Sun H, Gong G, Zheng Y, Huang M, Zhong S (2012) Dynamic characteristics of enzymes and transcriptome related to sugar metabolism and accumulation in sweet and non sweet watermelon fruits. In: Sari N, Solmaz I, Aras V (eds) Cucurbitaceae 2012, proceedings of the 10th Eucarpia meeting on genetics and breeding of Cucurbitaceae. Cukurova University, Adana, pp 103–111

Zhang H, Gong G, Guo S, Ren Y, Xu Y, Ling K-S (2011) Screening the USDA watermelon germplasm collection for drought tolerance at the seedling stage. HortScience 46:1245–1248

Zhang J, Guo S, Ren Y, Zhang H, Gong G, Zhou M, Wang G, Zong M, He H, Liu F, Xu Y (2017) High-level expression of a novel chromoplast phosphate transporter ClPHT4;2 is required for flesh color development in watermelon. New Phytol 213:1208–1221

Zohary D, Hopf M, Weiss E (2012) Domestication of plants in the Old World, 4th edn. Clarendon Press, Oxford, pp 9, 153–155

Caveats for Watermelon Whole Genome Sequencing

2

Purushothaman Natarajan, Padma Nimmakayala, Sudip Kumar Dutta, and Umesh K. Reddy

2.1 Introduction

A genome comprises the entire haploid set of chromosomes in a cell. In plants, there are significant variations in genome size and ploidy level, which can be challenging to sequence and assemble due to polyploidy and repetitive sequences. *Citrullus lanatus*, or watermelon, is an essential crop in the Cucurbitaceae family. While the diploid watermelon ($2n = 22$) is more prevalent, autopolyploids give rise to triploid ($3n = 3 \times = 33$) and tetraploid watermelon ($4n = 44$). The popular seedless watermelon is basically a triploid watermelon that arises from hybridization between the common diploid and tetraploid watermelon (Zhang et al. 2019). The initial estimation of Citrullus species' (*Citrullus vulgaris*) genome size using flow cytometry indicated it to be 490 Mb (Arumuganathan and Earle 1991). In a recent study, the k-mer-based analysis revealed the genome size of multiple watermelon species that fell between 420 and 490 Mb (Guo et al. 2013). Given its economic significance and economic relevance, sequencing the entire genome of watermelon is crucial for both biological research and plant breeding programs. With the emergence of high-throughput DNA sequencing technologies with less cost, multiple research groups worldwide have successfully sequenced the genomes of many watermelon accessions.

P. Natarajan · P. Nimmakayala · U. K. Reddy (✉)
Department of Biology, West Virginia State University, Institute, WV 25112, USA
e-mail: ureddy@wvstateu.edu

S. K. Dutta
ICAR-Research Complex for NEH Region, Sikkim Centre, Gangtok, India

2.2 Draft Genome Sequence

In 2013, the initial draft genome sequence of the watermelon cultivar '97103' ($2n = 22$) was assembled using Illumina sequencing technology (Guo et al. 2013). The Illumina GAII system was employed to sequence the long and short insert (paired-end, mate-pair) sequencing libraries. The sequencing coverage attained was 108.6X, based on the k-mer-based estimated genome size of 425 Mb. Subsequently, the quality-filtered reads were subjected to de novo assembly using a SOAP denovo genome assembler (Li et al. 2010). Following de novo assembly, the final assembly length was 353.5 Mb, which covered 83.2% of the watermelon genome, including 1793 scaffolds. The N50 size of the scaffolds and contigs was 2.38 Mb and 26.38 kb, respectively. A total of 11 chromosomes covering 330 Mb were constructed from 234 scaffolds. The accuracy of de novo assembly was assessed using one million ESTs, paired-end sequences of 667 BAC clones, and four completely sequenced BACs. Repeat sequences were masked using Repeat

S. Kr. Dutta et al. (eds.), *The Watermelon Genome*, Compendium of Plant Genomes,
https://doi.org/10.1007/978-3-031-34716-0_2

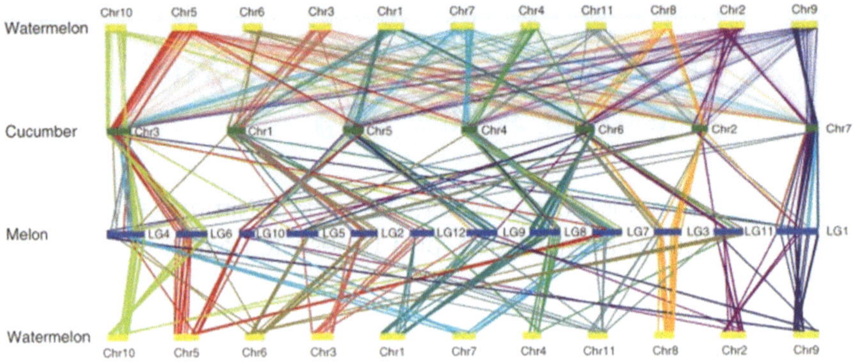

Fig. 2.1 Synteny plot shows the relationship between the genomes of watermelon, cucumber, and melon (Guo et al. 2013)

Masker (Chen 2004), and the ab initio gene predictions were performed using AUGUSTUS eukaryotic gene prediction software (Stanke and Waack 2003) and Glimmer HMM tools (Majoros et al. 2004).

The watermelon genome contained 159.8 Mb of transposable element repeats, corresponding to 45.2% of the total genome length. About 68% of these repeats were successfully annotated. Gene prediction data revealed the presence of 23,440 genes in the watermelon genome, with nearly 85% of these genes having either known homologs or functional classification. Furthermore, detailed synteny analysis of chromosomes from the different species of the Cucurbitaceae family revealed the relationship between the genome of watermelon and other cucurbit species (Fig. 2.1).

2.3 Improved Chromosome-Scale Genome Assembly

The watermelon cultivar '97103' draft genome assembled using Illumina sequencing reads was further improved by Guo et al. (2019) and generated chromosome-scale genome assembly. The assembly was produced using long-read sequencing (PacBio), optical mapping data (Bionano), and High-throughput Chromosome Conformation Capture (Hi-C) techniques. The assembly of 42 × data from the Pacbio with an N50 length of 10.8 kb produced 367 contigs. The N50 size of an assembly was significantly increased to 2.3 Mb using Pacbio data. Additionally, the Bionano optical map data were combined with Pacbio assembly to increase the contiguity, and this resulted in the improvement of the N50 length of assembly to 21.9 Mb. This assembly had a total size of 365.1 Mb with 149 scaffolds. To further refine the genome assembly, roughly 135.2 million Hi-C reads and genetic maps were used to generate chromosome-scale assembly. The ordering and orientation of the scaffolds into chromosomes were performed using Hi-C contact information data. The resulting version 2 assembly contained 31 scaffolds totaling 362.7 Mb, including 11 chromosomes. This improved assembly was achieved through the integration of PacBio, BioNano, and Hi-C data, which greatly enhanced the previous version 1 assembly (Table 2.1). A similar approach was employed by Renner et al. (2021) to assemble and annotate the Kordofan melon genome at the chromosome level using Pacbio, Illumina, and Hi-C technologies (Renner et al. 2021).

2.4 Resequencing of Watermelon Accessions

Guo et al. (2019) improved the '97103' watermelon reference genome assembly, enabling numerous resequencing studies to identify variants linked to important traits. Using a genome-wide association study (GWAS) approach, the

Table 2.1 Genome comparison between the two versions of the watermelon cultivar '97103' (Guo et al., 2019)

Watermelon '97103' genome	Version 1 assembly	Version 2 assembly
Total length	353,466,419	365,090,282
Maximum length	8,716,783	35,157,065
No. ordered scaffolds	125	31
No. oriented scaffolds	94	31
Total number of scaffolds	558	71
Total number of contigs	44,084	549
Scaffold N50	2,378,183	21,942,719
Contig N50	26,377	2,312,425

authors re-sequenced 414 watermelon accessions to identify genomic regions associated with fruit quality. GWAS analysis revealed 43 association signals, including eight overlapping with previously identified QTLs. One such QTL, QBRX2-1 (Ren et al. 2014), which had been previously reported, was associated with flesh sweetness and was found to harbor the sugar transporter gene, *ClTST2*. Additionally, two of the regions from chromosome 10 were strongly associated with flesh sweetness, in line with previous studies. The study also identified several other QTLs associated with flesh sweetness using the improved watermelon reference genome. RNA sequencing was employed to confirm the expression of these genes in both the flesh and rind, which uncovered an upregulation in their expression levels in watermelon fruits.

2.5 Comparison of Diploid and Tetraploid Watermelon

Garcia-Lozano et al. (2021) compared the transcriptome, methylome, and chromatin organization in diploid and tetraploid watermelon cultivars to study the effect of genome doubling. The study revealed that tetraploid watermelon had a transcriptome size per genome that was more than twice that of the diploid, with the majority of genes showing a dosage response greater than 1:1. In addition, the investigation detected more than 1688 hypomethylated areas in tetraploid watermelon located in intergenic regions, promoters, and near the transcriptional start sites of up-regulated genes (Fig. 2.2). Furthermore, high-throughput Chromosome Conformation Capture (Hi-C) was utilized to study the spatial genome organization and identify topologically associated domains, which showed a significant change in the A and B compartmental organization between diploid and tetraploid watermelon. The genes associated with these compartmental switches were related to plant development and stress responses in watermelons.

2.6 Genome Sequencing in Other Cucurbits

Sequencing and assembly of various cucurbit species, including economically important species within the Cucurbita genus, have been made possible by high-throughput sequencing techniques. Li et al. (2019) utilized reads from single-molecule real-time (SMRT) sequencing, 10X Genomics, and Hi-C to build the high-quality reference genome for the cucumber. Meanwhile, Sun et al. (2017) utilized high-throughput sequencing techniques and genetic maps to create the reference genomes for other cucurbit species, such as *C. maxima* and *C. moschata.* Illumina sequencing was used by Wu et al. (2017) to generate the high-quality reference genome sequence for bottle gourd (Wu et al. 2017). Several other cucurbit species are presently undergoing sequencing owing to their economic importance.

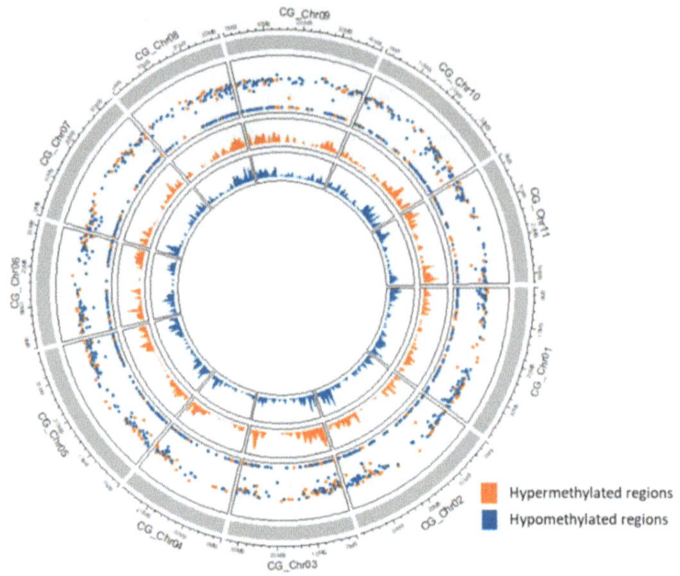

Fig. 2.2 Circos show the hyper- and hypomethylated regions identified from diploid and tetraploid watermelon (Garcia-Lozano et al. 2021)

2.7 Establishing Complete Reference Genomes (Telomere-To-Telomere Assembly)

The availability of various sequencing technologies has led to the improvement of draft or near-complete genome sequences into complete genome sequences that can serve as high-quality references for data analysis. A recent example is the improvement of the Arabidopsis thaliana reference genome (TAIR10) to produce a complete reference genome assembly of Col-0 with only two gaps, accomplished using reads from Oxford Nanopore Technologies, Pacific Biosciences, and Hi-C data. Wang et al. (2021) used this method to add 14.6 Mb of new sequences to the existing Arabidopsis genome, resulting in highly accurate chromosomes and the resolution of telomere-to-telomere manner for Chr3 and Chr5 (Wang et al. 2021). This same approach could be utilized to assemble or improve existing assemblies of cucurbit species, such as watermelon.

2.8 Pan-Genome for Watermelon

The pan-genome comprises the complete set of non-repetitive DNA sequences from a given species population, including both the core and dispensable genomes. The core genome consists of sequences shared among all individuals, while the dispensable genome contains unique sequences from two or more individuals (Bayer et al. 2020; Yao et al. 2015). A melon pan-genome was constructed by Sun et al. (2022) using resequencing data from 297 melon accessions, revealing significant genetic variations in melon groups for both favorable and unfavorable genes. This pan-genome included novel sequences of 168 Mb and 4325 genes, with 603 of 709 resistance gene analogs identified as core genes. Additionally, 106 variable genes were detected, including 55 absent in the melon genome and 13 presence–absence variations (PAVs) linked to length, shape, and width of the fruits. Various research teams are currently building a pan-genome for watermelon by combining reference genomes and resequencing data from multiple watermelon accessions, allowing for the identification of genes associated with key watermelon traits.

2.9 Conclusion

Watermelon is a vital crop species globally, and its extensive genomic resources enable the identification of mechanisms and markers for molecular breeding. The sequencing of the whole genome of several significant species in this group is aiding researchers worldwide in conducting GWAS and other studies. The creation of a pan-genome for watermelon, as well as the enhancement of the current reference genome using advanced genomics tools, would substantially benefit watermelon researchers and breeders.

References

Arumuganathan K, Earle E (1991) Nuclear DNA content of some important plant species. Plant Mol Biol Rep 9:208–218

Bayer PE, Golicz AA, Scheben A, Batley J, Edwards D (2020) Plant pan-genomes are the new reference. Nat Plants 6:914–920

Chen N (2004) Using Repeat Marker to identify repetitive elements in genomic sequences. Curr Protoc Bioinf 5:4.10.1–4.10.14

Garcia-Lozano M, Natarajan P, Levi A, Katam R, Lopez-Ortiz C, Nimmakayala P, Reddy UK (2021) Altered chromat in conformation and transcriptional regulation in watermelon following genome doubling. Plant J 106:588–600

Guo S, Zhang J, Sun H, Salse J, Lucas WJ, Zhang H, Zheng Y, Mao L, Ren Y, Wang Z (2013) The draft genome of watermelon (*Citrullus lanatus*) and resequencing of 20 diverse accessions. Nat Genet 45:51–58

Guo S, Zhao S, Sun H, Wang X, Wu S, Lin T, Ren Y, Gao L, Deng Y, Zhang J (2019) Resequencing of 414 cultivated and wild watermelon accessions identifies selection for fruit quality traits. Nat Genet 51:1616–1623

Li R, Zhu H, Ruan J, Qian W, Fang X, Shi Z, Li Y, Li S, Shan G, Kristiansen K (2010) De novo assembly of human genomes with massively parallel short read sequencing. Genome Res 20:265–272

Li Q, Li H, Huang W, Xu Y, Zhou Q, Wang S, Ruan J, Huang S, Zhang Z (2019) A chromosome-scale genome assembly of cucumber (*Cucumis sativus* L.) GigaSci 8:giz072

Majoros WH, Pertea M, Salzberg SL (2004) TigrScan and glimmerHMM: two open source ab initio eukaryotic gene-finders. Bioinformatics 20:2878–2879

Ren Y, Mcgregor C, Zhang Y, Gong G, Zhang H, Guo S, Sun H, Cai W, Zhang J, Xu Y (2014) An integrated genetic map based on four mapping populations and quantitative trait loci associated with economically important traits in watermelon (*Citrullus lanatus*). BMC Plant Biol 14:1–11

Renner SS, Wu S, Pérez-Escobar OA, Silber MV, Fei Z, Chomicki G (2021) A chromosome-level genome of a Kordofan melon illuminates the origin of domesticated watermelons. Proc Natl Acad Sci 118: e2101486118

Stanke M and Waack S (2003) Gene prediction with a hidden Markov model and a new intron submodel. Bioinf 19:ii215–ii225

Sun H, Wu S, Zhang G, Jiao C, Guo S, Ren Y, Zhang J, Zhang H, Gong G, Jia Z (2017) Karyotype stability and unbiased fractionation in the paleo-allotetraploid Cucurbita genomes. Mol Plant 10:1293–1306

Sun Y, Wang J, Li Y, Wang X, Xu WH, Wang YQ, Zhang PT, Zhang Y, Kong XD (2022) Pangenome analysis reveals the abundant gene presence/absence variations (PAVs) among different varieties of melon and their influence on traits. Front Plant Sci 420

Wang B, Yang X, Jia Y, Xu Y, Jia P, Dang N, Wang S, Xu T, Zhao X, Gao S (2021) High-quality Arabidopsis thaliana genome assembly with nanopore and HiFi long reads. Genomics, Proteomics Bioinf

Wu S, Shamimuzzaman M, Sun H, Salse J, Sui X, Wilder A, Wu Z, Levi A, Xu Y, Ling KS (2017) The bottle gourd genome provides insights into Cucurbitaceae evolution and facilitates mapping of a Papaya ring-spot virus resistance locus. Plant J 92:963–975

Yao W, Li G, Zhao H, Wang G, Lian X, Xie W (2015) Exploring the rice dispensable genome using a metagenome-like assembly strategy. Genome Biol 16:1–20

Zhang N, Bao Y, Xie Z, Huang X, Sun Y, Feng G, Zeng H, Ren J, Li Y, Xiong J (2019) Efficient Characterization of Tetraploid Watermelon. Plants 8:419

Watermelon Genetic Resources and Diversity

3

Nebahat Sari and İlknur Solmaz

3.1 Introduction

The Cucurbitaceae family includes approximately 118 genera and 825 species in which most produced watermelon, melon, cucumber, squash, and pumpkin. Fresh consumed dessert watermelon (*Citrullus lanatus* (Thunb.) Matsum. and Nakai), grown for fresh market with high economic importance, is one of the most important cultivated cucurbits with a world production of 102 million tons in 2020. Asia is the leading continent in watermelon production; China is the largest producer of watermelon with 59.3% (60.25 million tons) of the production and 47.4% of cultivated area. Turkiye is ranking in second with 3.4% of production (3.49 million tons) and 2.9% of area and India is the third highest watermelon producer country with 2.5% of production (2.78 million tons) and 3.2% of area of world production (Table 3.1). Iran ranks 4th in terms of production with 2.74 million tons (2.7% of the world's total). This is followed by Algeria (2.29 million tons—2.3% of the world's total), Brazil (2.18 million tons—2.2% of the world's total), USA (1.74 million tons—1.7% of the world's total), Senegal (1.68 million tons—1.6% of the world's total), Russian Federation (1.58 million tons—1.6% of the world's total), and Egypt (1.49 million tons—1.5% of the world's total). In terms of yield per unit area (hg/ha), the highest watermelon yield was obtained from Turkiye (447 hg/ha), USA (445 hg/ha), China (429 hg/ha) and this is followed by Algeria (375 hg/ha), and Egypt (331 hg/ha), respectively. The countries with the lowest watermelon yield per unit area within ten most producer countries are Russian Federation (158 hg/ha), Senegal (194 hg/ha), Brazil (222 hg/ha), India (253 hg/ha), and Iran (272 hg/ha) (FAOSTAT 2020).

Watermelon production in China has a long history. It was introduced in China and India around AD 1100 and 800, respectively (Zhao 2015). Although Turkiye is not the primary origin of watermelon, it is an important and rich genetic diversity center due to its location between three continents as an immigration road and Anatolia has climatologically different regions (Sari et al. 2006). Watermelon has been introduced to North America from Africa and Spain. A wide various varieties have been developed in the USA in the last 250 years and most of these varieties share a similar genetic basis (Levi et al. 2001a, b, 2016).

With the development of the first seedless watermelon cultivar by Kihara in 1951 as a result

N. Sari (✉) · İ. Solmaz
Department of Horticulture, Faculty of Agriculture, Cukurova University, 01330 Adana, Türkiye
e-mail: nebahat.sari0163@gmail.com

İ. Solmaz
e-mail: isolmaz@cu.edu.tr

S. Kr. Dutta et al. (eds.), *The Watermelon Genome*, Compendium of Plant Genomes,
https://doi.org/10.1007/978-3-031-34716-0_3

Table 3.1 Top ten producers of watermelon in the world (FAOSTAT 2020)

Rank	Country	Production (tons)	%	Area (ha)	%	Yield (hg/ha)
1	China	60,246,888	59.3	1,405,871	46.0	428,538
2	Turkiye	3,491,554	3.4	78,179	2.6	446,610
3	India	2,787,000	2.7	110,000	3.6	253,364
4	Iran	2,736,283	2.7	100,732	3.3	271,640
5	Algeria	2,286,809	2.3	61,000	2.0	374,887
6	Brazil	2,184,907	2.2	98,205	3.2	222,484
8	USA	1,741,579	1.7	39,133	1.3	445,041
9	Senegal	1,677,476	1.6	86,564	2.8	193,784
9	Russian Federation	1,584,319	1.6	100,082	3.3	158,301
10	Egypt	1,491,108	1.5	45,029	1.5	331,144
	Others	21,392,506	21.0	928,463	30.4	–
	Total	101,620,420	100.0	3,053,258	100.0	–

of hybridization of a tetraploid parent obtained by colchicine treatment with a diploid paternal parent, the triploid watermelon breeding studies also gained a different dimension (Solmaz et al. 2018). However, seeded (diploid) watermelons are still produced and consumed worldwide, especially in Asia, Africa, the Middle East, and South America. The American varieties "Crimson Sweet" and "Allsweet" have high flesh quality and small seeds. According to Levi et al. (2016), these varieties were developed at the Kansas State University by Charles V. Hall and launched in 1966 and 1972, respectively, perhaps leading diploid varieties and parents of most of today's seedless varieties.

The rate of genetic diversity in Cucurbits is very high, which facilitates the adaptation of Cucurbit species to various climates from tropical and subtropical regions to arid deserts and temperate regions, even mild climates. In cucurbit plants, the haploid chromosome number varies between 7 and 20 chromosomes, so diploids are 14–40. $n = 7$ (*Cucumis sativus*), $n = 11$ (*Citrullus* spp., *Momordica* spp., *Lagenaria* spp., *Sechium* spp., and *Trichosanthes*), $n = 12$ (*Benincasa hispida*, *Coccinia cordifolia*, *Cucumis* spp. other than *C. sativus*, and *Praecitrullus fistulosus*), contains various monoploid (x) chromosome numbers up to $n = 13$ (*Luffa* spp.) and $n = 20$ (*Cucurbita* spp.) (Rai et al. 2008). All Citrullus have 22 chromosomes ($2n = 22$, $x = 11$). In 1930, *C. vulgaris* was proposed to be separated into *lanatus* and *citroides* species by Bailey. The genus *Citrullus* has been taxonomically studied and divided into four species *C. lanatus* (syn. *C. vulgaris*), *C. ecirrhosus*, *C. colocynthis*, and *C. rehmii* (Levi et al. 2001a). *C. ecirrhosus* is closer and similar to *C. lanatus* than *C. colocynthis*. *Praecitrullus fistulosus* from India and Pakistan and *Acanthosicyos naudinianus* from the south of Africa are two other similar species. Gymnopetalum, Lagenaria, Momordica, Trichosanthes, and Melothria, which do not resemble watermelons, are other 22 chromosome members of the Cucurbitaceae family (Wehner 2008).

3.2 Systematic, Origin, and Domestication of Watermelon

The nomenclature and taxonomy of the genus Citrullus have been confusing (Paris 2015). According to Chomicki and Renner (2015), the systematic classification of watermelon species on the basis of order, family, and species is presented below:

Order: Cucurbitales
Family: Cucurbitaceae
Species: *Citrullus lanatus* (Thunb.) Matsumura and Nakai (Syn. *Citrullus vulgaris* Schrader; dessert and edible watermelon)*Citrullus amarus* Schrad. (Syn. *Citrullus lanatus* var. *citroides*) (citron melon)*Citrullus colocynthis* (L.) Schrad. (Colocynth, wild ancestor of watermelon)*Citrullus ecirrhosus* (Cogniaux) (Tendril-less melon)*Citrullus naudinianus* (Sond.) Hooker*Citrullus rehmii* (De Winter) *Citrullus mucosospermus* Fursa (Fursa) (Egusi melon).

Citrullus lanatus (Thunb.) Matsumura and Nakai are the best-known consumed watermelon, its leaves have 3–5 lobes or no lobes, flowers are monoecious or andromonoecious, annual, sweety fruits are from 1–3 to 70–80 kg, seed color is white, cream, green, red, red brown, brown, and black. Some accessions may have a secondary color of testa or hilum and different colors on the edge.

Citrullus amarus Schrad. is citron melon, also called as *Citrullus lanatus* var. *citroides*. *Citrullus amarus* Schrad. can be divided to two as wild (citron melon, tsamma melon, bitter melon, fodder melon) and cultivated (preserving melon, pastèque a confiture, gingérine, citre, jejerine, méréville, pateca, gila) form (Renner et al. 2017). Since the flesh is white, it is used in jam making in some countries (e.g. called pastèque a confiture in France). In recent years, citron watermelon has also been used as an alternative to Cucurbita rootstocks for watermelon (Solmaz et al. 2018; Kombo and Sari 2019; Hussein and Sari 2020).

Citrullus colocynthis (L.) Schrad. leaves are narrow lobed, small, and hairy, flowers are small, monoecious, perennial, fruits are small (< 10 cm) (Fig. 3.3).

Citrullus ecirrhosus (Cogniaux) is a wild perennial and monocieous watermelon (Renner et al. 2017) whose vegetative properties are similar to *C. colocynthis* (L.) Schrad. Its leaves are larger, and the degree of lobing is more deeper than *C. colocynthis* (L.) Schrad. and have no tendril (tendril-less melon).

Citrullus naudinianus (Sond.) Hooker (syn. *Acanthosicyos naudinianus*) is also a wild and perennial type of watermelon. Its leaves are densely covered with fine hairs; flower type is dioecious; fruits are elliptical, 6–12 × 4–8 cm; seeds are white (Jeffrey 1978).

Citrullus rehmii (De Winter) is a wild type of watermelon and is not cultivated (Paris 2015). It has annual and monoecious plants (Renner et al. 2017).

Citrullus mucosospermus Fursa (Fursa) is known as Egusi melon grown as oilseeds in the origin sub-Saharan Western Africa (Jarret et al. 1997; Achigan-Dako et al. 2006; Jensen et al. 2011; Akusu and Kiin-Kabari 2015). It is annual and has monoecious sex type (Renner et al. 2017).

Leaves and young fruits of some *Citrullus* species are given in Figs. 3.1 and 3.2, respectively.

Watermelons are grown in temperate and tropical regions of the world to serve as a source of water and food for animals and humans (Wehner 2008; Paris 2015). The origin of the watermelon is thought to be in South Africa because wild forms grow in these areas, and a rich maximum diversity has been found. It has been cultivated for over 4000 years in Africa and distributed around the world as trade and knowledge developed in central Africa. Citron (*Citrullus lanatus* var. *citroides*) grows in these wild areas and is thought to be related to the wild ancestor of watermelon. Livingstone (1857) reported that watermelons grew unusually in abundance in the Kalahari Desert after heavy rains. *Citrullus colocynthis* (L.) Schrad. is thought to be the wild ancestor of watermelon in North and West Africa. Although the fruits of Colocynth are small, their maximum diameter is 7.5 cm. The flesh is bitter, the seeds are small and brown. Watermelon has been grown in Africa and the Middle East for thousands of years and in China since at least 900 AD. Watermelon was brought to the New World in the 1500s. It was recorded that watermelons were brought to Europe by Spanish seafarers in

Fig. 3.1 Leaves of the some *Citrullus* spp.; *Citrullus lanatus* var. *lanatus* (Kar126), *Citrullus amarus* (Kar325), *Citrullus colocynthis* (Kar318), and *Citrullus rehmii* (Kar330)

Fig. 3.2 Young fruits of some *Citrullus* spp.; *Citrullus lanatus* var. *lanatus* (left), *Citrullus amarus* (middle), and *Citrullus colocynthis* (right)

Cordoba in 961 AD and Seville in 1158 AD. The spread of watermelons in Northern Europe was relatively slow, and cultivation of watermelon in Northern Europe did not become widespread until the end of the sixteenth century due to unsuitable climatic conditions for watermelons in the British Isles. Around this time, watermelons were brought to the New World and were recorded in Massachusetts colonies in the USA in 1629, when the plants were cultivated. The recognition of watermelon in other parts of the world has been through established trade routes (Wehner 2008). Watermelon was brought to North America from Africa and Spain, the number of varieties has increased rapidly in the last 250 years, but it has been determined that most of these varieties have similar genetic bases (Levi et al. 2001a, b).

Watermelon's primary origin center is South Africa, and its wild relatives are also found in West Africa. The secondary center is China, and related species can be found in India. The Middle Eastern regions as well as countries close to the Mediterranean can be good places to collect ancient landraces and wild accessions of *Citrullus* (Levi et al. 2017). For example, there is a history based on very old watermelons in Turkiye and there are many genetic resources, especially in Southeastern Anatolia and Aegean regions. Whitaker reported that *Citrullus colocynthis* was the ancestor of watermelon. Although *Citrullus colocynthis* is morphologically similar to *C. lanatus*, its fruits are bitter and its seeds are very small. Citron is thought to be an intermediate form between a more primitive and bitter form of *C. lanatus* and today's culture variety. Although the *Citrullus* species grows naturally in South and Central Africa, *C. colocynthis* also grows naturally in India. It is thought that watermelon cultivation started in ancient Egypt and India and spread from these countries to Asia and the Near East along the Mediterranean area (Levi et al. 2017).

3.3 Importance of Watermelon Gen Banks

The collection, maintaining, characterization, documentation, and distribution of plant genetic resources are under the responsibilities of plant gene banks for research, education, and breeding purposes (Anglin et al. 2018).

The facilities of National Gene Bank (NGB) for seed collection have been designed for the needs of long-term (− 18 °C), medium-term (0 °C), and short-term (4 °C) storage for multi-species crops and plants. Cryopreservation and in vitro conservation facilities also exist for specific species. Against the loss of genetic resources due to challenges of natural (floods, earthquakes, erosions, etc.) or man-made disasters (such as wars) a gene bank (Svalbard Global Seed Vault) between Norway and North Pole also keeps the seeds under natural conditions with a contract.

There is very important genetic diversity in watermelons in the World. The USDA-ARS, Plant Genetic Resources and Conservation Unit has one of the most important gene banks where watermelon genetic resources have been collected from around the world. There are 500 accessions of *Citrullus lanatus* (Thunb.) Matsum. and Nakai, 133 accessions of *Citrullus amarus* Schrad., 22 accessions of *C. colocynthis* (L.) Shrad., 1 accessions of *C. rehmii* (De Winter) genetic resources from watermelon in this gene bank (https://npgsweb.ars-grin.gov/gringlobal/search). Since Turkiye is an important center between Asia, Europe, and the Middle East, although watermelon is not the primary gene center, especially the Southeastern Anatolia, Aegean, Thrace, and Central Anatolia regions have significant watermelon biodiversity (Sari et al. 2006). Watermelon genetic resources have been collected by the National Plant Genetic Resources and Plant Diversity Research Program (NPGRRP) since 1964 by the Aegean Agricultural Research Institute. Later on, within the framework of the watermelon genetic resource collection program that started in 1993 at Cukurova University, around 400 watermelon genetic resources were collected and morphological (Sari and Solmaz 2018) and molecular characterizations (Solmaz and Sari 2009; Solmaz et al. 2010, 2016) were determined. In the genetic characterization studies, it was determined that watermelons are based on similar genetic bases. South Africa is counted as the homeland of watermelon, and a large number of *Citrullus* spp. has its type. The South African Development Society (SADC), the Center for Plant Genetic Resources (SPGRC), and the National Network of Plant Genetic Resources have prioritized the collection and preservation of watermelon genetic resources (Munyenyembe 2009; Mujaju et al. 2010). Similarly, the National Plant Genetic Resource Center of Namibia, the National Botanical Research Institute, and the Zambia Agricultural Research Institute have collected a large number of *Citrullus* species (McGregor 2012).

3.4 Diversity in Wild, Feral, and Cultivated Watermelons

Phenotypic diversity in watermelons can be distinguished from each other with the help of various descriptors. The most used of these descriptors is the UPOV (International Union for the Protection of New Varieties of Plants) descriptor (UPOV 2019). UPOV descriptor has 38 characters in total, excluding disease and pest resistance. One of these characters are ploidy level (diploid, triploid, and tetraploid), 3 are seedling (1—Cotyledon: size: small, medium, large; 2—Cotyledon: shape: narrow elliptic, medium elliptic, and broad elliptic; 3—Cotyledon: intensity of green color: light, medium, dark); 6 are leaves (1—Leaf blade: size: small, medium, and large; 2—Leaf blade: ratio length/width: low, medium, and high; 3—Leaf blade: color: yellowish green, green, grayish green, and bluish green; 4—Leaf blade: degree of lobing: absent or very weak, weak, medium, strong, and very strong; 5—Leaf blade: blistering: weak, medium, and strong; 6—Leaf blade: color of veins: green and yellow); 18 are fruits (1—Fruit: weight: very low, very low to low, low, low to medium, medium, medium to high, high, high to very high, very high; 2—Fruit: shape in longitudinal section: circular, broad elliptic, medium elliptic, and narrow elliptic; 3—Fruit: depression at base: absent or very shallow, shallow, medium, deep, and very deep; 4—Fruit: Shape of apical part: truncate, truncate to rounded, rounded, rounded to acute, and acute; 5—Fruit: Depression at apex: absent or very shallow, shallow, medium, deep, and very deep; 6—Fruit: Ground color of skin: yellow, very light green, very light green to light green, light green, light green to medium green, medium green, medium green to dark green, dark green, dark green to very dark green, and very dark green; 7—Fruit: Conspicuousness of veining: inconspicuous or very weakly conspicuous, weak, medium, strong, and very strong; 8—Fruit: Pattern of stripes: only one colored, one colored and veins, one colored, veins and marbled, one colored and marbled, two colored, veins and marbled and only veins; 9—Fruit: Width of stripes: very narrow, narrow, medium, broad, and very broad; 10—Fruit: Main color of stripes: yellow, very light green, light green, medium green, dark green, and very dark green; 11—Fruit: conspicuousness of stripes: Weak, medium, strong, and very strong; 12—Fruit: Margin of stripes: diffuse, medium, and sharp; 13—Fruit: Size of insertion of peduncle: small, medium, and large; 14—Fruit: Size of pistil scar: small, medium, and large; 15—Fruit: Grooving: absent or very weak, weak, medium, and strong; 16—Fruit: Waxy layer: absent or very weak, medium, and very strong; 17—Fruit: Thickness of pericarp: very thin, thin, medium, thick, and very thick; 18—Fruit: Main color of flesh: white, yellow, orange, pink, pinkish red, red, and dark red); 7 are seeds (1—Only triploid varieties: Seed coat: Size: small, medium, and large; 2—Only diploid and tetraploid varieties: Fruit: Number of seeds: none or few, medium and many; 3—Only diploid and tetraploid varieties: Seed: Length: very short, short, medium, long, and very long; 4—Only diploid and tetraploid varieties: Seed: Ratio length/width: very low, low, medium, high, and very high; 5—Only diploid and tetraploid varieties: Seed: Ground color of testa: white, cream, green, red, red brown, brown, and black; 6—Only diploid and tetraploid varieties: Seed: Area of over color: Absent or very small, small, medium, large, and very large; 7—Only diploid and tetraploid varieties: Seed: Patches at hilum: absent and present); 1 is about earlines/flowering time (Time of female flowering: early, medium, and late); 2 are resistance to disease (1—Resistance to *Fusarium oxysporum* f. sp. *niveum*: Races 0, 1, and 2: absent and present; 2—Resistance to *Colletotrichum orbiculare* Race 1: absent and present).

The excessive number of characters used for the identification in watermelons led to the development of a minimum descriptor list. While ECP-GR defined 12 characters as the minimum descriptor list (ECP/GR 2008), Sari and Solmaz (2018) used full UPOV criteria in their study with 310 Turkish watermelon germplasms, however they had chosen only 17 of these characters such as the minimum descriptor (Plant growth, Leaf blade: Length, Leaf blade: Width, Leaf: Degree of lobing, Inflorescence: Sex expression, Ovary: Size, Fruit: Weight, Fruit: Shape of longitudinal section, Fruit: Ground color of skin, Fruit: Stripes, Fruit: Width of stripes, Fruit: Thickness of rind, Fruit: Main color of flesh, Fruit: Number of seed, Seed: Size Seed, Ground color of testa, Seed: Secondary color of testa) for the Turkish watermelon genetic resources atlas.

European Cooperative Programme for Plant Genetic Resources (ECP/GR) descriptor is also provided as a minimum list of descriptors (ECP/GR 2008) for cultivated watermelon (*Citrullus lanatus*) with just 12 characters (1—Plant growth habit: ushy and runner; 2—Leaf blade: degree of secondary lobing: weak, intermediate, and strong; 3—Hermaphroditic flowers: absent and present; 4—Fruit weight [kg]: To be recorded at maturity stage; 5—Fruit shape: flattened, round, broad elliptical, elliptical, pyriform, and oblong; 6—Predominant (or ground) fruit skin color: light green, edium green, dark green, white, yellow, brown, other; 7—Secondary fruit skin color pattern: no secondary fruit skin color, solid, striped, spotted, mixed, other; 8—Fruit skin stripe color: light green, medium green, dark green, white, ellow, brown, other; 9—Flesh color: red, pink, canary yellow, salmon yellow, white, mixed, orange, green, other; 10—Thickness of pericarp as mm: measured at maturity stage; 11—Distribution of grooves: absent, at basal half, at apical half, on whole fruit; 12—Fruit bitterness: absent, slightly bitter, bitter).

Some leaf, fruit, and seed characteristics used in morphological characterization of watermelons are presented in Figs. 3.3, 3.4, 3.5, 3.6, 3.7, 3.8, 3.9, and 3.10, respectively.

Fig. 3.3 Leaf: degree of lobing; weak (left), medium (middle), strong (right)

Fig. 3.4 Fruit shapes of some watermelon cultigens: round (left), elliptical (second from left), broad elliptical (second from right), and cylindrical (right)

Fig. 3.5 Fruit: ground color of skin; white (left), yellow (middle), green (right)

Fig. 3.6 Green fruit skin colors: solid dark green (left), solid intermediate green (middle), solid light green (right)

Morphological characterization studies of *Citrullus* were started in 1977 by Oyulu. Since that time germplasms from various origins (Nigeria, Botswana, Sudan, USA, Germany, Netherlands, Romania, Yugoslavia, Israel, Turkiye, Libya, Taiwan, Korea, Namibia, Hungary, Brazil, Corsica, Kenya, Mali, EU, Morocco, Benin, Ivory Coast, France, Ghana, Nigeria, Togo, and Tunisia) comprising different *Citrullus* species as *C. lanatus*, *Citrullus* spp. *C. mucosospermus*, and *C. amarus* were characterized by Taylor (1985), El Mekki (1991, 1992), Krasteva (2000), Maggs-Kolling et al. (2000), Maggs-Kolling and Christiansen (2003), Nagy (2005), Wehner et al. (2001), da Silva et al. (2006), Goda (2007), Idehen et al. (2007), Laghetti and Hammer (2007), Sari et al. (2006, 2008), Szamosi et al. (2008, 2009), Huh et al.

Fig. 3.7 Fruit: width of stripes: absent (upper left), very narrow (upper middle), narrow (upper right), medium (below left), broad (below middle), and very broad (below right)

Fig. 3.8 Fruit skin and flesh color of some Turkish watermelon germplasms

Fig. 3.9 Seed characters of watermelon accessions

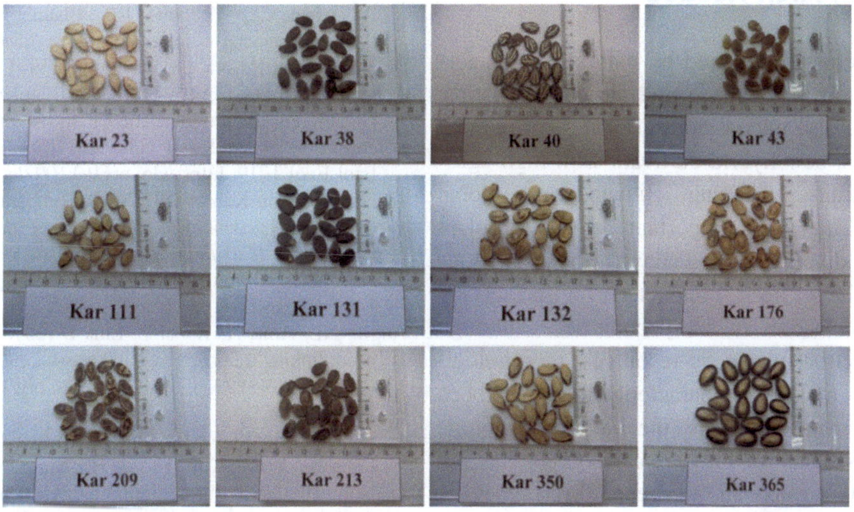

Fig. 3.10 Some Turkish large snack seed watermelon accessions

(2008, 2014), Solmaz and Sari (2009), Elbekkay et al. (2009, 2021), Gichimu et al. (2009), Jensen et al. (2011), Nantoumé et al. (2012), Solmaz et al. (2012), Achigan-Dako et al. (2015), Hakimi and El Madidi (2015), Gbotto et al. (2016), and Assefa et al. (2020).

Molecular characterization studies in the genus *Citrullus* were started by Zamir et al. (1984) Researchers used isozyme markers to reveal the genetic diversity of 44 accessions of *C. lanatus*, *C. colocynthis* from Israel. Isozymes were also applied by Navot and Zamir (1987) and Biles et al. (1989). Later on RAPD marker systems were exploited in genetic diversity studies for many years by several researchers as Zhang et al. (1994), Lee et al. (1996), Levi et al. (2000, 2001a), da Silva et al. (2006), Goda (2007), Levi and Thomas (2007), Mujaju et al. (2010), Solmaz et al. (2010), and Elbekkay et al. (2021). In parallel with the development of marker technology different marker systems as SSR (Levi et al. 2000; Kwon et al. 2007, 2010; Nimmakayala et al. 2010; Minsart et al. 2011; Mujaju et al. 2011; Gama et al. 2013; Nantoumé et al. 2013; Zhang et al. 2016a, b; Zhu et al. 2016; Solmaz et al. 2016; Mashilo et al. 2017), ISSR (Levi et al. 2004; Pandey et al. 2019), AFLP (Che et al. 2003; Levi et al. 2004;

Nimmakayala et al. 2010; Hwang et al. 2011a), Ms-AFLP (Nimmakayala et al. 2011), cpDNA (Dane and Lang 2004; Dane and Liu 2007; Achigan-Dako et al. 2021), EST-PCR (Levi et al. 2008), EST-SSR (Verma and Arya 2008; Hwang et al. 2011a, b; Kwon 2013; Mujaju et al. 2013), SRAP (Uluturk et al. 2011; Solmaz et al. 2016), HFO-TAG (Levi et al. 2013), CAPS (Liu et al. 2016), DAtTseq-based SNPs (Yang et al. 2016), SNPs (Nimmakayala et al. 2014; Joshi et al. 2019; Lee et al. 2019; Maragal et al. 2019; Wu et al. 2019; Cui et al. 2020).

3.5 Conclusions and Future Research Needed

Genetic erosion occurred by natural causes, forest fire, misuse of forest and agricultural fields, excessive and unconscious use of fertilizer and pesticides, new hybrid cultivars with high agronomic traits, changing climate, and global warming are some important threats for plant genetic diversity. Nowadays, many serious studies have been conducted for preserving genetic resources in many countries. Although the collection of genetic resources, morphological, and molecular characterizations has been done in many countries, the biotic and abiotic stress resistance of these rare resources has not been investigated in all genetic resources of every country yet.

Some of the watermelon genetic resources have been reported to be resistant to anthracnose, gummy stem blight, powdery mildew, downy mildew, Fusarium wilt which are the important fungal disease, *Acidovorax avenae* subsp. *citrulli* has bacterial disease, some accessions against viruses (Zucchini yellow mosaic virus, Squash vein yellowing virus, the watermelon strain of Papaya ringspot virus) and insects such as root-knot nematodes, whiteflies, and mites. It is recommended that the characterization of genetic resources collected by various gene banks of the world be completed, their biotic and abiotic stress resistance/tolerance status should be revealed and directed to new cultivar improvement.

Much progress has been made in describing morphological traits of watermelon. However, biotic and abiotic tolerance/resistance studies of all genetic resources have not been completed. These studies should be researched and used in new variety development studies.

References

Achigan-Dako EG, Enoch G, Fanou N, Kouke A, Avohou H, Vodouhe RS, Ahanchede A (2006) Evaluation agronomique de trois espèces de Egusi (Cucurbitaceae) utilisées dans l'alimentation au Béninet élaboration d'un modèle de prédiction du rendement. Biotechnol Agron Soc Environ 10:121–129

Achigan-Dako EG, Avohou ES, Linsoussi C, Ahanchede A, Vodouhe RS, Blattner FR (2015) Phenetic characterization of egusi-type (*C. mucosospermus*). Genet Resour Crop Evol 62:1159–1179

Achigan-Dako EG, Degbey H, Hale I, Blattner FR (2021) Georeferenced phylogenetic analysis of a global collection of wild and cultivated *Citrullus* species. Ecol Evol 1–19. https://doi.org/10.1002/ece3.7189

Akusu MO, Kiin-Kabari DB (2015) Comparative studies on the physicochemical and sensory properties of watermelon (*Citrullus lanatus*) and melon (*Citrullus vulgaris*) seed flours used in "egusi" soup preparation. J Food Res 4:1–6

Anglin NL, Amri A, Kehel Z, Ellis D (2018) A case of need: linking traits to genebank accessions. Biopreserv Biobank 16(5):337–349. https://doi.org/10.1089/bio.2018.0033

Assefa AD, Hur OS, Ro NY, Lee JE, Hwang AJ, Kim BS, Rhee JH, Yi JY, Kim JH, Lee HS, Sung JS, Kim MK, Noh JJ (2020) Fruit morphology, Citrulline and Arginine levels in diverse watermelon (*Citrullus lanatus*) germplasm collections. Plants 9:1054. https://doi.org/10.3390/plants9091054

Biles CL, Martyn RD, Wilson HD (1989) Isozymes and general proteins from various watermelon cultivars and tissue types. Hortic Sci 29:810–812

Che KP, Wang CY, Wang YG, Jin DM, Wang B, Xu Y, Kang GB, Zhang HY (2003) Genetic assessment of watermelon germplasm using the AFLP technique. Hortic Sci 38:81–84

Chomicki G, Renner SS (2015) Watermelon origin solved with molecular phylogenetics including Linnaean material: another example of museomics. New Phytol 205:526–532. https://doi.org/10.1111/nph.13163

Cui H, Ding Z, Zhu Q, Wu Y, Gao P (2020) Population structure and genetic diversity of watermelon (*Citrullus lanatus*) based on SNP of chloroplast genome. 3 Biotech 10:374. https://doi.org/10.1007/s13205-020-02372-5

Dane F, Lang P (2004) Sequence variation at cpDNA regions of watermelons and related wild species:

implications for evolution of *Citrullus* haplotypes. Am J Bot 91:1922–1929
Dane F, Liu J (2007) Diversity and origin of cultivated and citron type watermelon (*Citrullus lanatus*). Genet Resour Crop Evol 54:1255–1265
da Silva ML, Queiros MA, Ferreira MAJF, Busco GSC (2006) Morphological and molecular characterization of watermelon. Hortic Bras 24:405–409
ECP/GR (2008) https://www.ecpgr.cgiar.org/fileadmin/templates/ecpgr.org/upload/NW_and_WG_UPLOADS/Cucurbits_DescriptorLists.pdf
Elbekkay M, Laarayedh L, Lamari R, Hamza H, Ferchichi A (2009) Characterization of several local cultivars of watermelon collected from arid region in Tunisia. J Arid Land Stud 19:2005–2008
Elbekkay M, Hamza H, Neily MH, Djebali N, Ferchichi A (2021) Characterization of watermelon local cultivars from Southern Tunisia using morphological traits and molecular markers. Euphytica 217:74. https://doi.org/10.1007/s10681-021-02809-9
El Mekki MMD (1991) Effect of variety and some cultural practices on yield, quality, cracking and blossom-end rot of watermelon (*Citrullus lanatus*). M.Sc. thesis. University of Gezira, Faculty of Agricultural Science, Sudan
El Mekki MMD (1992) Blossum-end rot (BER) and cracking in watermelon *Citrullus lanatus* (Thunb.) Matsum & Nakai. Cucurb Genet Coop Rep 15:69–70
FAOSTAT (2020) http://www.fao.org/faostat/en/#data/QC (date: 24 Feb 2023)
Gama RNCS, Santos CAF, de Dias CS, Souza FF (2013) Molecular characterization of watermelon cultivars using microsatellite markers. Hortic Bras 31:522–527
Gbotto AA, Koffi KK, Fouha Bi ND, Doubi BIST, Tro HH, Baudooin JP, Zoro Bi IA (2016) Morphological diversity in oleaginous watermelon (*Citrullus mucosospermus*) from Nangui Abrogoua University germplasm collection. Afr J Biotechnol 15: 917–929
Gichimu BM, Owuor BO, Mwai GN, Dida MM (2009) Morphological characterization of some wild and cultivated watermelon (*Citrullus* spp.) accessions in Kenya. J Agric Biol Sci 4:10–18
Goda M (2007) Diversity of local genetic resources of watermelon *Citrullus lanatus* (Thunb.) Matsum and Nakai in Sudan. M.Sc. thesis, Swedish Biodiversity Center, No. 35
Hakimi F, El Madidi S (2015) Variability of agro-morphological traits in some Moroccan watermelon landraces (*Citrullus lanatus* Thunb Matsum & Nakai). Int J Curr Sci 17:90–96
Huh YC, Solmaz I, Sari N (2008) Morphological characterization of Korean and Turkish watermelon germplasm. In: Pitrat M (ed) Cucurbitaceae 2008. Proceedings of the IXth EUCARPIA meeting on genetics and breeding of Cucurbitaceae. INRA, Avignon, pp 327–333
Huh YC, Choi HS, Solmaz I, Sari N, Kim S (2014) Morphological characterization of Korean and Turkish watermelon germplasm. Korean J Agric Sci 41 (4):309–314. https://doi.org/10.7744/cnujas.2014.41.4.309
Hussein S, Sari N (2020) Effects of different rootstocks on seed yield and quality of triploid watermelon grown in greenhouse. In: Sari N et al (eds) Acta Horticulturae, vol 1282, pp 67–74
Hwang J, Kang J, Son B, Kim K, Park Y (2011a) Genetic diversity in watermelon cultivars and related species based on AFLPs and EST-SSRs. Not Bot Horti Agrobot 39:285–292
Hwang JH, Ahn SG, Oh JY, Choi YW, Kang JS, Park YH (2011b) Functional characterization of watermelon (*Citrullus lanatus*) EST-SSR by gel electrophoresis and high resolution melting analysis. Sci Hortic 130:715–724
Idehen EO, Kehinde OB, Ariyo OJ (2007) Numerical analysis of variation among Nigerian accessions of 'egusi' melon (*Citrullus lanatus* (Thunb.) Matsum & Nakai). J Am Sci 3:7–15
Jarret RL, Merrick LC, Holms T, Evans J, Aradhya MK (1997) Simple sequence repeats in watermelon [*Citrullus lanatus* (Thunb.) Matsum & Nakai]. Genome 40:433–441
Jeffrey C (1978) Cucurbitaceae. In: Flora Zambesiaca, vol 4
Jensen BD, Touré FM, Hamattal MA, Touré FA, Nantoume AD (2011) Watermelons in the sand of Sahara: cultivation and use of indigenous landraces in the Tombouctou region of Mali. Ethnobot Res Appl 9:151–162
Joshi V, Shinde S, Nimmakayala P, Abburi VL, Alaparthi SB, Lopez-Ortiz C, Levi A, Panicker G, Reddy UK (2019) Haplotype networking of GWAS hits for Citrulline variation associated with the domestication of watermelon. Int Mol Sci 20:5392. https://doi.org/10.3390/ijms20215392
Kombo MD, Sari N (2019) Rootstock effects on seed yield and quality in watermelon. Hortic Environ Biotechnol 60:303–312
Krasteva L (2000) Watermelon genetic resources in Bulgaria. Acta Hortic 510:253–256
Kwon YS (2013) Use of EST-SSR markers for genetic characterization of commercial watermelon varieties and hybrid seed purity testing. Seed Sci Technol 41:245–256
Kwon YS, Park EK, Lee WS, Yi SI, Bae KM, An JS, Kim HY (2007) Genetic assessment of watermelon (*Citrullus lanatus*) varieties using SSR markers developed from Cucurbit species. Korean J Genet 29:137–146
Kwon YS, Oh YH, Yi SI, Kim HY, An JM, Yang SG, Ok SH, Shin JS (2010) Informative SSR markers for commercial variety discrimination in watermelon (*Citrullus lanatus*). Genes Genom 32:115–122
Laghetti G, Hammer K (2007) The Corsican citron melon (*Citrullus lanatus* (Thunb.) Matsum & Nakai sub sp. *lanatus* var. *citroides* (Bailey) Mansf. ex Greb.) a traditional and neglected crop. Genet Resour Crop Evol 54:913–916

Lee SJ, Shin JS, Par KW, Hong YP (1996) Detection of genetic diversity using RAPD-PCR and sugar analysis in watermelon [*Citrullus lanatus* (Thunb.) Mansf.] germplasm. Theor Appl Genet 92:719–725

Lee KJ, Lee JR, Sebastin R, Shin MJ, Kim SH, Cho GT, Hyun DY (2019) Genetic diversity assessed by genotyping by sequencing (GBS) in watermelon germplasm. Genes 10:822. https://doi.org/10.3390/genes10100822

Levi A, Thomas CE (2007) DNA markers from different linkage regions of watermelon genome useful in differentiating among closely related watermelon genotypes. Hortic Sci 42:210–214

Levi A, Thomas CE, Keinath AP, Wehner TC (2000) Estimation of genetic diversity among *Citrullus* accessions using RAPD markers. Acta Hortic 510:385–390

Levi A, Thomas CE, Keinath AP, Wehner TC (2001a) Genetic diversity among watermelon (*Citrullus lanatus* and *Citrullus colocynthis*) accessions. Genet Resour Crop Evol 48:559–566

Levi A, Thomas CE, Wehner TC, Zhang X (2001b) Low genetic diversity indicates the need to broaden the genetic base of cultivated watermelon. Hortic Sci 36:1096–1101

Levi A, Thomas CE, Newma M, Reddy OUK, Zhang X, Xu Y (2004) ISSR and AFLP markers differ among American watermelon cultivars with limited genetic diversity. J Am Soc Hortic Sci 129:553–558

Levi A, Wechter P, Davis A (2008) EST-PCR markers representing watermelon fruit genes are polymorphic among watermelon heirloom cultivars sharing a narrow genetic base. Plant Genet Res 7:16–32

Levi A, Thies JA, Wechter WP, Harrison HF, Simmons AL, Reddy UK, Nimmakayala P, Fei Z (2013) High frequency oligonucleotides: targeting active gene (HFO-TAG) markers revealed wide genetic diversity among *Citrullus* spp. accessions useful for enhancing disease or pest resistance in watermelon cultivars. Genet Resour Crop Evol 60:427–440

Levi A, Coffey J, Massey L, Guner N, Oren E, Tadmor Y, Ling K (2016) Resistance to papaya ringspot virus-watermelon strain (PRSV-W) in the desert watermelon *Citrullus colocynthis*. Hortic Sci 51:4–7

Levi A, Jarret R, Kousik S, Wechter WP, Nimmakayala P, Reddy UK (2017) Genetic resources of watermelon. In: Grumet R et al (eds) Genetics and genomics of Cucurbitaceae. Plant genetics and genomics: crops and models. Springer International Publishing AG, p 87. https://doi.org/10.1007/7397_2016_34. Published Online: 23 Mar 2017

Liu S, Gao P, Zhu Q, Luan F, Davis AR, Wang X (2016) Development of cleaved amplified polymorphic sequence markers and a CAPS based genetic linkage map in watermelon (*Citrullus lanatus* [Thunb.] Matsum & Nakai) constructed using whole-genome re-sequencing data. Breed Sci 66:244–259. https://doi.org/10.1270/jsbbs.66.244

Livingstone JD (1857) Livingstone's missionary travels manuscript. https://livingstoneonline.org/in-his-own-words/missionary-travels-manuscript/livingstones-missionary-travels-manuscript-1857

Maggs-Kolling GL, Christiansen JL (2003) Variability in Namibian landraces of watermelon (*Citrullus lanatus*). Euphytica 132:251–258

Maggs-Kolling GL, Madsen S, Christiansen JL (2000) A phenetic analysis of morphological variation in *Citrullus lanatus* in Namibia. Genetic Resour Crop Evol 47:385–393

Maragal S, Lakshmana RES, Reddy DC (2019) Genetic analysis of fruit quality traits in prebred lines of watermelon derived from a wild accession of *Citrullus amarus*. Euphytica 215:199. https://doi.org/10.1007/s10681-019-2527-x

Mashilo J, Shimelis H, Odindo AO, Amelework B (2017) Genetic diversity and differentiation in citron watermelon [*Citrullus lanatus* var. *citroides*] landraces assessed by simple sequence repeat markers. Sci Hortic 21:99–106

McGregor C (2012) *Citrullus lanatus* germplasm of Southern Africa. Isr J Plant Sci 60:403–414

Minsart LA, Zoro BI, Dje Y, Baudoin JP, Jacquemart AL, Bertin P (2011) Set up of simple sequence repeat markers and first investigation of the genetic diversity of West-African watermelon (*Citrullus lanatus* ssp. *vulgaris* oleaginous type). Genet Resour Crop Evol 58:805–814

Mujaju C, Sehic J, Werlemark G, Garkava-Gustavsson L, Faith MH (2010) Genetic diversity in watermelon (*Citrullus lanatus*) landraces from Zimbabwe revealed by RAPD and SSR markers. Hereditas 147:142–153

Mujaju C, Sehic J, Werlemark G, Garkava-Gustavsson L, Andersen SB, Nybom H (2011) Genetic diversity among and within (*Citrullus lanatus*) landraces in Southern Africa. J Hortic Sci Biotechnol 86:353–358

Mujaju C, Sehic J, Nybom H (2013) Assessment of EST-SSR markers for evaluating genetic diversity in watermelon accessions from Zimbabwe. Am J Plant Sci 4:1448

Munyenyembe P (2009) Conservation of plant genetic resources of southern Africa: prospects and challenges. In: The 36th session of the food and agricultural organization (FAO) of the United Nations conference, Rome

Nagy J (2005) Description of the important cultivated watermelon varieties (in Hungarian). In: Nagy J (ed) Melon and watermelon. Szaktudas Kiado Haz Rt, Budapest, pp 114–131

Nantoumé AD, Traore S, Christiansen JL, Andersen SB, Jensen BD (2012) Traditional uses and cultivation of indigenous watermelons (*Citrullus lanatus*) in Mali. Int J Biodivers Conserv 4:461–471

Nantoumé AD, Andersen S, Jensen B (2013) Genetic differentiation of watermelon landrace types in Mali revealed by microsatellite markers. Genet Resour Crop Evol 2129–2141

Navot N, Zamir D (1987) Isozyme and seed protein phylogeny of the genus *Citrullus* (Cucurbitaceae). Plant Syst Evol 156:61–67

Nimmakayala P, Tomason YR, Jeong J, Ponniah SK, Karunathilake A, Levi A, Perumal R, Reddy UK (2010) Genetic reticulation and interrelationships among *Citrullus* species as revealed by joint analysis of shared AFLPs and species-specific SSR alleles. Plant Genet Res 8:16–25

Nimmakayala P, Vajja G, Gist RA, Tomason YR, Levi A, Reddy UK (2011) Effect of DNA methylation on molecular diversity of watermelon heirlooms and stability of methylation-specific polymorphisms across the genealogies. Euphytica 177:79–89

Nimmakayala P, Levi A, Abburi L, Abburi VL, Tomason YR, Saminathan T, Vajja VG, Malkaram S, Reddy R, Wehner TC, Mitchell SE, Reddy UK (2014) Single nucleotide polymorphisms generated by genotyping by sequencing to characterize genome-wide diversity, linkage disequilibrium and selective sweeps in cultivated watermelon. BMC Genom 15:767

Pandey A, Khan MK, Isik R, Turkmen O, Acar R, Seymen M, Hakki EE (2019) Genetic diversity and population structure of watermelon (*Citrullus* sp.) genotypes. 3 Biotech 9:210. https://doi.org/10.1007/s13205-019-1736-2

Paris HS (2015) Origin and emergence of the sweet dessert watermelon, *Citrullus lanatus*. Ann Bot 116 (2):133–148. https://doi.org/10.1093/aob/mcv077

Rai M, Pandey S, Kumar S (2008) Cucurbit research in India: a retrospect. In: Pitrat M (ed) Proceedings of the IXth EUCARPIA meeting on genetics and breeding of Cucurbitaceae, 21–24 May. INRA, Avignon (France), pp 285–293

Renner SS, Sousa A, Chomicki G (2017) Chromosome numbers, Sudanese wild forms and classification of the watermelon genus *Citrullus*, with 50 names allocated to seven biological species. Taxon 66 (6):1393–1405. https://doi.org/10.12705/666.7

Sari N, Solmaz I (2018) Turkey watermelon and melon genetic resources. 72 Tasarim Ltd. Şti. Press, Ankara, 280 p. ISBN: 978-605-2353-62-2

Sari N, Solmaz I, Yetisir H, Unlu H (2006) Watermelon genetic resources in Turkey and their characteristics. Acta Hortic 731:433–438

Sari N, Tan A, Yanmaz R, Yetisir H, Balkaya A, Solmaz I, Aykas L (2008) General status of cucurbit genetic resources in Turkey. In: Pitrat M (ed) Proceedings of the IXth EUCARPIA meeting on genetics and breeding of Cucurbitaceae. INRA, Avignon, pp 21–32

Solmaz I, Sari N (2009) Characterization of watermelon (*Citrullus lanatus*) accessions collected from Turkey for morphological traits. Genet Resour Crop Evol 56:173–188

Solmaz I, Sari N, Aka-Kacar Y, Yalcin-Mendi NY (2010) The genetic characterization of Turkish watermelon (*Citrullus lanatus*) accessions using RAPD markers. Genet Resour Crop Evol 57:763–771

Solmaz I, Sari N, Kartal E, Yetisir H (2012) Seed characteristics and seed-fruit correlation of Turkish watermelon germplasm. In: Sari N, Solmaz I, Aras N (eds) Proceedings of the Xth EUCARPIA meeting on genetics and breeding of Cucurbitaceae. Antalya, pp 340–345

Solmaz I, Aka Kacar Y, Sari N, Simsek O (2016) Genetic diversity within Turkish watermelon [*Citrullus lanatus* (Thunb.) Matsumura & Nakai] accessions revealed by SSR and SRAP markers. Turk J Agric For 40:407–419. https://doi.org/10.3906/tar-1511-26

Solmaz I, Sari N, Kombo MD, Simsek I, Hussein S, Namlı M (2018) Rootstock capacity in improving production and quality of triploid watermelon seeds. Turk J Agric For 42:298–308

Szamosi C, Solmaz I, Sari N (2008) Seed characteristics of Hungarian and Turkish watermelon genotypes. In: Proceedings of the 3rd Turkish seed congress, Capadocia, 25–28 June 2008, pp 149–154

Szamosi C, Solmaz I, Sari N, Barsony C (2009) Morphological characterization of Hungarian and Turkish watermelon (*Citrullus lanatus* (Thunb.) Matsum & Nakai) genetic resources. Genet Resour Crop Evol 56:1091–1105

Taylor FW (1985) The potential for the commercial utilization of indigenous plants in Botswana. In: Wickens GE, Goodin JR, Fields DV (eds) Plants for arid lands. George Allen & Unwin, London, pp 232–241

Uluturk ZI, Frary A, Doganlar S (2011) Determination of genetic diversity in watermelon (*Citrullus lanatus* (Thunb.) Matsum & Nakai) germplasm. Aust J Crop Sci 5:1832–1836

UPOV (2019) https://www.upov.int/edocs/tgdocs/en/tg142.pdf

Verma M, Arya L (2008) Development of EST-SSRs in watermelon (*Citrullus lanatus*) and their transferability to *Cucumis*. J Hortic Sci Biotechnol 83:732–736

Wehner TC (2008) Watermelon. In: Prohens J, Nuez F (eds) Handbook of plant breeding; vegetables I, pp 381–418

Wehner TC, Shetty NV, Elmstrom GW (2001) Breeding and seed production. In: Maynard DN (ed) Watermelons. Characteristics, production and marketing. ASHS Press, Alexandria, pp 21–73

Wu S, Wang X, Reddy U, Sun H, Bao K, Gao L, Mao L, Patel T, Ortiz C, Abburi VL, Nimmakayala P, Branham S, Wechter P, Massey L, Ling KS, Kousik C, Hammar SA, Tadmor Y, Portnoy V, Gur A, Katzir N, Guner N, Davis A, Hernandez AG, Wright CL, McGregor C, Jarret R, Zhang X, Xu Y, Wehner TC, Grumet R, Levi A, Fei Z (2019) Genome of 'Charleston Gray' the principal American watermelon cultivar and genetic characterization of 1,365 accessions in the U.S. National Plant Germplasm System watermelon collection. Plant Biotechnol J 17:2246–2258. https://doi.org/10.1111/pbi.13136

Yang X, Ren R, Ray R, Xu J, Li P, Zhang M, Liu G, Yao X, Kilan A (2016) Genetic diversity and population structure of core watermelon (*Citrullus lanatus*) genotypes using DArTseq-based SNPs. Plant Genet Resour 14:226–233

Zamir D, Navot N, Rudich J (1984) Enzyme polymorphism in *Citrullus lanatus* and *C. colocynthis* in Israel and Sinai. Plant Syst Evol 146:163–170

Zhang XP, Rhodes BB, Skorupska H (1994) RAPD molecular markers in watermelon. Cucurb Genet Coop Rep 17:116–119

Zhang H, Fan J, Guo S, Ren Y, Gong G, Zhang J (2016a) Genetic diversity population structure and formation of a core collection of 1197 *Citrullus* accessions. Hortic Sci 51:23–29

Zhang J, Guo S, Ren Y, Zhang H, Gong G, Zhou M, Wand G, Zong M, He H, Liu F, Xu Y (2016b) High-level expression of a novel chromoplast phosphate transporter ClPHT4; 2 is required for flesh color development in watermelon. New Phytol. https://doi.org/10.1111/nph.14257

Zhao R (2015) A history of food culture in China. SCPG Publishing Co., New York

Zhu H, Song P, Koo DH, Guo L, Li Y, Sun S, Weng Y, Yang L (2016) Genome wide characterization of simple sequence repeats in watermelon genome and their application in comparative mapping and genetic diversity analysis. BMC Genom 17:557. https://doi.org/10.1186/s12864-016-2870-4

4 The NLR Family of Disease Resistance Genes in Cultivated Watermelon and Other Cucurbits: Opportunities and Challenges

Andres Salcedo, Camilo H. Parada-Rojas, Rafael Guerrero, Madison Stahr, Kimberly N. D'Arcangelo, Cecilia McGregor, Chandrasekar Kousik, Todd Wehner, and Lina M. Quesada-Ocampo

4.1 Introduction

Watermelon (*Citrullus lanatus*) is a fruit crop in the family Cucurbitaceae, which also includes other important vegetables such as cucumber, pumpkin, gourds, and other types of melon. China is the largest producer and accounts for 62% of the world's watermelons (FAO 2019). The United States (US) is one of the top ten watermelon producers and consumers of watermelon, and most watermelon production in the US occurs in California and the southern states of Florida, Georgia, Texas and North Carolina due to the warm environment and sandy soils, which are ideal growing conditions for this crop (United States Department of Agriculture, National Agricultural Statistics Service 2020; Wehner 2008). Watermelon is an annual crop and is planted in mid-April–June in the southern US. Transplants are grown for 4–6 weeks in the greenhouse before being transplanted in the field. Watermelons are commonly affected by different pathogens in the greenhouse and field during a growing season, and integrated management approaches are necessary to achieve high yields and prevent losses to disease (Cohen et al. 2007).

There are many cultivars of watermelon grown in the US including seeded and seedless varieties. In the past, most watermelons produced were diploid seeded watermelons, but triploid seedless varieties have become popular in recent years (Levi et al. 2017). While these varieties have a higher profit margin, they also have increased production costs, different cultivation practices, and are often more susceptible to disease. For example, diploid cultivars are directly seeded, whereas more expensive triploid and hybrid cultivars are often transplanted from a greenhouse. From an economic study published in 2006, directly seeding an open-pollinating diploid cultivar costs \$0.01–\$0.02 per plant,

A. Salcedo · C. H. Parada-Rojas · M. Stahr · K. N. D'Arcangelo · L. M. Quesada-Ocampo (✉)
Department of Entomology and Plant Pathology, North Carolina State University, Raleigh, NC 27695, USA
e-mail: lmquesad@ncsu.edu

R. Guerrero
Department of Biological Sciences, North Carolina State University, Raleigh, NC 27695, USA

C. McGregor
Department of Horticulture and Institute of Plant Breeding, Genetics and Genomics, University of Georgia, Athens, GA 30602, USA

C. Kousik
USDA, ARS, U.S. Vegetable Laboratory, Charleston, SC 29414, USA

T. Wehner
Department of Horticultural Science, North Carolina State University, Raleigh, NC 27695, USA

S. Kr. Dutta et al. (eds.), *The Watermelon Genome*, Compendium of Plant Genomes,
https://doi.org/10.1007/978-3-031-34716-0_4

whereas a triploid variety costs $0.28 (Taylor et al. 2008). Nonetheless, consumer preference continues to shift the market toward seedless watermelon production, making breeding disease resistance into these preferred varieties a priority for private and public breeding programs (Marr and Gast 1991).

4.2 Important Diseases in Watermelon Production and Their Management

Commercial watermelon varieties are very susceptible to diseases caused by virtually all microbial pathogens (fungi, oomycetes, bacteria, viruses, nematodes), which can affect any part of the plant at any time during the production cycle. Diseases affecting watermelon fruits result in unmarketable or prone to decay products, and foliar diseases reduce fruit quality by exposing the fruits to sunscald or affecting ripening and sugar accumulation. Therefore, strategies to control diseases such as Fusarium wilt, gummy stem blight, bacterial fruit blotch, anthracnose, powdery mildew, *Phytophthora* fruit rot, and downy mildew (Fig. 4.1) are critical for researchers and product development specialists (Kousik et al. 2016).

Anthracnose caused by *Colletotrichum orbiculare* [(Berk. and Mont.) Arx] represents one of the most economically important diseases of watermelon worldwide (Kousik et al. 2016). This ascomycete fungus is capable of infecting all parts of the plant under favorable environmental conditions (Keinath 2017a) and is considered a threat for watermelon production in recent years (Keinath 2018; Kousik et al. 2016). Characteristic symptoms of anthracnose include foliar, stem, and fruit lesions. Dark water-soaked lesions render fruits unmarketable under high disease pressure (Keinath 2017a). *C. orbiculare* exhibits a hemibiotrophic lifestyle, characterized by a brief biotrophic phase followed by a switch to a destructive necrotrophic phase as the infection advances, and is considered one of the most destructive fungal pathogens (Dean et al. 2012; Gan et al. 2013).

Extended rain events, morning dews, and foliar wetness caused by overhead irrigation promote development of disease in the field. Conidia are mainly dispersed by rain splashing, but also by wind and farm equipment (Keinath 2017a). The fungus can overwinter by surviving on the debris of infected plants perpetuating its disease cycle (Rennberger et al. 2018). Watermelon growers in the US rely on routine fungicide applications of strobilurin or quinone outside inhibitor (QoI) fungicides (The Fungicide Resistance Action Committee, FRAC group 11) to manage anthracnose throughout the growing season. Fungicides can be applied upon symptoms detection on an interval of 5–10 days. If disease severity is high or environmental conditions are conducive for disease, growers apply fungicides on a shorter application interval (Quesada-Ocampo 2018a). Despite the efficacy of these fungicides in controlling anthracnose, repeated applications within a single growing season exert an economic toll on farmers' budgets and may impose strong selection pressures on pathogens toward fungicide resistance (Michelmore et al. 2017). Field management practices against anthracnose also include the use of host resistance. Multiple commercial varieties and watermelon wild relatives claim intermediate to high levels of resistance against *C. orbiculare* races (Patel 2019). However, highly desirable watermelon seedless cultivars remain untested for resistance to *C. orbiculare* despite this disease being a top research priority in watermelon (Kousik et al. 2016).

Gummy stem blight (GSB) is a disease caused by three ascomycete fungal *Stagonosporopsis* species: *S. citrulli*, *S. cucurbitacearum,* and *S. caricae* (Stewart et al. 2015). GSB is very devastating in transplant greenhouses and major watermelon-producing areas worldwide (Basım et al. 2016; Brewer et al. 2015; Mahapatra et al. 2020). In the southeastern US, GSB represents an important foliar pathogen capable of infecting all above-ground parts of the watermelon resulting in an average yield loss of 43% in unmanaged fields (Rennberger et al. 2018). Under hot and humid environments, GBS on watermelon induces symptoms that include necrotic lesions on the

Fig. 4.1 Foliar, stem, and fruit symptoms of several watermelon diseases. **a** Anthracnose caused by fungus *Colletotrichum orbiculare*. **b** Gummy stem blight caused by *Stagonosporopsis* spp. **c** Bacterial fruit blotch caused by the bacterium *Acidovorax citrulli* (photograph Gerald Holmes, Strawberry Center, California Polytechnic State University, San Luis Obispo). **d** *Fusarium* wilt caused by the fungus *Fusarium oxysporum* formae speciales *Niveum* (FON). **e** *Phytophthora* fruit rot caused by the oomycete *Phytophthora capsica*. **f** Downy mildew caused by the oomycete *Pseudoperonospora cubensis*. **g** Powdery mildew caused by the fungi *Podosphaera xanthii* (syn. *Sphaerotheca fuliginea*) and *Erysiphe cichoracearum*

leaf and petioles, stem canker, and extensive defoliation (Quesada-Ocampo 2018b). GSB can also result in fruit lesions, but fruit infection is not commonly observed in watermelon. The fungus can overwinter on plant debris, so deep-turning the debris after harvest is one cultural practice to manage GSB (Keinath 2002). Genetic resistance to the causal *Stagonosporopsis* spp. exists in wild watermelon germplasm (Gusmini et al. 2005); however, its introgression into commercially available cultivars remains challenging (Gusmini et al. 2017; Ren et al. 2019), and multiple attempts to deploy resistant breeding lines have failed under field conditions (Norton et al. 1995). Because of the lack of GSB-resistant watermelon cultivars commercially available, the most effective management option is chemical control (Avenot et al. 2012; Gusmini et al. 2017). Although multiple fungicides of various chemical classes are available for managing GSB, the causal species are considered a high-risk pathogen with potential to develop resistance to fungicides due to their short life cycle and frequent fungicide applications over a growing season (Keinath 2015a; Thomas et al. 2012).

Phytophthora fruit rot of watermelons, caused by the soilborne oomycete *Phytophthora capsici*, is an emerging disease in many watermelon growing regions in the US (Kousik et al. 2014a, 2016; McGrath 2017a, b). Most of the commercial watermelon varieties evaluated so far are highly susceptible to *Phytophthora* fruit rot (Kousik et al. 2014a, 2018b). *Phytophthora* fruit rot caused by *P. capsica* was first reported in 1940 (Wiant 1940), but the pathogen is able to infect more plant tissues including roots, crowns, and foliage in a wide number of plant families (Granke et al. 2012; Kousik et al. 2015; Mansfeld et al. 2020; Parada-Rojas and Quesada-Ocampo 2019; Quesada-Ocampo et al. 2015).

P. capsici produces three types of spores: a distinctive lemon-shaped sporangia, zoospores which discharge from sporangia in water, and thick cell wall oospores which allow the pathogen to overwinter and persist in soils for several years (Granke et al. 2012; Pavón et al. 2007). Signs of the disease are characterized by expanding lesions that produce fresh sporangia over a five-day period as a distinctive white "powdered sugar" layer on the surface of the fruit becomes visible to the naked eye (Hausbeck and Lamour 2004).

About 50% of US watermelon is grown in the southeastern US, where optimal conditions for development of Phytophthora fruit rot are prevalent (Kousik et al. 2017). *Phytophthora* fruit rot on watermelons can be a pre- or post-harvest problem. If the disease occurs early during fruit formation, it could result in total yield loss. Watermelon fruits resting on infested soil develop symptoms significantly reducing yields, and harvested fruits may rot during and after shipping leading to market rejections and revenue losses (Kousik et al. 2014a, 2016).

The most common control strategies include the application of fungicides, effective crop rotation schemes, soil solarization, planting in well-drained fields, water surveillance, and the avoidance of overhead irrigation (Granke et al. 2012; Kousik et al. 2011a, b, 2014a, 2017). Fungicides (mefenoxam, cyazofamid, fluopicolide, mandipropamid, and oxathiapiprolin) are generally effective in managing fruit rot (Jackson et al. 2012; Ji and Csinos 2015; Keinath 2007; Keinath and Kousik 2011). Unfortunately, during high disease pressure and favorable environmental conditions, their effectiveness can be significantly curtailed (Kousik et al. 2011a, b), and some isolates may possess fungicide insensitivity (Jackson et al. 2012; Keinath 2007; Kousik and Keinath 2008; Lamour and Hausbeck 2000; Parra and Ristaino 2001). The emergence of fungicide-resistant strains suggests and supports the need for constant development of new alternatives, including host resistance for managing *P. capsici*. Regular use of fungicides is not cost-effective, and host resistance is a viable alternative.

Fusarium wilt of watermelon, caused by the soilborne ascomycete fungus *Fusarium oxysporum* formae speciales *Niveum* (FON), is considered to be highly destructive and a limiting factor for watermelon production worldwide (Everts et al. 2014). FON produces three spore types: straight and tapered septate macroconidia, abundant aseptate elliptical microconidia, and thick-walled chlamydospores that can survive in soil for up to 5–10 years (Carlile 1956; Leslie and Summerell 2006). Plants are typically infected by overwintered chlamydospores germinating in the spring, but symptoms are not apparent until later in the summer (Keinath et al. 2017). Symptoms of *Fusarium* wilt can vary based on a number of factors, including environmental conditions, plant age, and density of the pathogen population in the soil. However, plants afflicted with *Fusarium* wilt can display symptoms such as leaves turning dull to chlorotic and eventually necrotic, a gummy exudate on the plant stems, and the more characteristic unilateral wilt and vascular discoloration, which results from pathogen growth and tyloses, gums, and gels produced within the plant xylem (Kleczewski and Egel 2011).

Since this disease was first identified in the states of Georgia and South Carolina in 1894 (Smith 1894), four separate races of FON have been described (Netzer 1976; Zhou et al. 2010). Races 0 and 1 are of lesser economic importance due to disease resistance available in many commercial varieties (Martyn and Netzer 1991). While there is currently no host resistance to race 3, it is also of lesser importance due to its limited geographic presence (Zhou et al. 2010). Therefore, race 2 is the most economically important as it is widely spread throughout the US and other watermelon-producing countries, and only low levels of partial resistance are available in diploid (seeded) and triploid (seedless) watermelon cultivars (Keinath and Hassell 2014). Grafting of watermelon plants onto resistant rootstocks can suppress the onset and development of disease caused by race 2 (Keinath and Hassell 2014), but is economically prohibitive, and thus not widely adopted by US watermelon growers (Taylors et al. 2006). Chemical management of *Fusarium* wilt has also become more limited over time due to the phasing out of methyl bromide (Sande et al. 2011), although some fungicides such as prothioconazole and pydiflumetofen can be used (Everts et al. 2014; Miller et al. 2020). While other cultural practices such as long-term crop rotation can be incorporated to help limit yield damages resulting from *Fusarium* wilt, continued research into host resistance would greatly benefit watermelon growers suffering from this destructive disease.

Downy mildew (DM) of watermelon, caused by the obligate oomycete pathogen, *Pseudoperonospora cubensis*, is the most devastating foliar disease to affect cucurbit production in the US (Savory et al. 2011). DM can rapidly defoliate watermelon vines reducing yield, fruit quality, and exposing fruit to sunscald (Salcedo et al. 2020). *P. cubensis* has two types of isolates, referred to as clade 1 and clade 2, that preferentially infect cucurbit crops and weeds. Clade 1 more frequently infects watermelon, pumpkin, and squashes, while clade 2 more frequently infects cucumber and cantaloupe (Wallace et al. 2020). Due to the obligate nature of *P. cubensis*, it cannot survive in areas with annual frosts and therefore overwinters in areas with hosts available year-round such as southern Florida, along the Gulf of Mexico, or within greenhouse production (Quesada-Ocampo et al. 2012). In these areas, *P. cubensis* inoculum increases on susceptible hosts, and annually asexual sporangia are aerially dispersed long distances moving northward along the east coast of the US during spring (Main et al. 2001; Nusbaum 1944; Rahman et al. 2021).

Favorable conditions for *P. cubensis* proliferation include those with high humidity and moisture from morning dew, rain, or overhead irrigation (lasting 6–12 h) and cool temperatures around 60 F (Cohen 1977). *P. cubensis* reproduces both sexually and asexually (Cohen and Rubin 2012), and the presence, morphology, and pigmentation of the sporangiophores and sporangia, as well as symptoms on the host, have been traditionally used for diagnosis (Crandall et al. 2018; Rahman et al. 2017; Savory et al. 2011). In watermelon, DM symptoms are variable compared to other cucurbit crops, and lesions frequently appear as yellow or brown spots with irregular shape, unrestricted by leaf veins with scarce sporulation (Crandall et al. 2018; Holmes et al. 2015), making diagnosis more difficult (Salcedo et al. 2021).

Cultural disease management practices of DM on watermelon and other cucurbits include planting early to escape high disease pressure, avoiding overhead irrigation, frequent scouting and removal of infected plants, and planting tolerant varieties when possible (D'Arcangelo et al. 2021; Quesada-Ocampo 2020a). Since there are no watermelon varieties with resistance to DM, growers regularly rely on fungicide applications each year for control of DM. In general, before the disease is present, fungicides are applied preventatively on a 7–10-day interval depending on cucurbit crop and then upon detection on a 5–7-day interval (Quesada-Ocampo 2020a).

Powdery mildew (PM), caused most frequently by the obligate biotrophic fungal pathogens *Podosphaera xanthii* (syn. *Sphaerotheca fuliginea*) and *Erysiphe cichoracearum*, is one of the most common, conspicuous, and widespread diseases and limiting factors for cucurbit crop production in both the field and greenhouse (Belanger et al. 2002; Keinath 2015b; Keinath and DuBose 2004; Pérez-García et al. 2009). Powdery mildew has been occurring more frequently on watermelon in recent years (Davis et al. 2001; Kousik et al. 2011a, b, 2016; Mercier et al. 2014). While both pathogens have been observed to cause the disease, *P. xanthii* is considered to be the main causal agent of powdery mildew on cucurbits (Pérez-García et al. 2009). Pathogens can infect cotyledons, stems, leaves, and fruit (Keinath et al. 2018; Keinath and DuBose 2004; Kousik et al. 2011a, b, 2018a), in some instances causing death at the seedlings stage on cucurbits (Jarvis 1992).

Both pathogens produce aerially dispersed, asexual conidia that can remain viable for 7–8 days, as well as sexual fruiting bodies called cleistothecia, which are able to overwinter in leaf debris each season, although reports of their occurrence remain rare outside of North America (McGrath 1994). Sources of initial PM primary inoculum include, but may not be limited to, conidia from greenhouse-grown cucurbits, alternative hosts, and overwintered cleistothecia. Characteristic symptoms of PM include a whitish, talcum-like, powdery appearance which develops on both the adaxial and abaxial surfaces of the leaf, petioles, stems, and in extreme cases, fruit (McGrath 2017a, b). Oftentimes, cucurbit crops are not affected by PM until after fruit initiation has occurred. While both causal agents induce identical symptoms on cucurbits, they can be easily distinguished from each other using light microscopy (Braun et al. 2002). Favorable conditions for infection and conidial survival include low light intensity, dense plant growth, and relatively high humidity. Dryness, or a lack of free moisture on the plant surface, is conducive for colonization, sporulation, and dispersal, although disease development can occur whether it is dry or not. Temperatures of 50–90 F are conducive for pathogen infection; however, PM development is hindered when daytime temperatures exceed 100 F (McGrath 2021).

A detailed description of the cucurbit powdery mildew pathogen and race classification system based on melon has been provided by Cohen et al. (2004). Physiological races of this pathogen have not been clearly classified for many cucurbits because of the lack of differentials and fully resistant germplasm. However, seven races of *P. xanthii* have been identified using melon (*Cucumis melo* L.) differentials (Cohen et al. 2004). The potential for existence of *P. xanthii* "races" based on watermelon differentials is still under discussion, and so far, there appear to be at least two races designated as 1W and 2W (Kousik and Ikerd 2014). However, many researchers have used the race classification based on melon differentials for evaluating watermelon and other cucurbit germplasms (Davis et al. 2007; Tetteh et al. 2010).

PM is managed using a combination of cultural and chemical control methods. Cultural control methods include the use of resistant cultivars, adequate plant spacing, establishing successive plantings up-wind of older plantings as older plants can serve as sources of conidia, and the removal of debris, weeds, and volunteer plants that may harbor the pathogen (Quesada-Ocampo 2020b). Application of fungicides was shown to reduce powdery mildew and increase marketable yield of watermelon (Keinath 2015b). Fungicide applications should be made every 7–10 days after the initial detection and onset of disease development. For best results, fungicide applications are necessary on the under-surface of the leaves and those in the low canopy (McGrath 2001). The development of fungicide resistance and resulting control failure have been a concern as reduced sensitivity to fungicides from several chemical groups has been detected with increased fungicide use (Keinath 2015b; Keinath et al. 2018; McGrath and Shishkoff 2001).

Tank mixes of fungicides with systemic movement and contact fungicides such as chlorothalonil have been effective in delaying the build-up of fungicide-resistant isolates (McGrath 2001). Additionally, biopesticides-containing ingredients such as botanical oils, bicarbonates lipopeptides, and hydrogen dioxide have been approved and registered for organic production in the US, with some having ranges of efficacy comparable to conventional fungicides (McGrath 2021). While resistant varieties are most desirable and are available in other cucurbits such as cucumbers and pumpkins, there are no commercially available PM-resistant watermelon varieties even though enormous efforts have been made to identify resistance to PM in watermelon (Ben-Naim and Cohen 2015; Davis et al. 2006, 2007; Tetteh et al. 2010, 2013a).

Bacterial fruit blotch (BFB) caused by the gram-negative, soilborne bacterium *Acidovorax citrulli* (syn. *A. avenae* subsp. *citrulli* and *Pseudomonas pseudoalcaligenes* subsp. *citrulli*; Willems et al. 1992; Schaad et al. 1978) was first recovered from watermelon plants in Georgia in 1965 (Webb and Goth 1965). Since then, BFB has led to sporadic, yet significant disease outbreaks in several countries with large-scale cucurbit production (Latin and Rane 1990; Palkovics et al. 2008; Yang et al. 2019), making it a serious economic threat.

Cucurbit hosts can be infected by at least two genetically distinct groups (group I and II) of *A. citrulli*; group I strains have been mainly found on non-watermelon hosts; however, strains from group II are more predominant and aggressive on watermelon (Eckshtain-Levi et al. 2014). Infected seedlings exhibit water-soaked spots primarily on cotyledons that can extend along the hypocotyl to affect true leaves. As the plant matures, foliar symptoms develop into brown to reddish-brown lesions along leaf veins that can be hard to distinguish from other foliar diseases like anthracnose (Keinath et al. 2017b). Symptoms on watermelon fruits typically appear as small, greasy, water-soaked spots on the upper surface of harvest-ready fruits that rapidly expand into dark green, irregularly shaped water blotches (Latin and Hopkins 1995). Under cool and dry growing conditions, lesions will instead gain an atypical appearance that resembles insect damage. These atypical lesions are more restricted in growth, developing brown cracks on the fruit rind as the fruit expands, and can exude amber-colored ooze (Keinath et al. 2017b).

Seedborne transmission is a key component for dissemination of BFB, as *A. citrulli* on seeds can retain the ability to infect seedlings after being stored for up to 35 years (Block and Shepherd 2008; Latin and Hopkins 1995). Seed becomes infested when *A. citrulli* infects the watermelon blossom of the parent plant (Walcott et al. 2003), which depending on the route of pathogen ingress can affect whether subsequent fruits become symptomatic and where the bacterium localizes in infected seed (Dutta et al. 2012). Seedlings grown from infected seeds can then serve as a secondary source of inoculum, which is often exacerbated in transplant greenhouses as the warm and humid environment is conducive for disease development and overhead irrigation allows for splash transmission, thus propagating diseased plants before they even enter the field (Latin and Hopkins 1995).

Management tactics for BFB are extremely limited and require strict adherence to integrated pest management regiments to try and exclude BFB in watermelon production. Only seed and transplant seedlings that test negative for *A. citrulli* are recommended for use, and frequent monitoring for removal of symptomatic and surrounding plants is common practice. Sanitizers and steam can be used to treat seeds and sterilize greenhouse spaces or equipment but cannot manage the disease alone (American Seed Trade Association 2009). In the field, copper-based bactericide sprays can be applied as a preventative measure but may be ineffective if environmental conditions are highly conducive for disease (Keinath et al. 2017b). For these reasons, the addition of resistant plant cultivars would greatly improve management options for BFB as no resistant varieties currently exist (Carvalho et al. 2013). However, until the challenges to breeding watermelon plants that are no longer highly susceptible in early and late stages of the plant life cycle are overcome, only tolerant varieties can be used (Bahar et al. 2009).

4.3 Resistance Genes and Pathogen Recognition

The ability of the host to suppress pathogen growth and reproduction is called disease resistance. The most widely observed form of resistance is non-host resistance, which occurs when a non-adapted pathogen fails to colonize the plant, establish infection, and produce disease. Non-host resistance relies on a complex combination of constitutive and induced defense components and plant immunity systems (Fonseca and Mysore 2019; Lee et al. 2017). Host resistance is mediated by resistance genes (R-genes) that act under different mechanisms. For example, the *Hm1* gene from maize encodes for a fungal detoxification enzyme (Johal and Briggs 1992), adult plant resistance (APR) genes Lr34/Yr18/Sr57/Pm38 and Lr67/Yr46/Sr55/Pm46 in wheat encode for a protein transporter (Krattinger et al. 2009), and recessive mutations on the *Mlo* gene that encodes for a plasma membrane-anchored protein similar to G-protein-coupled receptors confer durable broad-spectrum resistance to powdery mildew (PM) in cucumber and other species (Berg et al. 2017; Devoto et al. 1999).

In recent decades, our understanding of the molecular fundamentals of host resistance based on the innate plant immune system has shown that plants resist pathogen attacks with a series of specialized receptor proteins present in each plant cell (Dangl and Jones 2001; Jones and Dangl 2006) (Fig. 4.2). In a simplified and elegant zig-zag model proposed by Jones and Dangl (2006), plant immune receptor proteins are associated with two defense networks. In one of these networks, termed pathogen-associated molecular patterns (PAMPs)-triggered immunity (PTI) (Jones and Dangl 2006; Monaghan and Zipfel 2012), plant host transmembrane pattern recognition receptors (PRRs) are able to recognize conserved molecular pathogen markers called pathogen (or microbial) associated molecular patterns (PAMPs/MAMPs). These PAMPs/MAMPs include molecules such as fungal chitin, bacterial flagelin, elongation factor tu (EF Tu), lipopolysaccharides, and peptidoglucan (Dow et al. 2000; Erbs et al. 2008; Felix et al. 1993, 1999; Kunze et al. 2004) or

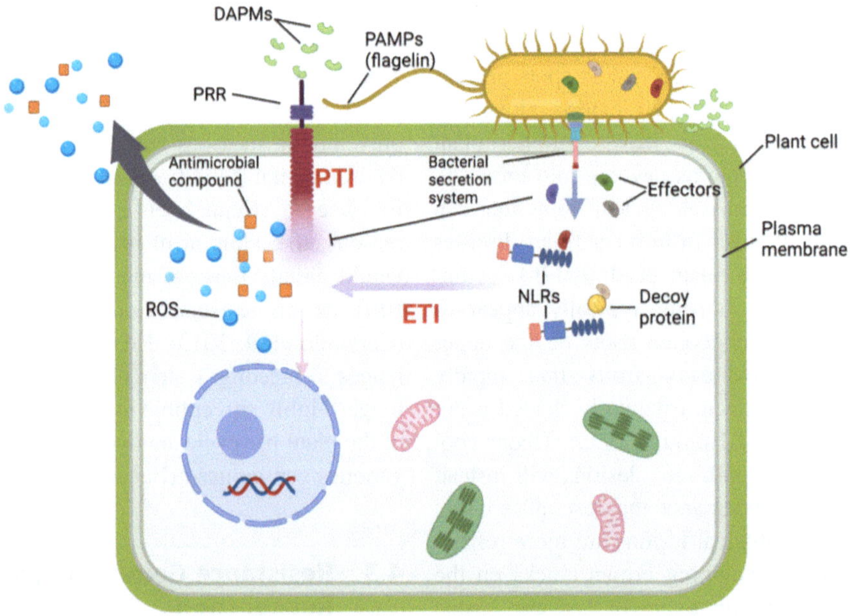

Fig. 4.2 Model of plant immunity. Pathogen-derived molecules (pathogen-associated molecular patterns, PAMPs) such as lipopolysaccharides, flagellin, and chitin and endogen molecules product of pathogen tissue (damage-associated molecular patterns, DAMPs) are released into the extracellular spaces and recognized by cell surface pattern recognition receptors (PRRs) initiating PAMP-triggered immunity (PTI). Pathogens deliver effectors to both the extracellular spaces (not shown here for simplicity) and to the plant cell interior to suppress PAMP/MAMP perception facilitating virulence. Effectors are recognized by intracellular nucleotide-binding (NB)-LRR receptors (NLRs), by direct interaction or by sensing effector-mediated alteration of a host virulence target or decoy proteins that mimic effector targets, inducing effector-triggered immunity (ETI)

endogen molecules that are the product of tissue damage or infection from the lytic activity of the pathogens called damage-associated molecular patterns (DAMPs) (Choi and Klessig 2016; Yamaguchi et al. 2010).

PRRs recognize PAMPs/MAMPs and DAMPs as potential threats activating an intracellular immune response that hamper the pathogen capacity to infect and proliferate in the host (Zhang and Zhou 2010). PTI is a major player in non-host resistance, allowing plants to resist many potential pathogens present in the environment (Fonseca and Mysore 2019). Plant-biotrophic or hemibiotrophic pathogen coevolution has selected for the production and secretion of diverse effector proteins that suppress PTI, resulting in diseases as consequence of effector-triggered susceptibility (Chisholm et al. 2006; Jones and Dangl 2006; Thordal-Christensen 2020). Plant hosts have evolved toward a second defense network of plant immunity based in cytoplasmic receptors capable of recognizing directly or indirectly pathogen protein effectors into the host cytoplasm, activating an immune response known as effector-triggered immunity (ETI) (Jones and Dangl 2006).

ETI response involves reactive oxygen species production, changes in hormone defense levels, accumulation of defense proteins, antimicrobial compounds, localized programed cell death at the site of infection, and sometimes a subsequent systemic acquired resistance, providing the host plant with broad-spectrum resistance and activating in all tissues a plant defense response to future pathogen challenges (Bari and Jones 2009; Gururani et al. 2012; Heath 2000; Spoel and Dong 2012). The strict distinction between PTI and ETI is often not clear, and recent models of plant immunity and cloning of PRRs seem to be functioning in a gene-for-gene fashion (Kanyuka and Rudd 2019).

One of the major objectives of plant breeding toward disease resistance has been the introgression of resistance to biotic stress into commercial cultivars using major dominant disease resistance genes (R-genes) (Dangl and Jones 2001; Hammond-Kosack and Jones 1997; REX Consortium 2016). The monogenic resistance and dominant nature of *R*-genes led to the inference that they encode proteins that recognize specific dominant pathogen gene products called *Avr* (avirulence) genes in a gene-for-gene complement. R-protein-mediated recognition of Avr proteins activates host defense mechanisms impairing pathogen proliferation (Flor 1971; Keen 1990; Van Der Biezen and Jones 1998). This model agrees with *R*-genes coding for intracellular receptors capable of recognizing pathogen effectors that once recognized by the immune system become avirulence factors (Ellis et al. 2000).

The most abundant *R*-genes cloned and characterized to date correspond to intracellular receptor proteins known as NLRs for their similarity to intracellular NOD-like receptors in animal immune systems (Maekawa et al. 2011). These receptors confer resistance to a wide variety of taxonomically unrelated plant pests (virus, bacteria, fungi, oomycetes, nematodes, and insects) (Rossi et al. 1998; Sekhwal et al. 2015). NLRs are modular proteins with a conserved central nucleotide-binding site (NBS) domain containing approximately 300 amino acids, a carboxy terminal domain with variable leucine-rich repeats (LRR) involved in specific recognition of pathogen-derived effector protein and signaling (Jones and Dangl 2006; Martin et al. 2003; McHale et al. 2006), as well as an amino terminal domain that divides the NLRs into functional subfamilies differentiated by the presence of a particular domain in the protein amino terminus: the TIR domain (TNL), defined by the presence of rToll/interleukin-1 receptor (TIR); the CC domain (CNL) that includes the coiled-coil (CC) domain, and CCr-NLRs/RNLs that contain the resistance to powdery mildew8 (RPW8) domain (Collier et al. 2011; Maekawa et al. 2011; Meyers et al. 1999, 2003).

NLRs encode one of the most ancient, diverse, and largest gene families known in plants (Jacob et al. 2013). TNLs are the most common NLR and seem to be restricted to dicots, while CNLs occur in both monocots and dicots (Shao et al. 2016). NLR number varies considerably among plant species, being generally few in Cucurbitaceae species (Lin et al. 2013) to thousands in hexaploid wheat (Andersen et al. 2020). NLRs are generally arranged in clusters of tandem duplications and diversified for unequal crossing over, rearrangements, and gene conversion events. These clusters are the incubators of new resistant specificities through domain swaps between homologous NLRs and become large reservoirs of multiple NLR copies that can be targets of beneficial mutations and new specificities (Baragan and Weigel 2020; Friedman and Baker 2007; Hulbert et al. 2001). Furthermore, in addition to variation on coding sequence, NLRs have variation in copy number, genomic locations, presence–absence between and within species, haplotypes, and cultivars (Baragan and Weigel 2020; Lee and Chae 2020; Stam et al. 2019).

Current models suggest that R-proteins interact with pathogen effectors through direct and indirect recognition (Dodds et al. 2006; Van Der Biezen and Jones 1998). Under direct interaction, pathogen effector molecules are detected through direct binding to the NLR protein. By contrast, indirect interaction models, where the R-proteins monitor or guard a host conserved defense protein (guardee) that are the target of pathogen effector proteins (guard model), occur more frequently. Changes in the integrity or modification of the guardee protein after effector interaction results in the activation of the defense response (Dangl and Jones 2001; DeYoung and Innes 2006). In a variation of this model, host proteins targeted by pathogen effectors play no specific role in defense and mimic a true defense component acting as a decoy for the pathogen effector. Again, under this model changes in the integrity or modification of the decoy protein after effector interaction perceived by an R-protein results in the activation of the defense response (van der Hoor and Kamoun 2008). In an extension of the decoy model, NLRs can act as a decoy themselves by integrating domains in their structure (NLR-IDs), which mimic targets of effector proteins serving as a sensor (Cesari et al. 2013; Kroj et al. 2016). NLR-IDs function with other genetically linked NLR that in several cases form a heterocomplex receptor in which each member has specific attributes: the NLR-ID acts as sensor (receptor) detecting a specific pathogen effector or effectors activity on a host target, while the other NLR acts as signaling molecule (transducer or helper) inducing the defense response (Cesari et al. 2014; Zhai et al. 2014). In this context, NLRs can work as NLR pairs and helpers, and sensors can connect with others forming a NLR network that could increase the robustness of the immune response (Adachi et al. 2019).

4.4 Host Resistance Genes in Watermelon: A Historical Perspective

Search for disease resistance in watermelon has involved the massive screening of watermelon accessions (PIs, commercial lines, breeding lines) under different environments. Frequently, wild-type accessions tend to be more resistant than cultivated species, and it is expected that most resistance be mediated by major resistance (R) genes (qualitative resistance) or multiple genes with minor effect (quantitative resistance). Major genes confer complete resistance and are easily manipulable for research and breeding purposes but frequently are overcome by new virulent pathogens. In contrast, quantitative disease resistance only reduces pathogen multiplication and severity of disease symptoms, but having lower selection pressure against pathogen variants results in more durable resistance than major *R*-genes (Parlevliet 2002; Poland et al. 2009; Zhang et al. 2013a, b). Quantitative resistance is associated with genomic regions (QTLs or quantitative trait loci) that contribute with a variable effect to explain the resistant phenotype observed (Young 1996). The development of next-generation sequencing (NGS) technologies and genotyping platforms combined with mapping populations has increased the identification and mapping of QTLs controlling quantitative resistance in cucurbits (Grumet et al. 2020).

Anthracnose: Differential cultivars of several cucurbit species have been used to identify seven races of *C. orbiculare* grouped into three vegetative compatibility groups (Wasilwa et al. 1993).

Four races cause anthracnose in watermelon, but commercial cultivars are resistant to only three (Patel 2019). Several watermelon commercial varieties have resistance to *C. orbiculare* race 1 (Jang et al. 2019). Anthracnose-resistant cultivars developed by Layton in 1937 were released in 1949 and are being widely used to control the disease. They were developed by using resistant accessions from Africa crossed with elite cultivars such as Iowa Belle. Unfortunately, the resistant cultivars began showing disease symptoms in 1958, indicating that new races of the pathogen were overcoming the resistance (Goode 1958). Among the most common watermelon cultivars, Garrison is susceptible to all races, while Charleston Gray was resistant to races 1, 3, and 4 (Winstead et al. 1959). Resistance to race 2 is available in PI 189225, PI 271775, PI 271778, and PI 512385 (Boyhan et al. 1994a, b; Sowell et al. 1980).

Inheritance of resistance to anthracnose races 1 and 3 is controlled by a single dominant gene, *Ar-1* (Layton 1937; Winstead et al. 1959). Using a combination of genotyping by sequencing (GBS) and bulked-segregant analysis sequencing (BSA-seq) data, Jang et al. (2019) identified that resistance to anthracnose race 1 is associated with a non-synonymous substitution in the LRR domain of the NLR annotated as *Cla001017*, and this substitution is conserved between *Cucurbitaceae* and *Fabaceae*. Inheritance of resistance to race 2 is also controlled by a single dominant gene, *Ar-2-1* (Suvanprakorn and Norton 1980). In contrast to single dominant inheritance of anthracnose resistance, race 2 resistance in the line R309 (*Citrullus colocynthis*) is quantitative involving multiple genes (Love and Rhodes 1988).

Gummy Stem Blight: Numerous sources of resistance to gummy stem blight in watermelon have been described (Boyhan et al. 1994a, b; Gusmini et al. 2005; Sowell 1975; Sowell and Pointer 1962). The *Citrullus amarus* accession PI 189225 has been identified as a source of resistance to gummy stem blight for more than 50 years and determined by the single recessive gene, *db* (Norton 1979). PI 189225 and PI 271778 were used to breed the cultivars AU-Jubilant and AU-Producer, the first watermelon cultivars reported to be resistant to the disease (Norton et al. 1986). However, these cultivars proved to not be resistant under field conditions (Song et al. 2002), and currently no cultivars with field level resistance are available to growers.

Recent studies suggest that resistance to gummy stem blight resistance from PI 189225 and other sources is more complex and controlled by more than one gene (Gusmini et al. 2017; Lee et al. 2021; St Amand and Wehner 1995). It has also been suggested that different genes are responsible for resistance in stems and leaves (Ren et al. 2019). The slow progress in the development of GSB-resistant cultivars in watermelon is affected by a combination of complex inheritance of resistance and the presence of sources of resistance in wild watermelon relatives with poor fruit quality. However, multiple crossing and intercrossing of wild plant introductions with elite cultivars have proved to be beneficial to produce high-quality germplasm with resistance to GSB (Rivera-Burgos et al. 2021).

Hassan et al. (2019) measured expression of NLR genes in infected and non-infected plants of PI 189225 (resistant) and Charleston Gray (susceptible) at 12, 24, and 72 h post inoculation. Six genes (*Cla001821*, *Cla019863*, *Cla020705*, *Cla012430*, *Cla012433*, and *Cla012439*) showed increased expression levels in PI 189225 compared to Charleston Gray. PI 189225 is the source of a major QTL (*Qgsb8.1*; phenotypic variation explained PVE ~ 32%) on chromosome 8 of the watermelon genome associated with gummy stem blight resistance of leaves (Ren et al. 2019). The *Qgsb8.1* locus was narrowed down to a 571.27 kb region on the *C. lanatus* cv 93107v1 draft genome (Guo et al. 2013). This region contains 19 genes, from which two genes, *Cla001017* (NLR) and *Cla001019* (pathogenesis related gene), were suggested as candidate *R*-genes based on differential gene expression (Ren et al. 2019). Lee et al. (2021) also identified QTLs associated with gummy stem blight in PI 189225. The authors evaluated leaf lesions and stem blight symptoms separately and identified co-localized QTLs on chromosome 8 (*Qgsb8.2* syn. *qSB-8.1*, *qLL1-8.1*; PVE ~ 10%). *Qgsb8.2* stretches a region of 872 kb on the 93107v1 draft genome (Guo et al. 2013). Three receptor-like kinase genes (*Cla022133*, *Cla022184*, and *Cla022195*) and an LRR-RLK gene (*Cla022196*) were suggested as candidate genes.

Gimode et al. (2021) identified three QTLs (*Qgsb3.1* syn. *ClGSB3.1*, PVE » 6%; *Qgsb5.1* syn. *ClGSB5.1*, PVE » 10%; and *Qgsb7.1* syn. *ClGSB7.1*, PVE ~ 21%) associated with gummy stem blight resistance using the wild watermelon relative *C. amarus* accession PI 482276 as a resistance source. *Qgsb7.1* stretches approximately 6 kb in the Charleston Gray draft genome (Wu et al. 2019) and is syntenic to a region associated with gummy stem blight resistance in melon (Hu et al. 2018). In this genomic region, the receptor-like protein (RLP) Avr9/Cf-9 with an extracytoplasmic LRR domain and designated as *ClCG07G013230* was identified as a candidate *R*-gene based on homology with a candidate orthologous gene in melon and its association with a non-synonymous point mutation in a watermelon mapping population (Gimode et al. 2021).

Phytophthora Fruit Rot of Watermelon: Deployment of host resistance presents as a desirable strategy to incorporate in an integrated disease management approach against *Phytophthora* fruit rot (Michelmore et al. 2017). However, the majority of the commercial watermelon varieties evaluated are highly susceptible to *Phytophthora* fruit rot with large lesions and abundant sporulation (Kousik et al. 2014a, 2018b). Sources of resistance to *Phytophthora* fruit rot of watermelon have been identified (Kousik et al. 2012a, b), and resistant germplasm lines for use in watermelon breeding programs have been developed (Kousik et al. 2014a). The fruit of the resistant germplasm lines was resistant at all ages, whereas fruit of susceptible lines is susceptible at all stages of development with larger fruit rot lesions and abundant sporangia (Kousik et al. 2018b). These resistant germplasm lines belong to *Citrullus mucocospermus* and can be easily crossed with cultivated watermelon (*C. lanatus*). Several lines of these resistant germplasm are resistant to 20 isolates of *P. capsici* collected from different states and crops in the US (Kousik and Ikerd 2014). However, it has been a challenge to incorporate resistance to *Phytophthora* fruit rot into usable germplasm lines (Kousik et al. 2012a, b). The United States Department of Agriculture (USDA) recently released four *Phytophthora* fruit rot-resistant watermelon germplasm lines USVL489-PFR, USVL203-PFR, USVL020-PFR, USVL782-PFR, and USVL531-MD (Kousik et al. 2014b, 2019). To date, these resistant lines represent important sources of resistance for breeding programs in the USDA and watermelon seed industry. Ongoing studies have indicated that resistance to *Phytophthora* fruit rot in watermelon may be controlled by multiple genes and is complex as observed in *P. capsici*-pepper interactions (Lefebvre and Palloix 1996; Minamiyama et al. 2007; Wang et al. 2015a, b). A recombinant inbred line (RIL) population from a cross of a *Phytophthora* fruit rot-resistant and susceptible line is being developed for fine mapping and further studies similar to those done for *P. capsici*-pepper interactions (Jiang et al. 2015; Ogundiwin et al. 2005; Sy et al. 2008; Truong et al. 2012). Though resistant accessions have been identified and germplasm lines developed (Kousik et al. 2012a, b, 2014a), the genetics behind the resistance to *Phytophthora* fruit rot in watermelon has not yet been clearly elucidated. Interestingly, smaller lesions with lower intensity of sporulation were observed on a susceptible watermelon cultivar "Sugar Baby" grafted onto the fruit rot-resistant germplasm lines used as rootstocks (Kousik et al. 2012a, b). Further studies are needed to determine if grafting can be a viable alternative in combination with fungicides to control *Phytophthora* fruit rot.

Fusarium Wilt: Resistance to *Fusarium oxysporum* f. sp. *Niveum* (FON) race 1 in the cultivar Calhoun Gray is controlled by a single dominant gene, *Fo-1* (Henderson et al. 1970; Netzer 1976).

A major QTL (*Qfon1-1.1* syn. *Fo-1.1*, *Qfon1.1*, *Qfon1*, *FON-1*)[1] associated with this resistance was mapped to the tip of chromosome 1 of the watermelon genome (Branham et al. 2018; Fall et al. 2018; Lambel et al. 2014; Meru and McGregor 2016; Ren et al. 2015) and is found in PI 296341-FR (Martyn and Netzer 1991; Ren et al. 2015). Lambel et al. (2014) identified additional minor QTLs on chromosomes 1 (*Qfon1-1.2* syn. *Fo-1.2*), 3 (*Qfon1-3* syn. *Fo-1.3*), 4 (*Qfon1-4* syn. *Fo-1.4*), 9 (*Qfon1-9* syn. *Fo-1.5*), and 10 (*Qfon1-10* syn. *Fo-1.6*). A major QTL for resistance to FON-1 from *C. amarus* line USVL246-FR2 was also mapped on chromosome 9 (Branham et al. 2020). The latter QTL has some overlap with *Qfon1-9* described by Lambel et al. (2014), but it remains to be determined whether they represent the same locus. Branham et al. (2018) identified an additional QTL associated with FON race 1 resistance using QTL-Seq on chromosome 11 (*Qfon1-11* syn. *Fo-1.11*). *Qfon1-11* overlaps with a FON race 2 resistance QTL (*Qfon2-11* syn. *Qfon11*) identified by Meru and McGregor (2016) in a *C. lanatus* background.

QTLs associated with resistance to FON race 2 in *C. amarus* PI 296341-FR (Martyn and Netzer 1991) have been identified on chromosome 9 (*Qfon2-9* syn. *Qfon2.1*; PVE = 13.7%), and a resistant QTL associated with the susceptible parent, 97103, was identified on chromosome 10 (*Qfon2-10* syn. *Qfon2.2*; PVE = 12.5%) (Ren et al. 2015). *Qfon2-9* (syn. *qFon2-9*) was also determined to be the major QTL (PVE = 43%) associated with FON race 2 resistance in *C. amarus* USVL246-FR2 (Branham et al. 2017). Four minor QTLs were also identified in the latter study, including the same QTL on chromosome 10 (*Qfon2-10* syn. *qFon2-10*) identified by Ren et al. (2015). The three other novel QTLs were on chromosomes 2 (*Qfon2-2* syn. *qFon2-2*), 5 (*Qfon2-5* syn. *qFon2-5*), and 8 (*Qfon2-8.1* syn. *qFon2-8*). Seven QTLs associated with FON race 2 resistance were identified in a *C. amarus* population from a cross between USVL252-FR2 and PI 244019$^{PRSV\text{-}R}$ (Branham et al. 2020). *Qfon2-1* (syn. *qFon2-1*, PVE ~ 18%) overlaps with the major QTL for race 1 resistance (*Qfon1-1*); however, it is not known whether this is due to pleiotropy or linkage (Branham et al. 2020). *Qfon2-2*, *Qfon2-5,* and *Qfon2-8.1* previously identified in USVL246-FR2 (Branham et al. 2017) were also confirmed in the USVL252-FR2 population, and a QTL overlapping with *Qfon1-11* (Branham et al. 2018) and *Qfon2-11* (Meru and McGregor 2016) was also detected (Branham et al. 2020). Two novel minor QTLs (*Qfon2-6* syn. *qFon2-6* and *Qfon2-8.2* syn. *qFon2-8.2*) were also identified in USVL252-FR2 and PI 244019$^{PRSV\text{-}R}$ population (Branham et al. 2020).

Downy Mildew: Early screening carried out by Winstead et al. (1957) using watermelon introductions and breeding lines found two highly resistant lines (PI 179660 and PI 179875) to downy mildew and several moderately resistant lines after the screening of 300 accessions. Downy mildew frequently causes losses to watermelon (Holmes et al. 2015); however, currently there are no watermelon varieties or hybrids resistant to downy mildew, and most of the disease control relies on preventive fungicide applications.

Powdery Mildew of Watermelon: Using watermelon differentials, Kousik and Ikerd (2014) identified the potential presence of at least two *P. xanthii* races. The designated race 1W has been used if the powdery mildew isolate can infect watermelon leaves and the melon differential Iran H but not the melon differential PMR-45. The designation race 2W has been used if the powdery mildew isolate can infect leaves of both the melon differential Iran H and PMR-45. Sources of resistance to race 1W or 2W have been identified (Ben-Naim and Cohen 2015; Davis et al. 2006; Kousik and Ikerd 2016; Tetteh et al. 2010; Thomas et al. 2005), and the mode of inheritance has been determined for some sources (Ben-Naim and Cohen 2015; Kim et al. 2015; Tetteh et al. 2013a, b). In one study, it was determined that resistance to race 1W in watermelon PI 189225 was controlled by three complementary and partially dominant genes (Ben-Naim and Cohen 2015),

[1] Note that naming of Fusarium QTLs follows the format, trait name and race-chromosome number.QTL order, rather than the more traditional, trait name.chromosome number.QTL order.

whereas resistance to race 2W seems to be controlled by a single dominant gene (Tetteh et al. 2013b). Other studies with different PIs have indicated multigenic control of resistance (Davis et al. 2006) or a combination of recessive and dominant genes (Tetteh et al. 2013a).

Genome-wide Association (GWAS) studies using historic powdery mildew (PM) screening (Davis et al. 2007; Tetteh et al. 2010) and genotyping by sequencing (GBS) data identified the presence of SNPs on the watermelon genome associated with resistance to PM race 2W in stem and leaves, some of them localized on chromosome 2 (Wu et al. 2019). A major QTL designated *pmr2.1*, which explained 80% of the variation, was detected on chromosome 2 in a study by Kim et al. (2015) in the Indian watermelon variety "ArkaManik". The major QTL detected by Kim et al. (2015) overlapped with the peak detected on chromosome 2 by the GWAS study (Wu et al. 2019). In a more recent study by Mandal et al. (2020), using a combination of RNAseq and whole genome resequencing (WGS) using PM-resistant USVL531-MDR and susceptible parent (USVL677-PMS) identified a candidate NLR resistance gene-designed *ClaPMR2*. This gene is homologous to the *Arabidopsis thaliana* PM resistance protein *AtRPW8* and increased its expression > twofold in the resistant line USVL531-MDR compared to the susceptible line USVL677-PMS eight days after inoculation with PM. Whole genome resequencing also confirmed the introgression breakpoints in four advanced PM resistance RIL lines developed from a cross of USVL531-MDR and USVL77-PMS. The transcriptome data further revealed a complex regulatory network associated with the introgressed junctions mediated by PM resistance R-proteins that may involve multiple signal regulators and transducers, carbohydrate metabolism, cell wall modifications, and the hormone-signaling pathway (Mandal et al. 2020). A cleaved amplified polymorphic sequence (CAPS) marker developed based on a SNP on *ClaPMR2* that co-segregated with the resistance locus in USVL531-MDR effectively identified resistant and susceptible individuals in an F_2 population developed from a cross of USVL531-MDR and USVL677-PMS with 99% accuracy.

Grafting of PM susceptible variety "Mickey Lee" on resistant watermelon and bottle gourd lines demonstrated the utility of grafting on PM-resistant rootstocks as a means of using host resistance to manage PM (Kousik et al. 2018c). Grafting on some of the PM-resistant rootstocks provided significant resistance to the scion (the upper part of the graft). However, the level of resistance provided by the various PM-resistant rootstocks to the susceptible scion varied. A distinctly different metabolomic profile was observed in some of the PM-resistant germplasm lines that were used as rootstocks compared to susceptible scion, indicating that melatonin was the most significantly translocated metabolite from the PM-resistant rootstock to the PM susceptible scion (Mahmud et al. 2015). Further studies in which the PM susceptible watermelon line USVL677 was transformed with the melatonin biosynthetic gene (serotonin N-acetyltransferease, SNAT) from the resistant line indicated that the hormone melatonin played a role in the defense mechanism (Mandal et al. 2018). Transformed watermelon plants overexpressing SNAT displayed higher levels of melatonin and a reduced development of PM compared to the susceptible USVL677-PMS. This reduction on PM symptoms was also observed when melatonin was applied exogenously in other cucurbits (Mandal et al. 2018).

Overall, the research on resistance to PM in watermelon is advancing rapidly. PM-resistant seedless and seeded watermelon varieties are now being developed and released by various seed companies. Watermelon varieties such as the seedless "Suprema", commercial pollinizers ("SP5" and Lion), and the highly resistant line USVL531-MDR were found to be resistant to PM including a South Carolina prevailing strain of *P. xanthii* (Kousik et al. 2019). The USDA Vegetable Laboratory (Charleston, SC) has been developing watermelon germplasm lines with high levels of resistance to powdery mildew and recently released four lines with broad resistance to six PM isolates from different states in the US (Kousik et al. 2018a).

Bacterial Fruit Blotch (BFB): Currently, there is a lack of commercial resistant cultivars to BFB. Large-scale screening in laboratories, greenhouses, and fields suggests that foliar resistance is multigenic (Branham et al. 2019). However, few studies have been conducted to identify fruit-resistant germplasm (Carvalho et al. 2013). Recently, Daley and Wehner (2021) through a three-year germplasm screening of 1452 *Citrullus* spp. cultigens (PIs and commercial cultivars) identified BFB fruit resistance under field conditions at the immature fruit stage. Several African PIs such as PI 494819 (*C. lanatus*, Zambia), PI 596659 (*C. amarus* Schrad., South Africa), PI 596670 (*C. amarus*, South Africa), PI 490384 [*C. mucosospermus* (Fursa) Fursa, Mali], and PI 596656 (*C. amarus*, South Africa) had the lowest average disease ratings compared with the commercial check, Charleston Gray, being a potential source of resistance for breeding programs, and model to understand the genetics of resistance mechanism.

4.5 Genomics Approaches to Accelerate Resistance Gene Discovery

4.5.1 Resistance Genes in the Watermelon Genome

Cultivated watermelon genomes (365 Mb) were among the first Cucurbitaceae genomes being sequenced, assembled, and annotated (Guo et al. 2013, 2019) together with several cucurbit genomes, including cucumber Chinese long (Huang et al. 2009; Li et al. 2019), cucumber inbred line Gy14 (Yang et al. 2012), wild cucumber accession PI 183967 (*C. sativus* var. *hardwickii*) (Qi et al. 2013), double-haploid muskmelon line DHL92 (Garcia-Mas et al. 2012), *Cucurbita maxima* and *Cucurbita moschata* (Sun et al. 2017), *Cucurbita pepo* (Montero-Pau et al. 2018), silver seed gourd *Cucurbita argyrosperma* (Barrera-Redondo et al. 2019), bottle gourd (*Lagenaria siceraria*) (Wu et al. 2017a, b), wax gourd *Benincasa hispida* (Xie et al. 2019), the Kordofan melon *Citrullus lanatus* subsp. *Cordophanus*, the possible progenitor of the cultivated watermelon (Renner et al. 2021), and *Cucumis hystrix* a wild perennial relative of cucumber (Qin et al. 2021). Genomic resources are integrated with tools for comparative genomics and transcriptomic analysis in a family-wide cucurbit genomics database (CuGenDB; http://cucurbitgenomics.org) (Zheng et al. 2019; Yu et al. 2023). The availability of such genomic resources is useful for breeding programs because it is the foundation for the systematic analysis of the *R*-gene repertoire in the Cucurbitaceae (Hassan et al. 2019).

The initial watermelon genome reference was obtained from the diploid Asian cultivar “97103” ($2n = 2x = 22$), using Illumina short reads (Guo et al. 2013). Over 85% of genes on the watermelon genome were functionally identified, and 23,440 protein-coding genes were initially predicted, which is close to the number of genes found in cucumber (Huang et al. 2009), but with a comparatively lower number (44 vs. 78) of genes coding for NLR genes. Conserved domain and motif analysis of NLRs in watermelon found seven canonical (CC)-NLRs, 11 canonical (TIR)-NLRs, and three (RWP8) NLR-encoding genes distributed throughout all chromosomes (Hassan et al. 2019). No evidence for sequence exchange between different homologous NLRs in the watermelon genome suggests independent evolution of those genes and a general reduction of their variability. In addition to canonical NLRs, clusters of receptor-like genes coding for potential transmembrane pattern recognition receptors (PRRs) were identified, including 162 genes coding for receptor-like kinases with intracellular kinase and extracellular LRR domains as well as 35 genes coding for receptor-like proteins lacking a kinase domain (Guo et al. 2013). The first watermelon reference genome assembly was improved by (Guo et al. 2019), using PacBio long reads combined with BioNano optical and Hi-C chromatin interaction maps, increasing the N50 values and reducing the number of predicted genes to 22,596 compared with the first assembly. This new assembly consists of 149 scaffolds with an N50 size of 21.9 Mb. In addition, 414 watermelon accessions collected at different geographic locations were resequenced to investigate the diversity present in seven species of the *Citrullus* genus.

Genetic analysis of agricultural traits like fruit quality and disease resistance using association mapping and QTL analysis is limited using as reference the genome of unique watermelon cultivar, especially when the variation in gene presence/absence and copy numbers in the cultivar "97103" is compared with American germplasm. This propelled the de novo sequencing of the main American watermelon cultivar "Charleston Gray", which found that a significant number of genomic variations overlapped with QTLs of economic importance such as weight and shape. The Charleston gray assembly consists of 2034 scaffolds, with a N50 of 7.47 Mb and 22,546 predicted protein-coding genes (Wu et al. 2019).

Watermelon cultivars share a narrow genetic base resulting from years of breeding and selection focused on quality traits such as sugar content, flesh color, and rind pattern. During watermelon domestication, many resistance genes were lost, and little is known about NLR content and features in resistant and susceptible watermelon lines/cultivars and their role in plant defense to several pathogens (Levi et al. 2017). Resequencing of representative watermelon accessions showed that the semi-cultivated *Citrullus* species *C. colocynthis*, *C. amarus*, and *Citrullus ecirrhosus*, which can be intercrossed with *C. lanatus*, display the highest nucleotide diversity and are valuable to identify and characterize new sources of disease and pest resistance including NLR genes (Levi et al. 2017; Simmons et al. 2019).

4.5.2 Genome Survey of NLR-Encoding Genes in Watermelon and Other Cucurbitaceae

The analysis and inventory of the complete set of NLRs genes or NLR complement (NLRome) in watermelon genomes will reveal the diversity of *R*-genes in this important species and is a valuable resource for mapping-based cloning of resistance genes (Zhang et al. 2020) and breeding for disease resistance (Bent 2016; Giolai et al. 2017). *R*-genes or *R*-gene analogs also are valuable genomic resources for genome-wide mapping, to develop high-density genetic maps, diagnostic markers, and for QTL co-localization studies (Sekhwal et al. 2015).

Phylogenetic analysis of NLRs in sequenced genomes of Cucurbitaceae species indicates that those genes are ancient, being present before divergence with other dicots like *Arabidopsis*, diversified by duplication, frequent gene loss, mutations, and deficient duplications in extant NLR lineages (Wan et al. 2013). Cucurbitaceae genomes have a relatively low NLR number and low NLR lineages, and frequent gene loss has been pointed out as the major cause of this reduction (Lin et al. 2013). The repetitive nature of NLRs, complex genomic organization, variation (mutations, indels), and low expression between and within species make the annotation and curation challenging or time-consuming; hence, it is expected that many NLRs be misannotated by gene prediction algorithms based in genomic data (Meyers et al. 2003; Sekhwal et al. 2015). Moreover, it is estimated the approximately half of *R*-genes annotated in Cucurbitaceae genomes are pseudogenes (Lin et al. 2013). Truncated pseudogenes could confer resistance to pathogens since the absence of one NLR protein domain does not necessarily imply a lack of function (Nishimura et al. 2017). The ultimate goal of an NLRome survey is to capture a substantial fraction of intra- and interspecific haplotype diversity, understand NLR evolution (Van de Weyer et al. 2019), and ultimately discover their specificities and roles as sensor or helpers. However, it is unlikely that a single reference genome would be enough to capture all NLR diversity including presence/absence polymorphisms, haplotypes, and domain diversity (Bittner-Eddy et al. 2000; Botella et al. 1998); therefore, the assembly of more genomes from different accessions or varieties of the same species would help us to reach this goal. Nevertheless, depending on the species, it is not feasible to do an inventory of the NLR repertoire and determine orthology relationships using solely genome comparisons (Baragan and Weigel 2020).

To overcome these constraints, strategies for NLR surveys in cucurbits had included the use of degenerated primers based in the conserved NBS domain (Gharaei and Ghayeb Zamharir 2017; Lin et al. 2013; Reddy et al. 2019; Wan et al. 2013) when genomic information was limited. However, these experimental strategies frequently fail to detect all NLR complements or create inaccurate inventories (Giolai et al. 2016; Wan et al. 2013). With the reduction in the cost of whole genome sequencing, automated gene callers and manual curation have been used for identification of NLR-encoding genes using computational homology-based methods and sequenced alignment from curated databases, but it has been labor-intensive and insufficiently accurate (Steuernagel et al. 2015).

In recent years, the application of targeted enrichment technologies to reduce genome complexity has allowed the enrichment of genomic regions of interest and variant identification without whole genome sequencing (Mamanova et al. 2010; Zhang et al. 2020). These technologies have been successfully applied in the capture of NLRs by exhaustive sequencing of the NLRome in genomic DNAs in a method known as *R*-gene enrichment sequencing (RenSeq) (Jupe et al. 2013). RenSeq is based on the use of a RNA bait library (overlapping biotinylated RNA probes from 60 to 120-mer complementary to full or partially annotated NLRs from the same or a related species) that hybridize to target NLR loci present in a fragmented genomic DNA of the plant of interest (Jupe et al. 2014). Target DNA captured is amplified and subsequently sequenced using long-read high-throughput sequencing platforms (Giolai et al. 2017).

Baits for RenSeq are designed from reference gene models available; however, as we mentioned previously, NLR identification and annotation can be very time-consuming, and the automated gene callers misannotate or fail to identify a considerable percentage of NLR complements (Andolfo et al. 2014; Meyers et al. 2003). Computational approaches for NLR prediction have been using several pipelines and bioinformatic methods based on their conserved domains and motifs, but a standardized method and consistent annotation criteria without limitations have not been developed (Sekhwal et al. 2015). Some of the computational approaches such as RGAugury use protein sequences to detect conserved domains and motifs found in *R*-genes and use a integrative approach with databases like BLAST, InterProScan, pfam_scan, nCoil, and Phobius to generate a list of candidate genes including membrane-associated receptors (RLPs or RLKs) and NLRs (Li et al. 2016). The plant resistance gene database PRGdb uses transcriptomic, proteomic, and hidden Markov models (HMM) to automatically predict and annotate NLRs (Osuna-Cruz et al. 2018). An improvement in the automatic and accurate identification was made with automated NLR annotation tools such as NLR-Parser (Steuernagel et al. 2015). NLR-Parser uses motif alignment and a search tool (MAST) to identify NLR sequences based on curated amino acid motifs found in plant NLR proteins independent of gene annotation. NLR-Annotator, an expanded version of NLR-Parser, identifies unannotated (ab initio) NLRs from whole genomic sequence data with the possibility to identify novel domain structures (Steuernagel et al. 2020). NLR annotations in genome assemblies can be made also with NLGenomeSweeper (Toda et al. 2020), a pipeline based on the identification of the complete NB-ARC domain.

NLR automatic annotation tools like NLR-Parser have been successfully applied to improve the prediction on NLR annotation in a wide variety of species (Mondragón-Palomino et al. 2017; Parada-Rojas and Quesada-Ocampo 2019; Stam et al. 2019) and wild genotypes (Van de Weyer et al. 2019) with high sensitivity and specificity compared with previously reported annotations based on other bioinformatics tools or manual procedures (Kourelis et al. 2020). When NLR-Parser was used against the predicted proteins from watermelon sequenced genomes (97103, Charleston Gray) it yielded 56

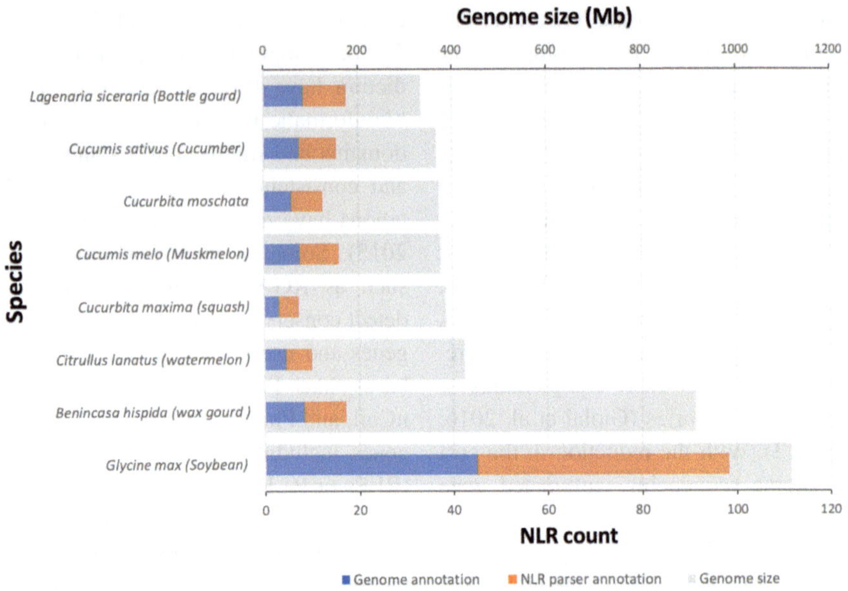

Fig. 4.3 Genome size and intracellular receptor proteins (NLRs) counts reported with automated gene callers (genome annotation) and estimated (this book chapter) using NLR-Parser annotation tool for selected Cucurbitaceae genomes. For comparative purposes, soybean (*Glycine max*, Fabaceae) NLRs were included

and 50 putative NLRs (compared with 44 NLRs previously reported), of which 20 and 18 were complete, 22 and 26 were members of the CNL class, and 17 and 14 of the TNL class, respectively. As previously mentioned, the relative number of NLRs per genome in cucurbits compared with other diploid species continues to be lower, and watermelon together with *C. maxima* displayed the lowest NLR number among genomes sequenced (Fig. 4.3).

The phylogenetic relationship of watermelon NLRs was estimated using complete NLRs identified with NLR-Parser in other cucurbits genomes sequenced http://cucurbitgenomics.org (Zheng et al. 2019) and the reference plant NLR database (RefPlantNLR) (Kourelis et al. 2020), a collection of 415 experimentally validated plant NLRs belonging to 11 orders of flowering plants (Fig. 4.4). Amino acid alignments of the highly conserved NBS domain were used to characterize NLRs, minimizing the variability of other domains. The unrooted phylogenetic tree separated cucurbit NLRs families and showed a reduced diversity among the cucurbits with less representation of the CNL family. Clades showed a mostly monophyletic origin for NLRs from all cucurbit species with an evident orthologous relationship previously documented (Lin et al. 2013).

The practical value with implications for breeding for characterization of NLRomes is the functional analysis of all NLRs or NLR candidates based on presence–absence variation (PAV) or haplotype diversity on resistant genotypes. NLRs can be expressed in heterologous plant systems such as tobacco or other plant models to determine their hypersensitive response and possible role in ETI-mediated disease resistance (Wang et al. 2015a, b; Xu et al. 2018) and for engineering of broad or specific disease resistance or introgression on susceptible backgrounds (Monteiro and Nishimura 2018; Narusaka et al. 2013a; Prigozhin and Krasileva 2021). Pan-NLRomes also can help to determine plant protein interactors since indirect interaction of effectors with NLRs seems to be the norm; various alternative single or multi-protein recognition targets (interactome) would be interacting with NLRs being potential targets for breeding (Weßling et al. 2014). NLRomes can help to determine what pathogen effectors

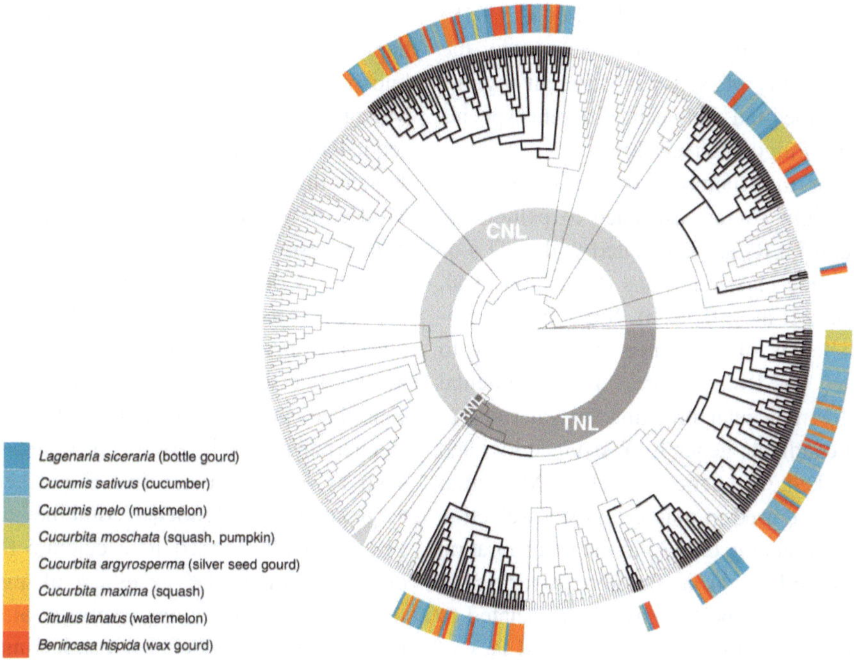

Fig. 4.4 Phylogenetic comparison of Cucurbitaceae NLR genes with plant NLR database (RefPlantNLR) collection. Complete NLRs for three major families of NLRs (CNL, TNL, and RNL) identified by the NLR-Parser tool (Steuernagel et al. 2015) on cucurbits transcriptome and curated NLRs (Kourelis et al. 2020) were considered. Phylogenetic tree was generated by RAxM (Stamatakis 2014) after alignment of NB-ARC domains using MUSCLE (Edgar 2004). Cucurbitaceous species are shown in color code, and their clades are bolded

interact with which NLRs (Baragan and Weigel 2020); this has important application for target breeding since breeders can select or design NLR haplotypes with recognition specificities for optimal response to a pathogen race without yield penalty (Harris et al. 2013). The interaction of NLRs with pathogen effectors can be exploited to accelerate disease resistance breeding and NLR gene cloning. Pan-NLRomes could be tested against the potential effector complement(s) of several races of a pathogen or pan-effectoromes. NLR candidates are screened in a heterologous system for transient complementation tests with pathogen effectors. These functional assays avoid redundancy in cloning efforts and benefit breeding by detecting resistance specificities in germplasm, finding better strategies for *R*-gene pyramiding, or selecting NLR haplotypes with a broad spectrum of resistance (Vleeshouwers and Oliver 2014).

4.6 Future Perspectives

Watermelon is prone to numerous and devastating diseases that negatively impact its production, which is in a steady increase in demand. Watermelon disease control has been limited by few phytosanitary alternatives available and the lack of resistant germplasm that slows the progress in breeding for resistance. The relatively low number of NLRs and variability in cucurbits would explain to some extent the widespread susceptibility to pathogens on watermelon (Lin et al. 2013; Song et al. 2019; Wroblewski et al. 2007), and it is also more likely that other types of *R*-genes are involved in resistance observed (Berg et al. 2021). Cucurbit NLRomes are experimental model systems for understanding plant immunity and NLR networks (Wu et al. 2017a, b) and offer a great opportunity to study its regulation, fitness cost (Richard et al. 2018) and adaptation to multiple environments by the specific variability of NLRs in different regions of the genome and between different species (Morata and Puigdomènech 2017).

It is clear that members of the watermelon NLRome might confer resistance to pathogens, since several NLRs are associated with QTLs or resistance gene cluster regions (Chovelon et al. 2021; Zhang et al. 2013a, b). Since NLRs provide high levels of resistance and are easy to manipulate, NLRomes are a fundamental resource for molecular breeding in watermelon-providing candidate genes for resistance loci and closely linked or gene markers. Molecular markers can be used in marker-assisted selection to speed the breeding or help in the incorporation of multiple NLRs in a single cultivar (gene stacking) to overcome the caveats associated to monogenic resistant (race specificity and lack of durability) (Marone et al. 2013). An alternative approach is the development of transgenic plants that express NLRs in Cucurbitaceae (Liu et al. 2016; Narusaka et al. 2013b) taking advantage of the important progress that has been made to improve the transformation protocols in this plant family (Nanasato and Tabei 2020).

An important question to answer when NLRomes are generated is to determine the number of individuals needed to capture a representative fraction of the species pan-genome. This depends greatly on the diversity of accessions considered (Baragan and Weigel 2020). Genotyping of watermelon germplasm collected throughout the world revealed a clear distant relationship and genetic diversity of *C. amarus* (used as major source of disease resistance alleles in watermelon breeding) from *C. lanatus* and *C. mucosospermus*, which will be useful to identity potential genes/haplotypes for resistant to major watermelon diseases and cultivar development by germplasm interspecific hybridization (Guo et al. 2019; Levi et al. 2013; Thies et al. 2010).

RenSeq and derivative applications for *R*-gene validations might take advantage of diversity panels of watermelon wild relatives for rapid gene cloning of typical NLRs. An obvious strategy is to improve the bait design and capture efficiency for RenSeq after discovering new NLR haplotypes on wild relatives using baits based on known NLRs found in the reference genome. Mutagenesis resistance gene enrichment and sequencing (MutRenSeq) combines RenSeq sequencing data with mutagenesis in the genome. MutRenSeq includes the generation by chemical mutagenesis and screening of susceptible mutants. Comparison RenSeq data sequence between resistant parental and multiple independently derived mutants helps to identify *R*-genes after screening for mutations in a single candidate gene associated with susceptibility (Steuernagel et al. 2016). New developments such as associated genetics *R*-gene enrichment sequencing (AgRenSeq) combine association genetics with RenSeq to accelerate the *R*-gene cloning without the necessity of mapping or mutagenized populations (Arora et al. 2019). In this strategy, RenSeq is conducting disease-resistant diversity panels such as wild relatives to different pathogen races. Resistance or susceptibility phenotypic data observed on diversity panels is associated with the presence–absence of sub-sequences (*k*-mers) using accounting algorithm identifying NLRs for resistance.

References

Adachi H, Derevnina L, Kamoun S (2019) NLR singletons, pairs, and networks: evolution, assembly, and regulation of the intracellular immunoreceptor circuitry of plants. Curr Opin Plant Biol 50:121–131

American Seed Trade Association (2009) Bacterial fruit blotch: a commercial growers guide

Andersen EJ, Nepal MP, Purintun JM, Nelson D, Mermigka G, Sarris PF (2020) Wheat disease resistance genes and their diversification through integrated domain fusions. Front Genet 11:898

Andolfo G, Jupe F, Witek K, Etherington GJ, Ercolano MR, Jones JDG (2014) Defining the full tomato NB-LRR resistance gene repertoire using genomic and cDNA RenSeq. BMC Plant Biol 14:120

Arora S, Steuernagel B, Gaurav K, Chandramohan S, Long Y, Matny O, Johnson R, Enk J, Periyannan S, Singh N, Asyraf Md Hatta M, Athiyannan N, Cheema J, Yu G, Kangara N, Ghosh S, Szabo LJ, Poland J, Bariana H, Wulff BBH (2019) Resistance gene cloning from a wild crop relative by sequence capture and association genetics. Nat Biotechnol 37:139–143

Avenot HF, Thomas A, Gitaitis RD, Stevenson Langston Jr DB, Stevenson KL (2012) Molecular characterization of boscalid- and penthiopyrad-resistant isolates of *Didymella bryoniae* and assessment of their sensitivity to fluopyram. Pest Manag Sci 68:645–651

Bahar O, Kritzman G, Burdman S (2009) Bacterial fruit blotch of melon: screens for disease tolerance and role of seed transmission in pathogenicity. Eur J Plant Pathol 123:71–83

Baragan AC, Weigel D (2020) Plant NLR diversity: the known unknowns of pan-NL Romes. Plant Cell 33:814–831

Bari R, Jones JDG (2009) Role of plant hormones in plant defence responses. Plant Mol Biol 69:473–488

Barrera-Redondo J, Ibarra-Laclette E, Vázquez-Lobo A, Gutiérrez-Guerrero YT, Sánchez de la Vega G, Pinero D, Montes-Hernández S, Lira-Saade R, Eguiarte LE (2019) The genome of *Cucurbita argyrosperma* (silver-seed gourd) reveals faster rates of protein-coding genes and long noncoding RNA turnover and neofunctionalization within Cucurbita. Mol Plant 12:506–520

Basım E, Basım H, Abdulai M, Baki D, Öztürk N (2016) Identification and characterization of *Didymella bryoniae* causing gummy stem blight disease of watermelon (*Citrullus lanatus*) in Turkey. Crop Prot 90:150–156

Belanger RR, Dik AJ, Bushnell WR, Carver TLW (2002) The powdery mildews: a comprehensive treatise. American Phytopathological Society Press, St. Paul, Minnesota

Ben-Naim Y, Cohen Y (2015) Inheritance of resistance to powdery mildew race 1W in watermelon. Phytopathology 105:1446–1457

Bent A (2016) Resistance from relatives. Nat Biotechnol 34:620–621

Berg JA, Appiano M, Bijsterbosch G, Visser RGF, Schouten HJ, Bai Y (2017) Functional characterization of cucumber (*Cucumis sativus* L.) Clade V MLO genes. BMC Plant Biol 17:80

Berg JA, Hermans FWK, Beenders F, Abedinpour H, Vriezen WH, Visser RGF, Bai Y, Schouten HJ (2021) The amino acid permease (AAP) genes CsAAP2A and SlAAP5A/B are required for oomycete susceptibility in cucumber and tomato. Mol Plant Pathol 22:658–672

Bittner-Eddy PD, Crute IR, Holub EB, Beynon JL (2000) RPP13 is a simple locus in *Arabidopsis thaliana* for alleles that specify downy mildew resistance to different avirulence determinants in *Peronospora parasitica*. Plant J 21:177–188

Block CC, Shepherd LM (2008) Long-term survival and seed transmission of *Acidovorax avenae* subsp. *citrulli* in melon and watermelon seed. Plant Health Prog 9:36

Botella MA, Parker JE, Frost LN, Bittner-Eddy PD, Beynon JL, Daniels MJ, Holub EB, Jones JDG (1998) Three genes of the Arabidopsis *RPP1* complex resistance locus recognize distinct *Peronospora parasitica* virulence determinants. Plant Cell 10:1847–1860

Boyhan GE, Norton JD, Abrahams BR (1994a) Screening for resistance to anthracnose (race 2) gummy stem blight and root knot nematode in watermelon germplasm. Cucurbit Genet Coop Rep 17:1488–1499

Boyhan GE, Norton JD, Abrahams BR, Wen HH (1994b) A new source of resistance to anthracnose (race 2) in watermelon. Hortic Sci 29:111–112

Branham SE, Levi A, Farnham MW, Wechter WP (2017) A GBS-SNP-based linkage map and quantitative trait loci (QTL) associated with resistance to Fusarium oxysporum f. sp.niveum race 2 identified in Citrullus lanatus var.citroides. Theor Appl Genet 130:319–330

Branham SE, Patrick Wechter W, Lambel S, Massey L, Ma M, Fauve J, Farnham MW, Levi A (2018) QTL-seq and marker development for resistance to *Fusarium oxysporum* f. sp. *Niveum* race 1 in cultivated watermelon. Mol Breed 38:139

Branham SE, Levi A, Katawczik ML, Wechter WP (2019) QTL mapping of resistance to bacterial fruit blotch in *Citrullus amarus*. Theor Appl Genet 132:1463–1471

Branham SE, Wechter WP, Ling KS, Chanda B, Massey L, Zhao G, Guner N, Bello M, Kabelka E, Fei Z, Levi A (2020) QTL mapping of resistance to *Fusarium oxysporum* f. sp. *Niveum* race 2 and Papaya ringspot virus in *Citrullus amarus*. Theor Appl Genet 133:677–687

Braun U, Cook TA, Inman AJ, Shin HD (2002) The taxonomy of powdery mildew fungi. In: Belanger RR, Bushnell WR, Dik AJ, Carver TLW (eds) The powdery mildews. American Phytopathological Society (APS) Press, St. Paul, Minnesota, pp 13–15

Brewer MT, Rath M, Li HX (2015) Genetic diversity and population structure of cucurbit gummy stem blight fungi based on microsatellite markers. Phytopathology 105:815–824

Carlile MJ (1956) A study of the factors influencing non-genetic variation in a strain of *Fusarium oxysporum*. J Gen Microbiol 14:643–654

Carvalho FCQ, Santos LA, Dias RCS, Mariano RLR, Souza EB (2013) Selection of watermelon genotypes for resistance to bacterial fruit blotch. Euphytica 190:169–180

Cesari S, Thilliez G, Ribot C, Chalvon V, Michel C, Jauneau A, Rivas S, Alaux L, Kanzaki H, Okuyama Y, Morel JB, Fournier E, Tharreau D, Terauchi R, Kroj T (2013) The rice resistance protein pair RGA4/RGA5 recognizes the *Magnaporthe oryzae* effectors AVR-Pia and AVR1-CO39 by direct binding. Plant Cell 25:1463–1481

Cesari S, Bernoux M, Moncuquet P, Kroj T, Dodds P (2014) A novel conserved mechanism for plant NLR protein pairs: the "integrated decoy" hypothesis. Front Plant Sci 5:606

Chisholm ST, Coaker G, Day B, Staskawicz BJ (2006) Host-microbe interactions: shaping the evolution of the plant immune response. Cell 124:803–814

Choi HW, Klessig DF (2016) DAMPs, MAMPs and NAMPs in plant innate immunity. BMC Plant Biol 16:232

Chovelon V, Feriche-Linares R, Barreau G, Chadoeuf J, Callot C, Gautier V, Le Paslier MC, Berad A, Faivre-Rampant P, Lagnel J, Boissot N (2021) Building a cluster of NLR genes conferring resistance to pests and pathogens: the story of the Vat gene cluster in cucurbits. Hortic Res 8:72

Cohen Y (1977) The combined effects of temperature, leaf wetness, and inoculum concentration on infection of cucumbers with *Pseudoperonospora cubensis*. Can J Bot 55:1478–1487

Cohen Y, Rubin AE (2012) Mating type and sexual reproduction of *Pseudoperonospora cubensis*, the downy mildew agent of cucurbits. Eur J Plant Pathol 132:577–592

Cohen R, Burger Y, Katzir N (2004) Monitoring physiological races of *Podosphaera xanthii* (syn. *Sphaerotheca fuliginea*), the causal agent of powdery mildew in cucurbits: factors affecting race identification and the importance for research and commerce. Phytoparasitica 32:174–183

Cohen R, Burger Y, Carmela H, Koren A, Edelstein E (2007) Introducing grafted cucurbits to modern agriculture: the Israeli experience. Plant Dis 91:916–923

Collier SM, Hamel LP, Moffett P (2011) Cell death mediated by the N-terminal domains of a unique and highly conserved class of NB-LRR protein. MPMI 24:918–931

Crandall SG, Rahman A, Quesada-Ocampo LM, Martin FN, Bilodeau GJ, Miles TD (2018) Advances in diagnostics of downy mildews: lessons learned from other oomycetes and future challenges. Plant Dis 102:265–275

Daley J, Wehner TC (2021) Screening for bacterial fruit blotch resistance in watermelon fruit. Crop Sci 61:1228–1240

Dangl JL, Jones JDG (2001) Plant pathogens and integrated defence responses to infection. Nature 411:826–833

D'Arcangelo KN, Adams ML, Kerns JP, Quesada-Ocampo LM (2021) Assessment of fungicide product applications and program approaches for control of downy mildew on pickling cucumber in North Carolina. Crop Prot 140:105412

Davis AR, Bruton BD, Pair SD, Thomas CE (2001) Powdery mildew: an emerging disease of watermelon in the United States. Cucurbit Genet Coop Rep 28:24–32

Davis AR, Levi A, Wehner T, Pitrat M (2006) PI 525088-PMR, a melon race 1 powdery mildew-resistant watermelon line. Hortic Sci 41:1527–1528

Davis AR, Levi A, Tetteh A, Wehner T, Russo V, Pitrat M (2007) Evaluation of watermelon and related species for resistance to race 1W powdery mildew. J Am Soc Hortic Sci 132:790–795

Dean R, Van Kan JAL, Pretorius ZA, Hammond-Kosack KE, Di Pietro A, Spanu PD, Rudd JJ, Dickman M, Kahmann R, Ellis J, Foster GD (2012) The top 10 fungal pathogens in molecular plant pathology: top 10 fungal pathogens. Mol Plant Pathol 13:414–430

Devoto A, Piffanelli P, Nilsson I, Wallin E, Panstruga R, von Heijne G, Schulze-Lefert P (1999) Topology, subcellular localization, and sequence diversity of the Mlo family in plants. J Biol Chem 274:34993–35004

DeYoung BJ, Innes RW (2006) Plant NBS-LRR proteins in pathogen sensing and host defense. Nat Immunol 7:1243–1249

Dodds PN, Lawrence GJ, Catanzariti AM, Teh T, Wang CIA, Ayliffe MA, Kobe B, Ellis JG (2006) Direct protein interaction underlies gene-for-gene specificity and co-evolution of the flax resistance genes and flax rust a virulence genes. PNAS 103:8888–8893

Dow M, Newman MA, von Roepenack E (2000) The induction and modulation of plant defense responses by bacterial lipopolysaccharides. Annu Rev Phytopathol 38:241–261

Dutta B, Avci U, Hahn MG, Walcott RR (2012) Location of *Acidovorax citrulli* in infested watermelon seeds is influenced by the pathway of bacterial invasion. Phytopathology 102:461–468

Eckshtain-Levi N, Munitz T, Živanović M, Traore SM, Spröer C, Zhao B, Welbaum G, Walcott R, Sikorski J, Burdman S (2014) Comparative analysis of type III secreted effector genes reflects divergence of *Acidovorax citrulli* strains into three distinct lineages. Phytopathology 104:1152–1162

Edgar RC (2004) MUSCLE: a multiple sequence alignment method with reduced time and space complexity. BMC Bioinform 5:113

Ellis J, Dodds P, Pryor T (2000) Structure, function and evolution of plant disease resistance genes. Curr Opin Plant Biol 3:278–284

Erbs G, Silipo A, Aslam S, De Castro C, Liparoti V, Flagiello A, Pucci P, Lanzetta R, Parrilli M, Molinaro A, Newman MA, Cooper RM (2008) Peptidoglycan and muropeptides from pathogens *Agrobacterium* and *Xanthomonas* elicit plant innate immunity: structure and activity. Chem Biol 15:438–448

Everts KL, Egel DS, Langston D, Zhou XG (2014) Chemical management of *Fusarium* wilt of watermelon. Crop Prot 66:114–119

Fall LA, Clevenger J, McGregor C (2018) Assay development and marker validation for marker assisted selection of *Fusarium oxysporum* f. sp. *Niveum* race 1 in watermelon. Mol Breed 38:130

FAO (2019) Watermelon production report (FAOSTAT statistical database). Food and Agriculture Organization of the United Nations

Felix G, Regenass M, Boller T (1993) Specific perception of sub nanomolar concentrations of chitin fragments by tomato cells: induction of extracellular alkalinization, changes in protein phosphorylation, and establishment of a refractory state. Plant J 4:307–316

Felix G, Duran JD, Volko S, Boller T (1999) Plants have a sensitive perception system for the most conserved domain of bacterial flagellin: plants perceive a conserved domain of bacterial flagellin. Plant J 18:265–276

Flor HH (1971) Current status of the gene-for-gene concept. Annu Rev Phytopathol 9:275–296

Fonseca JP, Mysore KS (2019) Genes involved in non host disease resistance as a key to engineer durable resistance in crops. Plant Sci 279:108–116

Friedman AR, Baker BJ (2007) The evolution of resistance genes in multi-protein plant resistance systems. Curr Opin Gen Dev 17:493–499

Gan P, Ikeda K, Irieda H, Narusaka M, O'Connell RJ, Narusaka Y, Takano Y, Kubo Y, Shirasu K (2013) Comparative genomic and transcriptomic analyses reveal the hemibiotrophic stage shift of *Colleototrichum* fungi. New Phytol 197:1236–1249

Garcia-Mas J, Benjak A, Sanseverino W, Bourgeois M, Mir G, Gonzalez VM, Henaff E, Camara F, Cozzuto L, Lowy E, Alioto T, Capella-Gutierrez S, Blanca J, Canizares J, Ziarsolo P, Gonzalez-Ibeas D, Rodriguez-Moreno L, Droege M, Du L et al (2012) The genome of melon (*Cucumis melo* L.). PNAS 109:11872–11877

Gharaei F, Ghayeb Zamharir M (2017) Study of new NBS-LRR genes analogues in cucurbits native types in Iran. Iran J Genet Plant Breed 6:8–15

Gimode W, Bao K, Fei Z, McGregor C (2021) QTL associated with gummy stem blight resistance in watermelon. Theor Appl Genet 134:573–584

Giolai M, Paajanen P, Verweij W, Percival-Alwyn L, Baker D, Witek K, Jupe F, Bryan G, Hein I, Jones JDG, Clark MD (2016) Targeted capture and sequencing of gene-sized DNA molecules. Biotechniques 61:6

Giolai M, Paajanen P, Verweij W, Witek K, Jones JDG, Clark MD (2017) Comparative analysis of targeted long read sequencing approaches for characterization of a plant's immune receptor repertoire. BMC Genom 18:564

Goode MJ (1958) Physiological specialization in *Colletotrichum lagenarium*. Phytopathology 48:79–83

Granke LL, Quesada-Ocampo L, Lamour K, Hausbeck MK (2012) Advances in research on *Phytophthora capsici* on vegetable crops in the United States. Plant Dis 96:1588–1600

Grumet R, Fei Z, Levi A, Mazourek M, McCreight JD, Schultheis J, Weng Y, HausbeckM KS, Ling KS, Linares-Ramirez A, McGregor C, Quesada-Ocampo L, Reddy U, Smart C, Wechter P, Wehner T, Wessel-Beaver L, Wintermantel WM (2020) The CucCAP project: leveraging applied genomics to improve disease resistance in cucurbit crops. Acta Hortic 1294:91–104

Guo S, Zhang J, Sun H, Salse J, Lucas WJ, Zhang H, Zheng Y, Mao L, Ren Y, Wang Z, Min J, Guo X, Murat F, Ham BK, Zhang Z, Gao S, Huang M, Xu Y, Zhong S et al (2013) The draft genome of watermelon (*Citrullus lanatus*) and resequencing of 20 diverse accessions. Nat Genet 45:51–58

Guo S, Zhao S, Sun H, Wang X, Wu S, Lin T, Ren Y, Gao L, Deng Y, Zhang J, Lu X, Zhang H, Shang J, Gong G, Wen C, He N, Tian S, Li M, Liu J et al (2019) Resequencing of 414 cultivated and wild watermelon accessions identifies selection for fruit quality traits. Nat Genet 51:1616–1623

Gururani MA, Venkatesh J, Upadhyaya CP, Nookaraju A, Pandey SK, Park SW (2012) Plant disease resistance genes: current status and future directions. Physiol Mol Plant Pathol 78:51–65

Gusmini G, Song R, Wehner TC (2005) New sources of resistance to gummy stem blight in watermelon. Crop Sci 45:582–588

Gusmini G, Rivera-Burgos LA, Wehner TC (2017) Inheritance of resistance to gummy stem blight in watermelon. Hortic Sci 52:1477–1482

Hammond-Kosack KE, Jones JDG (1997) Plant disease resistance genes. Annu Rev Plant Physiol Plant Mol Biol 48:575–607

Harris CJ, Slootweg EJ, Goverse A, Baulcombe DC (2013) Step wise artificial evolution of a plant disease resistance gene. PNAS 110:21189–21194

Hassan M, Rahim M, Jung HJ, Park JI, Kim HT, Nou IS (2019) Genome-wide characterization of NBS-encoding genes in watermelon and their potential association with gummy stem blight resistance. Int J Mol Sci 20:902

Hausbeck MK, Lamour KH (2004) *Phytophthora capsici* on vegetable crops: research progress and management challenges. Plant Dis 88:1292–1303

Heath MC (2000) Hypersensitive response-related death. Plant Mol Biol 44:321–334

Henderson WR, Jenkins SF, Rawlings JO (1970) The inheritance of *Fusarium* wilt resistance in watermelon, *Citrullus lanatus* (Thumb.) Mansf. J Am Soc Hortic Sci 95:276–282

Holmes GJ, Ojiambo PS, Hausbeck MK, Quesada-Ocampo L, Keinath AP (2015) Resurgence of cucurbit downy mildew in the United States: a watershed event for research and extension. Plant Dis 99:428–441

Hu Z, Deng G, Mou H, Xu Y, Chen L, Yang J, Zhan M (2018) A re-sequencing-based ultra-dense genetic map reveals a gummy stem blight resistance-associated gene in *Cucumis melo*. DNA Res 25:1–10

Huang S, Li R, Zhang Z, Li L, Gu X, Fan W, Lucas WJ, Wang X, Xie B, Ni P, Ren Y, Zhu H, Li J, Lin K, Jin W, Fei Z, Li G, Staub J, Kilian A et al (2009) The genome of the cucumber, *Cucumis sativus* L. Nat Genet 41:1275–1281

Hulbert SH, Webb CA, Smith SM, Sun Q (2001) Resistance gene complexes: evolution and utilization. Annu Rev Phytopathol 39:285–312

Jackson KL, Yin J, Ji P (2012) Sensitivity of *Phytophthora capsici* on vegetable crops in Georgia to mandipropamid, dimethomorph and cyazofamid. Plant Dis 96:1337–1342

Jacob F, Vernaldi S, Maekawa T (2013) Evolution and conservation of plant NLR functions. Front Immunol 4:297

Jang YJ, Seo M, Hersh CP, Rhee SJ, Kim Y, Lee GP (2019) An evolutionarily conserved non-synonymous SNP in a leucine-rich repeat domain determines anthracnose resistance in watermelon. Theor Appl Genet 132:473–488

Jarvis WR (1992) Managing diseases in greenhouse crops. American Phytopathological Society Press, St. Paul, Minnesota

Ji P, Csinos AS (2015) Effect of oxathiapiprolin on asexual life stages of *Phytophthora capsici* and disease development on vegetables. Annu Appl Biol 166:229–235

Jiang L, Sanogo S, Bosland PW (2015) Using recombinant inbred lines to monitor changes in the race structure of *Phytophthora capsici* in Chile Pepper in New Mexico. Plant Health Prog 16:235–240

Johal G, Briggs S (1992) Reductase activity encoded by the HM1 disease resistance gene in maize. Sci 258:985–987

Jones JDG, Dangl JL (2006) The plant immune system. Nature 444:323–329

Jupe F, Witek K, Verweij W, Śliwka J, Pritchard L, Etherington GJ, Maclean D, Cock PJ, Leggett RM, Bryan GJ, Cardle L, Hein I, Jones JDG (2013) Resistance gene enrichment sequencing (RenSeq) enables reannotation of the NB-LRR gene family from sequenced plant genomes and rapid mapping of resistance loci in segregating populations. Plant J 76:530–544

Jupe F, Chen X, Verweij W, Witek K, Jones JDG, Hein I (2014) Genomic DNA library preparation for resistance gene enrichment and sequencing (RenSeq) in plants. Methods Mol Biol 1127:291–303

Kanyuka K, Rudd JJ (2019) Cell surface immune receptors: the guardians of the plant's extracellular spaces. Curr Opin Plant Biol 50:1–8

Keen NT (1990) Gene-for-gene complementarity in plant-pathogen interactions. Annu Rev Genet 24:447–463

Keinath AP (2002) Survival of *Didymella bryoniae* in buried watermelon vines in South Carolina. Plant Dis 86:32–38

Keinath AP (2007) Sensitivity of populations of *Phytophthora capsici* from South Carolina to mefenoxam, dimethomorph, zoxamide and cymoxanil. Plant Dis 91:743–748

Keinath AP (2015a) Baseline sensitivity of *Didymella bryoniae* to cyprodinil and fludioxonil and field efficacy of these fungicides against isolates resistant to pyraclostrobin and boscalid. Plant Dis 99:815–822

Keinath AP (2015b) Efficacy of fungicides against powdery mildew on watermelon caused by *Podosphaera xanthii*. Crop Prot 75:70–76

Keinath AP (2017a) Anthracnose. In: Keinath AP, Wintermantel WM, Zitter TA (eds) Compendium of cucurbit diseases and pests. American Phytopathological Society Press, St. Paul, Minnesota, pp 54–56

Keinath AP (2017b) Bacterial Fruit Blotch. In: Keinath AP, Wintermantel WM, Zitter TA (eds) Compendium of cucurbit diseases and pests. American Phytopathological Society Press, St. Paul, Minnesota, pp 71–73

Keinath AP (2018) Minimizing yield and quality losses in watermelon with multi-site and strobilurin fungicides effective against foliar and fruit anthracnose. Crop Prot 106:72–78

Keinath AP, DuBose VB (2004) Evaluation of fungicides for prevention and management of powdery mildew on watermelon. Crop Prot 23:35–42

Keinath AP, Hassell RL (2014) Suppression of Fusarium wilt caused by *Fusarium oxysporum* f. sp. *Niveum* race 2 on grafted triploid watermelon. Plant Dis 98:1326–1332

Keinath AP, Kousik CS (2011) Sensitivity of isolates of *Phytophthora capsici* from the eastern United States to fluopicolide. Plant Dis 95:1414–1419

Keinath AP, Rennberger G, Kousik CS (2018) First report of resistance to boscalid in *Podosphaera xanthii*, cucurbit powdery mildew in South Carolina. Plant Health Prog 19:220–221

Keinath AP, Wintermantel WM, Zitter TA (2017) Fusarium Wilts. In: Keinath AP, Wintermantel WM, Zitter TA (eds) Compendium of cucurbit diseases and pests. American Phytopathological Society Press, St. Paul, Minnesota, pp 32–40

Kim KH, Hwang JH, Han DY, Park M, Kim S, Choi D, Kim Y, Lee GP, Kim ST, Park YH (2015) Major quantitative trait loci and putative candidate genes for powdery mildew resistance and fruit-related traits revealed by an intraspecific genetic map for watermelon (*Citrullus lanatus* var. Lanatus). PLoS ONE 10: e0145665

Kleczewski NM, Egel DS (2011) A diagnostic guide for *Fusarium* wilt of watermelon. Plant Health Prog 12:27

Kourelis J, Sakai T, Adachi H, Kamoun S (2020) Ref plant NLR: a comprehensive collection of experimentally validated plant NLRs. BioRxiv 2020.07.08.193961

Kousik CS, Ikerd JL (2014) Evidence for cucurbit powdery mildew pathogen races based on watermelon differentials. In Harvey M, Weng Y, Day B, Grumet R (eds) Proceedings of Cucurbitaceae 2014. J Am Soc Hortic Sci, Alexandria, VA

Kousik CS, Ikerd JL (2016) Evaluation of watermelon varieties for tolerance to powdery mildew and *Phytophthora* fruit rot, 2014. Plant Dis Manag Rep 10: V082

Kousik CS, Keinath AP (2008) First report of insensitivity to cyazofamid among isolates of *Phytophthora capsici* from the southeastern United States. Plant Dis 92:979–979

Kousik CS, Adams ML, Jester WR, Hassell R, Harrison HF, Holmes GJ (2011a) Effect of cultural practices and fungicides on *Phytophthora* fruit rot of watermelon in the Carolinas. Crop Prot 30:888–894

Kousik CS, Donahoo RS, Webster CG, Turechek WW, Adkins ST, Roberts PD (2011b) Outbreak of cucurbit powdery mildew on watermelon fruit caused by *Podosphaera xanthii* in Southwest Florida. Plant Dis 95:1586–1586

Kousik CS, Ikerd JL Hassell R (2012a) Effect of grafting onto resistant rootstocks on the development of *Phytophthora* fruit rot on susceptible scions, 2011. Plant Dis Manag Rep 6:V012

Kousik CS, Ikerd JL, Wechter P, Harrison H, Levi A (2012b) Resistance to *Phytophthora* fruit rot of watermelon caused by *Phytophthora capsici* in U.S. plant introductions. Hortic Sci 47:1682–1689

Kousik CS, Ikerd JL, Harrison HF (2014a) Development of pre- and postharvest *Phytophthora* fruit rot on watermelons treated with fungicides in the field. Plant Health Prog 15:145–150

Kousik CS, Ling KS, Adkins S, Webster CG, Turechek WW (2014b) *Phytophthora* fruit rot-resistant watermelon germplasm lines USVL489-PFR, USVL782-PFR, USVL203-PFR, and USVL020-PFR. Hortic Sci 49:101–104

Kousik CS, Parada C, Quesada-Ocampo L (2015) First report of *Phytophthora* fruit rot on bitter gourd (*Mormodica charantia*) and sponge gourd (*Luffa cylindrica*) caused by *Phytophthora capsici*. Plant Health Prog 16:93–94

Kousik CS, Brusca J, Turechek WW (2016) Diseases and disease management strategies take top research priority in the watermelon research and development group members survey (2014 to 2015). Plant Health Prog 17:53–58

Kousik CS, Ji P, Egel DS, Quesada-Ocampo LM (2017) Fungicide rotation programs for managing *Phytophthora* fruit rot of watermelon in Southeastern United States. Plant Health Prog 18:28–34

Kousik CS, Ikerd JL, Mandal MK (2018a) Evaluation of watermelon varieties for tolerance to powdery mildew and *Phytophthora* fruit rot, 2015. Plant Dis Manag Rep 12:V076

Kousik CS, Ikerd JL, Turechek WW (2018b) Development of *Phytophthora* fruit rot caused by *Phytophthora capsici* on resistant and susceptible watermelon fruit of different ages. Plant Dis 102:370–374

Kousik CS, Ikerd J, Mandal M, Adkins S, Turechek WW (2018c) Watermelon germplasm lines USVL608-PMR, USVL255-PMR, USVL313-PMR and USVL585-PMR with broad resistance to powdery mildew. Hortic Sci 53:1212–1217

Kousik CS, Ikerd JL, Mandal M (2019) Relative susceptibility of commercial watermelon varieties to powdery mildew. Crop Prot 125:104910

Krattinger SG, Lagudah ES, Spielmeyer W, Singh RP, Huerta-Espino J, McFadden H, Bossolini E, Selter LL, Keller B (2009) A putative ABC transporter confers durable resistance to multiple fungal pathogens in wheat. Science 323:1360–1363

Kroj T, Chanclud E, Michel-Romiti C, Grand X, Morel JB (2016) Integration of decoy domains derived from protein targets of pathogen effectors into plant immune receptors is widespread. New Phytol 210:618–626

Kunze G, Zipfel C, Robatzek S, Niehaus K, Boller T, Felix G (2004) The N terminus of bacterial elongation factor Tu elicits innate immunity in *Arabidopsis* plants. Plant Cell 16:3496–3507

Lambel S, Lanini B, Vivoda E, Fauve J, Patrick Wechter W, Harris-Shultz KR, Massey L, Levi A (2014) A major QTL associated with *Fusarium oxysporum* race 1 resistance identified in genetic populations derived from closely related watermelon lines using selective genotyping and genotyping-by-sequencing for SNP discovery. Theor Appl Genet 127:2105–2115

Lamour KH, Hausbeck MK (2000) Mefenoxam insensitivity and the sexual stage of *Phytophthora capsici* in Michigan cucurbit fields. Phytopathology 90:396–400

Latin RX, Hopkins DL (1995) Bacterial fruit blotch of watermelon: the hypothetical exam question becomes reality. Plant Dis 79:761–765

Latin RX, Rane KK (1990) Bacterial fruit blotch of watermelon in Indiana. Plant Dis 74:331

Layton DV (1937) The parasitism of *Colletotrichum lagenarium* (Pass.) Ells. and Halst. Iowa Agr Expt Sta Ann Bull 223

Lee RRQ, Chae E (2020) Variation patterns of NLR clusters in *Arabidopsis thaliana* genomes. Plant Commun 1:100089

Lee HA, Lee HY, Seo E, Lee J, Kim SB, Oh S, Choi E, Choi E, Lee SE, Choi D (2017) Current understandings of plant nonhost resistance. MPMI 30:5–15

Lee ES, Kim DS, Kim SG, Huh YC, Back CG, Lee YR, Siddique MI, Han K, Lee HE, Lee J (2021) QTL mapping for gummy stem blight resistance in watermelon (*Citrullus* spp.). Plants 10:500

Lefebvre V, Palloix A (1996) Both epistatic and additive effects of QTLs are involved in polygenic induced resistance to disease: a case study, the interaction pepper—*Phytophthora capsici* Leonian. Theor Appl Genet 93:503–511

Leslie JF, Summerell BA (2006) 12: morphological characters. In: Leslie JF, Summerell BA (eds) The Fusarium laboratory manual. Wiley, Ames, IA, pp 113–120

Levi A, Thies JA, Wechter WP, Harrison HF, Simmons AM, Reddy UK, Nimmakayala P, Fei Z (2013) High frequency oligonucleotides: targeting active gene (HFO-TAG) markers revealed wide genetic diversity among *Citrullus* spp. accessions useful for enhancing disease or pest resistance in watermelon cultivars. Genet Resour Crop Evol 60:427–440

Levi A, Jarret R, Kousik S, Patrick Wechter W, Nimmakayala P, Reddy UK (2017) Genetic resources of watermelon. In Grumet R, Katzir N, Garcia-Mas J (eds) Genetics and genomics of Cucurbitaceae. Plant genetics and genomics: crops and models, vol 20. Springer, Cham, pp 87–110

Li P, Quan X, Jia G, Xiao J, Cloutier S, You FM (2016) RGAugury: a pipeline for genome-wide prediction of resistance gene analogs (RGAs) in plants. BMC Genom 17:852

Li Q, Li H, Huang W, Xu Y, Zhou Q, Wang S, Ruan J, Huang S, Zhang Z (2019) A chromosome-scale genome assembly of cucumber (*Cucumis sativus* L.). GigaScience 8:giz072

Lin X, Zhang Y, Kuang H, Chen J (2013) Frequent loss of lineages and deficient duplications accounted for low copy number of disease resistance genes in Cucurbitaceae. BMC Genom 14:335

Liu L, Gu Q, Ijaz R, Zhang J, Ye Z (2016) Generation of transgenic watermelon resistance to Cucumber mosaic virus facilitated by an effective *Agrobacterium*-mediated transformation method. Sci Hortic 205:32–38

Love SL, Rhodes BB (1988) Single gene control of anthracnose resistance in *Citrullus*? Cucurbit Gen Coop Rep 11:64–67

Maekawa T, Kufer TA, Schulze-Lefert P (2011) NLR functions in plant and animal immune systems: so far and yet so close. Nat Immunol 12:817–826

Mahapatra S, Rao ES, Sandeepkumar GM, Sriram S (2020) *Stagonosporopsis cucurbitacearum* the causal agent of gummy stem blight of watermelon in India. Australas Plant Dis 15:7

Mahmud I, Kousik C, Hassell R, Chowdhury K, Boroujerdi AF (2015) NMR spectroscopy identifies metabolites translocated from powdery mildew resistant rootstocks to susceptible watermelon scions. J Agric Food Chem 63:8083–8091

Main CE, Keever T, Holmes GJ, Davis JM (2001) Forecasting long-range transport of downy mildew spores and plant disease epidemics. APSnet Feature Articles. Online. https://doi.org/10.1094/APSnetFeature-2001-0501

Mamanova L, Coffey AJ, Scott CE, Kozarewa I, Turner EH, Kumar A, Howard E, Shendure J, Turner DJ (2010) Target-enrichment strategies for next-generation sequencing. Nat Methods 7:111–118

Mandal MK, Suren H, Ward B, Boroujerdi A, Kousik C (2018) Differential roles of melatonin in plant-host resistance and pathogen suppression in cucurbits. J Pineal Res 65:e12505

Mandal MK, Suren H, Kousik C (2020) Elucidation of resistance signaling and identification of powdery mildew resistant mapping loci (ClaPMR2) during watermelon-*Podosphaera xanthii* interaction using RNA-Seq and whole-genome resequencing approach. Sci Rep 10:14038

Mansfeld BN, Colle M, Zhang C, Lin YC, Grumet R (2020) Developmentally regulated activation of defense allows for rapid inhibition of infection in age-related resistance to *Phytophthora capsici* in cucumber fruit. BMC Genom 21:628

Marone D, Russo M, Laidò G, De Leonardis A, Mastrangelo A (2013) Plant nucleotide binding site–leucine-rich repeat (NBS-LRR) genes: active guardians in host defense responses. J Pineal Res 14:7302–7326

Marr CW, Gast KLB (1991) Reactions by consumers in a farmers' market to prices for seedless watermelon and ratings of eating quality. HortTechnology 1:105–106

Martin GB, Bogdanove AJ, Sessa G (2003) Understanding the functions of plant disease proteins. Annu Rev Plant Biol 54:23–61

Martyn RD, Netzer D (1991) Resistance to races 0, 1, and 2 of *Fusarium* wilt of watermelon in *Citrullus* sp. PI 296341-FR. Hortic Sci 26:429–432

McGrath MT (1994) Heterothallism in *Sphaerotheca fuliginea*. Mycologia 86:517–523

McGrath MT (2001) Fungicide resistance in cucurbit powdery mildew: experiences and challenges. Plant Dis 85:236–245

McGrath MT (2017a) Phytophthora fruit rot. In: Keinath P, Wintermantel WM, Zitter TA (eds) Compendium of cucurbit diseases and pests. American Phytopathological Society Press, St. Paul, Minnesota, pp 102–104

McGrath MT (2017b) Powdery mildew. In Keinath P, Wintermantel WM, Zitter TA (eds) Compendium of cucurbit diseases and pests. American Phytopathological Society Press, St. Paul, Minnesota, pp 62–64

McGrath MT (2021) Cucurbit powdery mildew. Cornell vegetables factsheets. https://www.vegetables.cornell.edu/pest-management/disease-factsheets/cucurbit-powdery-mildew. Accessed 23 May 2021

McGrath MT, Shishkoff N (2001) Resistance to triadimefon and benomyl: dynamics and impact on managing cucurbit powdery mildew. Plant Dis 85:147–154

McHale L, Tan X, Koehl P, Michelmore RW (2006) Plant NBS-LRR proteins: adaptable guards. Genome Biol 7:212

Mercier J, Muscara MJ, Davis AR (2014) First report of *Podosphaera xanthii* race 1W causing powdery mildew of watermelon in California. Plant Dis 98:158–158

Meru G, McGregor CE (2016) A genetic locus associated with resistance to *Fusarium oxysporum* f. sp. *Niveum* race 2 in *Citrullus lanatus* type watermelon. J Am Soc Hortic Sci 141:617–622

Meyers BC, Dickerman AW, Michelmore RW, Sivaramakrishnan S, Sobral BW, Young ND (1999) Plant disease resistance genes encode members of an ancient and diverse protein family within the nucleotide-binding super family. Plant J 20:317–332

Meyers BC, Kozik A, Griego A, Kuang H, Michelmore RW (2003) Genome-wide analysis of NBS-LRR–encoding genes in *Arabidopsis*. Plant Cell 15:809–834

Michelmore R, Coaker G, Bart R, Beattie G, Bent A, Bruce T, Cameron D, Dangl J, Dinesh-Kumar S, Edwards R, Eves-van den Akker S, Gassmann W, Greenberg JT, Hanley-Bowdoin L, Harrison RJ, Harvey J, He P, Huffaker A, Hulbert S et al (2017) Foundational and translational research opportunities to improve plant health. MPMI 30:515–516

Miller NF, Standish JR, Quesada-Ocampo LM (2020) Sensitivity of *Fusarium oxysporum* f. sp. *Niveum* to prothioconazole and pydiflumetofen in vitro and efficacy for *Fusarium* wilt management in watermelon. Plant Health Prog 21:13–18

Minamiyama Y, Tsuro M, Kubo T, Hirai M (2007) QTL Analysis for resistance to *Phytophthora capsici* in pepper using a high density SSR-based map. Breed Sci 57:129–134

Monaghan J, Zipfel C (2012) Plant pattern recognition receptor complexes at the plasma membrane. Curr Opin Plant Biol 15:349–357

Mondragón-Palomino M, Stam R, John-Arputharaj A, Dresselhaus T (2017) Diversification of defensins and NLRs in *Arabidopsis* species by different evolutionary mechanisms. BMC Evol Biol 17:255

Monteiro F, Nishimura MT (2018) Structural, functional and genomic diversity of plant NLR proteins: an evolved resource for rational engineering of plant immunity. Annu Rev Phytopathol 56:243–267

Montero-Pau J, Blanca J, Bombarely A, Ziarsolo P, Esteras C, Martí-Gómez C, Ferriol M, Gómez P, Jamilena M, Mueller L, Picó B, Cañizares J (2018) *De novo* assembly of the zucchini genome reveals whole-genome duplication associated with the origin of the *Cucurbita* genus. Plant Biotechnol J 16:1161–1171

Morata J, Puigdomènech P (2017) Variability among Cucurbitaceae species (melon, cucumber and watermelon) in a genomic region containing a cluster of NBS-LRR genes. BMC Genom 18:138

Nanasato Y, Tabei Y (2020) A method of transformation and current progress in transgenic research on cucumbers and *Cucurbita* species. Plant Biotechnol 37:141–146

Narusaka M, Kubo Y, Hatakeyama K, Imamura J, Ezura H, Nanasato Y, Tabei Y, Takano Y, Shirasu K, Narusaka Y (2013a) Interfamily transfer of dual NB-LRR genes confers resistance to multiple pathogens. PLoS ONE 8:e55954

Narusaka M, Kubo Y, Hatakeyama K, Imamura J, Ezura H, Nanasato Y, Tabei Y, Takano Y, Shirasu K, Narusaka Y (2013b) Breaking restricted taxonomic functionality by dual *resistance* genes. Plant Signal Behav 8:e24244

Netzer D (1976) Physiological races and soil population level of fusarium wilt of watermelon. Phytoparasitica 4:131–136

Nishimura MT, Anderson RG, Cherkis KA, Law TF, Liu QL, Machius M, Nimchuk ZL, Yang L, Chung EH, El Kasmi F, Hyunh M, Osborne Nishimura E, Sondek JE, Dangl JL (2017) TIR-only protein RBA1 recognizes a pathogen effector to regulate cell death in *Arabidopsis*. Proc Natl Acad Sci USA 114:E2053–E2062

Norton JD (1979) Inheritance of resistance to gummy stem blight caused by *Didymella bryoniae* in watermelon. Hortic Sci 14:630–632

Norton JD, Cosper RD, Smith DA, Rymal KS (1986) 'AU-jubilant' and 'AU-producer' watermelons. Hortic Sci 21:1460–1461

Norton JD, Boyhan GE, Smith DA, Abrahams BR (1995) 'AU-sweet scarlet' watermelon. Hortic Sci 30:393–394

Nusbaum CJ (1944) The seasonal spread and development of Cucurbit downy mildew in the Atlantic coastal States. Plant Dis Rep 28:82–85

Ogundiwin EA, Berke TF, Massoudi M, Black LL, Huestis G, Choi D, Lee S, Prince JP (2005) Construction of 2 intraspecific linkage maps and identification of resistance QTLs for *Phytophthora capsici* root-rot and foliar-blight diseases of pepper (*Capsicum annuum* L.). Genome 48:698–711

Osuna-Cruz CM, Paytuvi-Gallart A, Di Donato A, Sundesha V, Andolfo G, Aiese Cigliano R, Sanseverino W, Ercolano MR (2018) PRGdb 3.0: a comprehensive platform for prediction and analysis of plant disease resistance genes. Nucleic Acids Res 46:D1197–D1201

Palkovics L, Petróczy M, Kertész B, Németh J, Cs B, Zs M, Hevesi M (2008) First report of bacterial fruit blotch of watermelon caused by *Acidovorax avenae* subsp. *citrulli* in Hungary. Plant Dis 92:834–834

Parada-Rojas CH, Quesada-Ocampo LM (2019) Characterizing sources of resistance to *Phytophthora* blight of pepper caused by *Phytophthora capsici* in North Carolina. Plant Health Prog 20:112–119

Parlevliet JE (2002) Durability of resistance against fungal, bacterial and viral pathogens; present situation. Euphytica 124:147–156

Parra G, Ristaino JB (2001) Resistance to mefenoxam and metalaxyl among field isolates of *Phytophthora capsici* causing *Phytophthora* blight of bell pepper. Plant Dis 85:1069–1075

Patel TK (2019) Studies on anthracnose resistance in watermelon germplasm. Ph.D. thesis, North Carolina State University, Raleigh, North Carolina

Pavón CF, Babadoost M, Lambert KN (2007) Quantification of *Phytophthora capsici* oospores in soil by sieving-centrifugation and real-time polymerase chain reaction. Plant Dis 92:143–149

Pérez-García A, Romero D, Fernández-Ortuño D, López-Ruiz F, De Vicente A, Torés JA (2009) The powdery mildew fungus *Podosphaera fusca* (synonym *Podosphaera xanthii*) a constant threat to cucurbits. Mol Plant Pathol 10:153–160

Poland JA, Balint-Kurti PJ, Wisser RJ, Pratt RC, Nelson RJ (2009) Shades of gray: the world of quantitative disease resistance. Trends Plant Sci 14:21–29

Prigozhin DM, Krasileva KV (2021) Analysis of intraspecies diversity reveals a subset of highly variable plant immune receptors and predicts their binding sites. Plant Cell 33:998–1015

Qi J, Liu X, Shen D, Miao H, Xie B, Li X, Zeng P, Wang S, Shang Y, Gu X, Du Y, Li Y, Lin T, Yuan J, Yang X, Chen J, Chen H, Xiong X, Huang K et al (2013) A genomic variation map provides insights into the genetic basis of cucumber domestication and diversity. Nat Genet 45:1510–1515

Qin X, Zhang Z, Lou Q, Xia L, Li J, Li M, Zhou J, Zhao X, Xu Y, Yang S, Yu X, Cheng C, Huang S, Chen J (2021) Chromosome-scale genome assembly of *Cucumis hystrix*—a wild species interspecifically cross-compatible with cultivated cucumber. Hortic Res 8:40

Quesada-Ocampo LM, Granke LL, Olsen J, Gutting HC, Runge F, Thines M, Lebeda A, Hausbeck MK (2012) The genetic structure of *Pseudoperonospora cubensis* populations. Plant Dis 96:1459–1470

Quesada-Ocampo LM, Vargas AM, Naegele RP, Francis DM, Hausbeck MK (2015) Resistance to crown and root rot caused by *Phytophthora capsici* in a tomato advanced backcross of *Solanum habrochaites* and *Solanum lycopersicum*. Plant Dis 100:829–835

Quesada-Ocampo LM (2018a) Anthracnose of cucurbits. NC State Extension Publications. https://content.ces.ncsu.edu/anthracnose-of-cucurbits. Accessed 24 May 2021

Quesada-Ocampo LM (2018b) Gummy stem blight of cucurbits. NC State Extension Publications. Vegetable Pathology Factsheets. https://content.ces.ncsu.edu/gummy-stem-blight-and-phoma-blight-on-cucurbits. Accessed 24 May 2021

Quesada-Ocampo LM (2020a) Cucurbit downy mildew. NC State Extension Publications. Cucurbit Downy Mildew Vegetable Pathology Factsheets. https://content.ces.ncsu.edu/cucurbit-downy-mildew. Accessed 23 May 2021

Quesada-Ocampo LM (2020b) Cucurbit powdery mildew. NC State Extension Publications. Cucurbit Downy Mildew Vegetable Pathology Factsheets. https://content.ces.ncsu.edu/cucurbit-downy-mildew. Accessed 23 May 2021

Rahman A, Miles TD, Martin FN, Quesada-Ocampo LM (2017) Molecular approaches for biosurveillance of the cucurbit downy mildew pathogen, *Pseudoperonospora cubensis*. Can J Plant Pathol 39:282–296

Rahman A, Standish JR, D'Arcangelo KN, Quesada-Ocampo LM (2021) Clade-specific biosurveillance of *Pseudoperonospora cubensis* using spore traps for precision disease management of cucurbit downy mildew. Phytopathology 111:312–320

Reddy AC, Lavanya B, Tejaswi T, Sreenivasa E, Lakshmana RDC (2019) Isolation and characterization of NBS-encoding disease resistance gene analogs in watermelon against *Fusarium* wilt. Curr Sci 117:617

Ren Y, Di Jiao, Gong G, Zhang H, Guo S, Zhang J, Xu Y (2015) Genetic analysis and chromosome mapping of resistance to *Fusarium oxysporum* f. sp. *Niveum* (FON) race 1 and race 2 in watermelon (*Citrullus lanatus* L.). Mol Breed 35:183

Ren R, Xu J, Zhang M, Liu G, Yao X, Zhu L, Hou Q (2019) Identification and molecular mapping of a gummy stem blight resistance gene in wild watermelon (*Citrullus amarus*) Germplasm PI 189225. Plant Dis 104:16–24

Renner SS, Wu S, Pérez-Escobar OA, Silber MV, Fei Z, Chomicki G (2021) A genome of a Kordofan melon illuminates the origin of domesticated watermelons. PNAS 118:e2101486118

Rennberger G, Gerard P, Keinath AP (2018) Factors influencing the occurrence of foliar pathogens in commercial watermelon fields in South Carolina based on stratified cluster sampling. Plant Dis 103:484–494

REX Consortium (2016) Combining selective pressures to enhance the durability of disease resistance genes. Front Plant Sci 7

Richard MMS, Gratias A, Meyers BC, Geffroy V (2018) Molecular mechanisms that limit the costs of NLR-mediated resistance in plants: how plants limit the costs of NLR resistance. Mol Plant Pathol 19:2516–2523

Rivera-Burgos LA, Silverman E, Nebahat S, Wehner TC (2021) Evaluation of resistance to gummy stem blight in a population of recombinant inbred lines of watermelon × citron. Hortic Sci 56:1–9

Rossi M, Goggin FL, Milligan SB, Kaloshian I, Ullman DE, Williamson VM (1998) The nematode resistance gene Mi of tomato confers resistance against the potato aphid. Proc Natl Acad Sci USA 95:9750–9754

Salcedo A, Hausbeck M, Pigg S, Quesada-Ocampo LM (2020) Diagnostic guide for cucurbit downy mildew. Plant Health Prog 21:166–172

Salcedo AF, Purayannur S, Standish JR, Miles T, Thiessen L, Quesada-Ocampo LM (2021) Fantastic downy mildew pathogens and how to find them: advances in detection and diagnostics. Plants 10:435

Sande D, Mullen J, Wetzstein M, Houston J (2011) Environmental impacts from pesticide use: a case study of soil fumigation in Florida tomato production. Int J Environ Res Public Health 8:4649–4661

Savory EA, Granke LL, Quesada-Ocampo LM, Varbanova M, Hausbeck MK, Day B (2011) The cucurbit downy mildew pathogen *Pseudoperonospora cubensis*: cucurbit downy mildew. Mol Plant Pathol 12:217–226

Schaad NW, Sowell G, Goth RW, Colwell RR, Webb RE (1978) *Pseudomonas pseudoalcaligenes* subsp. *citrulli* subsp. Int J Syst Bact 28:117–125

Sekhwal M, Li P, Lam I, Wang X, Cloutier S, You F (2015) Disease resistance gene analogs (RGAs) in plants. Int J Mol Sci 16:19248–19290

Shao ZQ, Xue JY, Wu P, Zhang YM, Wu Y, Hang YY, Wang B, Chen JQ (2016) Large-scale analyses of angiosperm nucleotide-binding site-leucine-rich repeat genes reveal three anciently diverged classes with distinct evolutionary patterns. Plant Physiol 170:2095–2109

Simmons AM, Jarret RL, Cantrell CL, Levi A (2019) *Citrullus ecirrhosus*: wild source of resistance against *Bemisia tabacci* (Hemiptera: Aleyrodidae) for cultivated watermelon. J Econ Entomol 112:2425–2432

Smith EF (1894) The watermelon disease of the South. Proc Am Adv Sci 42:289–290

Song H, Guo Z, Hu X, Qian L, Miao F, Zhang X, Chen J (2019) Evolutionary balance between LRR domain loss and young NBS–LRR genes production governs disease resistance in *Arachis hypogaea* cv. Tifrunner. BMC Genom 20:844

Song R, Gusmini G, Wehner T (2002) Screening the watermelon germplasm collection for resistance to gummy stem blight. In: XXVI international horticultural congress: advances in vegetable breeding, pp 63–68

Sowell G (1975) Additional source of resistance to gummy stem blight in watermelon. Plant Dis Rep 59:413–415

Sowell G, Pointer GR (1962) Gummy stem blight resistance of introduced watermelons. Plant Dis Rep 46:883–885

Sowell G, Rhodes BB, Norton JD (1980) New sources of resistance to watermelon anthracnose. J Am Soc Hortic Sci 105:197–199

Spoel SH, Dong X (2012) How do plants achieve immunity? Defence without specialized immune cells. Nat Rev Immunol 12:89–100

St Amand PC, Wehner TC (1995) Greenhouse, detached-leaf and field testing methods to determine cucumber resistance to gummy stem blight. J Am Soc Hortic Sci 120:673–680

Stam R, Silva-Arias GA, Tellier A (2019) Subsets of NLR genes show differential signatures of adaptation during colonization of new habitats. New Phytol 224:367–379

Stamatakis A (2014) RAxML version 8: a tool for phylogenetic analysis and post-analysis of large phylogenies. Bioinformatics 30:1312–1313

Steuernagel B, Jupe F, Witek K, Jones JDG, Wulff BBH (2015) NLR-parser: rapid annotation of plant NLR complements. Bioinformatics 31:1665–1667

Steuernagel B, Periyannan SK, Hernández-Pinzón I, Witek K, Rouse MN, Yu G, Hatta A, Ayliffe M, Bariana H, Jones JDG, Lagudah ES, Wulff BBH (2016) Rapid cloning of disease-resistance genes in plants using mutagenesis and sequence capture. Nat Biotechnol 34:652–655

Steuernagel B, Witek K, Krattinger SG, Ramirez-Gonzalez RH, Schoonbeek H, Yu G, Baggs E, Witek AI, Yadav I, Krasileva KV, Jones JDG, Uauy C, Keller B, Ridout CJ, Wulff BBH (2020) The NLR-annotator tool enables annotation of the intracellular immune receptor repertoire. Plant Physiol 183:468–482

Stewart JE, Turner AN, Brewer MT (2015) Evolutionary history and variation in host range of three *Stagonosporopsis* species causing gummy stem blight of cucurbits. Fungal Biol 119:370–382

Sun H, Wu S, Zhang G, Jiao C, Guo S, Ren Y, Zhang J, Zhang H, Gong G, Jia Z, Zhang F, Tian J, Lucas WJ, Doyle JJ, Li H, Fei Z, Xu Y (2017) Karyotype stability and unbiased fractionation in the paleo-allotetraploid cucurbita genomes. Mol Plant 10:1293–1306

Suvanprakorn K, Norton JD (1980) Inheritance of resistance to race 2 anthracnose in watermelon. J Am Soc Hortic Sci 105:862–865

Sy O, Steiner R, Bosland PW (2008) Recombinant inbred line differential identifies race-specific resistance to *Phytophthora* root rot in *Capsicum annuum*. Phytopathology 98:867–870

Taylor M, Bruton B, Fish W, Roberts W (2008) Cost benefit analyses of using grafted watermelon transplants for *Fusarium* wilt disease control. Acta Hortic 782:343–350

Taylors M, Bruton B, Fish W (2006) Cost benefit analysis of using grafted watermelons for disease control and the fresh-cut market. In: Cucurbitaceae 2006, Asheville, North Carolina, 17–21 Sept 2006, pp 277–285

Tetteh AY, Wehner TC, Davis AR (2010) Identifying resistance to powdery mildew race 2W in the USDA-ARS watermelon germplasm collection. Crop Sci 50:933–939

Tetteh AY, Wehner TC, Davis AR (2013a) Inheritance of resistance to the new race of powdery mildew in watermelon. Crop Sci 53:880–887

Tetteh AY, Wehner TC, Davis AR (2013b) Inheritance of resistance to powdery mildew race 2 in *Citrullus lanatus* var. Lanatus. Hortic Sci 48:1227–1230

Thies JA, Ariss JJ, Hassell RL, Olson S, Kousik CS, Levi A (2010) Grafting for management of southern root-knot nematode, *Meloidogyne incognita* in watermelon. Plant Dis 94:1195–1199

Thomas CE, Levi A, Caniglia E (2005) Evaluation of U.S. plant introductions of watermelon for resistance to powdery mildew. Hortic Sci 40:154–156

Thomas A, Langston DB, Stevenson KL (2012) Baseline sensitivity and cross-resistance to succinate-dehydrogenase-inhibiting and demethylation-inhibiting fungicides in *Didymella bryoniae*. Plant Dis 96:979–984

Thordal-Christensen H (2020) A holistic view on plant effectors-triggered immunity presented as an iceberg model. Cell Mol Life Sci 77:3963–3976

Toda N, Rustenholz C, Baud A, Le Paslier MC, Amselem J, Merdinoglu D, Faivre-Rampant P (2020) NL genome sweeper: a tool for genome-wide NBS-LRR resistance gene identification. Genes 11:333

Truong HTH, Kim KT, Kim DW, Kim S, ChaeY PJH, Oh DG, Cho MC (2012) Identification of isolate-specific resistance QTLs to *Phytophthora* root rot using an intraspecific recombinant inbred line population of pepper (*Capsicum annuum*): *Phytophthora* root rot resistance QTLs. Plant Pathol 61:48–56

USDA National Agricultural Statistics Service (2020) Watermelon US production (NASS—quick stats)

Van de Weyer AL, Monteiro F, Furzer OJ, Nishimura MT, Cevik V, Witek K, Jones JDG, Dangl JL, Weigel D, Bemm F (2019) A species-wide inventory of NLR genes and alleles in *Arabidopsis thaliana*. Cell 178:1260-1272.e14

Van Der Biezen EA, Jones JDG (1998) Plant disease-resistance proteins and the gene-for-gene concept. Trends Biochem Sci 23:454–456

van der Hoor RAL, Kamoun S (2008) From guard to decoy: a new model for perception of plant pathogen effectors. Plant Cell 20:2009–2017

Vleeshouwers VGAA, Oliver RP (2014) Effectors as tools in disease resistance breeding against biotrophic, hemibiotrophic and necrotrophic plant pathogens. MPMI 27:196–206

Walcott RR, Gitaitis RD, Castro AC (2003) Role of blossoms in watermelon seed infestation by *Acidovorax avenae* subsp. *citrulli*. Phytopathology 93:528–534

Wallace EC, D'Arcangelo KN, Quesada-Ocampo LM (2020) Population analyses reveal two host-adapted clades of *Pseudoperonospora cubensis* the causal agent of cucurbit downy mildew on commercial and wild cucurbits. Phytopathology 110:1578–1587

Wan H, Yuan W, Bo K, Shen J, Pang X, Chen J (2013) Genome-wide analysis of NBS-encoding disease resistance genes in *Cucumis sativus* and phylogenetic study of NBS-encoding genes in Cucurbitaceae crops. BMC Genom 14:109

Wang GF, Ji J, El-Kasmi F, Dangl JL, Johal G, Balint-Kurti PJ (2015a) Correction: molecular and functional analyses of a maize autoactive NB-LRR protein identify precise structural requirements for activity. PLoS Pathog 11:e1004830

Wang P, Liu X, Guo J, Liu C, Fu N, Shen H (2015b) Identification and expression analysis of candidate genes associated with defense responses to *Phytophthora capsici* in pepper line "PI 201234." Int J Mol Sci 16:11417–11438

Wasilwa LA, Corell JC, Morelock TE, McNew RE (1993) Re examination of races of the cucurbit anthracnose pathogen *Colletotrichum orbiculare*. Phytopathology 83:1190–1198

Webb RE, Goth RW (1965) A seedborne bacterium isolated from watermelon. Plant Dis Rep 49:818–821

Wehner TC (2008) Watermelon. In: Prohens J, Nuez F (eds) Vegetables I. Springer, New York, pp 381–418

Weßling R, Epple P, Altmann S, He Y, Yang L, Henz SR, McDonald N, Wiley K, Bader KC, Gläßer C, Mukhtar MS, Haigis S, Ghamsari L, Stephens AE, Ecker JR, Vidal M, Jones JDG, Mayer KFX, Loren V, van Themaat E et al (2014) Convergent targeting of a common host protein-network by pathogen effectors from three kingdoms of life. Cell Host Microb 16:364–375

Wiant JS, Tucker CM (1940) A rot of winter queen watermelon caused by Phytophthora capsici. J Agric Res 60:73–88.

Willems A, Goor M, Thielemans S, Gillis M, Kersters K, De Ley J (1992) Transfer of several phytopathogenic *Pseudomonas* species to *Acidovorax* as *Acidovorax avenae* subsp. Avenae subsp. Nov, comb. Nov, *Acidovorax avenae* subsp. *citrulli Acidovorax avenae* subsp. Cattleyae and *Acidovorax konjaci*. Int J Syst Bacteriol 42:107–119

Winstead NN, Aycock R, Jenkins JM (1957) Reactions of watermelon plant introductions and varieties to downy mildew in North Carolina. Plant Dis Rep 41:620–622

Winstead NN, Goode JM, Barham WS (1959) Resistance in watermelon to *Colletotrichum lagnerium* races 1, 2 and 3. Plant Dis Rep 43:570–577

Wroblewski T, Piskurewicz U, Tomczak A, Ochoa O, Michelmore RW (2007) Silencing of the major family of NBS-LRR-encoding genes in lettuce results in the loss of multiple resistance specificities: silencing of NBS-LRR genes in lettuce. Plant J 51:803–818

Wu CH, Abd-El-Haliem A, Bozkurt TO, Belhaj K, Terauchi R, Vossen JH, Kamoun S (2017a) NLR network mediates immunity to diverse plant pathogens. Proc Natl Acad Sci USA 114:8113–8118

Wu S, Shamimuzzaman M, Sun H, Salse J, Sui X, Wilder A, Wu Z, Levi A, Xu Y, Ling KS, Fei Z (2017b) The bottle gourd genome provides insights into Cucurbitaceae evolution and facilitates mapping of a *Papaya ring-spot virus* resistance locus. Plant J 92:963–975

Wu S, Wang X, Reddy U, Sun H, Bao K, Gao L, Mao L, Patel T, Ortiz C, Abburi VL, Nimmakayala P, Branham S, Wechter P, Massey L, Ling K, Kousik C, Hammar SA, Tadmor Y, Portnoy V et al (2019) Genome of 'Charleston Gray', the principal American watermelon cultivar and genetic characterization of 1,365 accessions in the U.S. National Plant Germplasm System watermelon collection. Plant Biotechnol J 17:2246–2258

Xie D, Xu Y, Wang J, Liu W, Zhou Q, Luo S, Huang W, He X, Li Q, Peng Q, Yang X, Yuan J, Yu J, Wang X, Lucas WJ, Huang S, Jiang B, Zhang Z (2019) The wax gourd genomes offer insights into the genetic diversity and ancestral cucurbit karyotype. Nat Commun 10:5158

Xu Y, Liu F, Zhu S, Li X (2018) The maize NBS-LRR gene ZmNBS25 enhances disease resistance in rice and *Arabidopsis*. Front Plant Sci 9:1033

Yamaguchi Y, Huffaker A, Bryan AC, Tax FE, Ryan CA (2010) PEPR2 is a second receptor for the Pep1 and Pep2 peptides and contributes to defense responses in *Arabidopsis*. Plant Cell 22:508–522

Yang L, Koo DH, Li Y, Zhang X, Luan F, Havey MJ, Jiang J, Weng Y (2012) Chromosome rearrangements during domestication of cucumber as revealed by high-density genetic mapping and draft genome assembly: chromosome evolution and domestication in *Cucumis*. Plant J 71:895–906

Yang YQ, Lan B, Sun Q, Chen HF, Huang YH, Chen J, Huang JH, Guo YH, Li XM (2019) First report of *Acidovorax avenae* subsp. *citrulli* causing bacterial fruit blotch of watermelon in Jiangxi, China. Plant Dis 103:2468–2468

Young ND (1996) QTL mapping and quantitative disease resistance in plants. Ann Rev Phytopathol 34:479–501

Yu J, Wu S, Sun H, Wang X, Tang X, Guo S, Zhang Z, Huang S, Xu Y, Weng Y, Mazourek M, McGregor C, Renner SS, Branham S, Kousik C, Wechter WP, Levi A, Grumet R, Zheng Y, Fei Z (2023) CuGenDBv2: an updated database for cucurbit genomics. Nucleic Acids Res 6:51(D1):D1457–D1464.

Zhai C, Zhang Y, Yao N, Lin F, Liu Z, Dong Z, Wang L, Pan Q (2014) Function and interaction of the coupled genes responsible for *Pik*-h encoded rice blast resistance. PLoS ONE 9:e98067

Zhang J, Zhou JM (2010) Plant immunity triggered by microbial molecular signatures. Mol Plant 3:783–793

Zhang SP, Liu MM, Miao H, Zhang SQ, Yang YH, Xie BY, Wehner TC, Gu XF (2013a) Chromosomal mapping and QTL analysis of resistance to downy mildew in *Cucumis sativus*. Plant Dis 97:245–251

Zhang Y, Lubberstedt T, Xu M (2013b) The genetic and molecular basis of plant resistance to pathogens. JGG 40:23–35

Zhang J, Zhang P, Dodds P, Lagudah E (2020) How target-sequence enrichment and sequencing (TEnSeq) pipelines have catalyzed resistance gene cloning in the wheat-rust patho system. Front Plant Sci 11:678

Zheng Y, Wu S, Bai Y, Sun H, Jiao C, Guo S, Zhao K, Blanca J, Zhang Z, Huang S, Xu Y, Weng Y, Mazourek M, Reddy U, Ando K, McCreight JD, Schaffer AA, Burger J, Tadmor Y et al (2019) Cucurbit genomics database (CuGenDB): a central portal for comparative and functional genomics of cucurbit crops. Nucleic Acids Res 47:D1128–D1136

Zhou XG, Everts KL, Bruton BD (2010) Race 3, a new and highly virulent race of *Fusarium oxysporum* f. sp. *Niveum* causing *Fusarium* wilt in watermelon. Plant Dis 94:92–98

5 Genetics and Genomics of Fruit Quality Traits of Watermelon

Cecilia McGregor, Samikshya Rijal, Samuel Josiah, and Lincoln Adams

5.1 Introduction

The non-bitter, dessert watermelon (*Citrullus lanatus*) is one of seven species in the genus *Citrullus* that includes *C. mucosospermus* (including egusi melon), *C. amarus* (citron), *C. colocynthis*, *C. ecirrhosus*, *C. rehmii,* and *C. naudinianus* (Chomicki and Renner 2015). For most of the twentieth century, it was widely believed that watermelon was domesticated from *C. amarus* in southern Africa (Bailey 1930). As more genomic evidence emerged in the 2010s, the probable center of domestication and closest relative of watermelon shifted from the citron melons of southern Africa to *C. mucosospermus* of western Africa (Chomicki and Renner 2015; Guo et al. 2013). Recent evidence, however, points to northeastern Africa, in the area of present-day Sudan as the most likely center of origin (Renner et al. 2021). This hypothesis is based on archeological evidence as well as recently discovered non-bitter, white fleshed watermelons belonging to *C. lanatus* susp. *cordophanus* or Kordofan melon (Renner et al. 2021).

C. McGregor (✉) · S. Josiah
Department of Horticulture, University of Georgia, Athens, GA 30602, USA
e-mail: cmcgre1@uga.edu

C. McGregor · S. Rijal · L. Adams
Institute of Plant Breeding, Genetics and Genomics, University of Georgia, Athens, GA 30602, USA

Watermelon is an important annual horticultural crop which is commercially produced in 118 countries around the world. Global production in 2019 exceeded 100 million tons, with China responsible for approximately 60% of global watermelon production. Turkey is the second largest producer but represents a considerably smaller share of the market than China, with 3.8% of global production. Other top producers include India, Brazil, Algeria, Iran, Russia, and the USA each responsible for between 1.6 and 2.5% of global production (FAOSTAT 2020).

Watermelon fruit is usually sold whole or precut and consumed fresh in the form of slices or cut-up pieces (Wehner 2008). The fruit flesh and rind can also be used to make jams, candied or pickled rinds or processed into juice (Wehner 2008), while specialized cultivars are used for the production of edible seeds (Zhang and Jiang 1990; Gusmini et al. 2004). Fruit quality traits are important in watermelon for a variety of reasons, many related to consumer preference. Due to its function as a dessert item, internal fruit quality characteristics such as sugar content, flavor, color, and texture are very important in order to maintain a delicious, highly desirable product for the consumer (Wehner 2008). Regional consumer preferences for specific fruit sizes and rind patterns drive selection for specific phenotypes in different parts of the world (Wu et al. 2019; Guo et al. 2013). As globalization expands the markets of watermelon growers, rind

S. Kr. Dutta et al. (eds.), *The Watermelon Genome*, Compendium of Plant Genomes,
https://doi.org/10.1007/978-3-031-34716-0_5

thickness has become an increasingly important trait for watermelon. Thick rinds allow watermelons to survive the shipping process, but this must be balanced against consumer preference and fruit size, as consumers prefer a rind that is a small portion of the total fruit (Wehner 2008; Wu et al. 2019).

The Mendelian inheritance of fruit traits in watermelon have been studied for more than a hundred years (Porter 1937; Porter et al. 1940; Weetman 1937; Poole and Grimball 1945; McKay 1936; Poole 1944) and gene lists have been maintained since 1944 (Wehner 2012). With the advent of molecular biology and genomics, it became possible to determine the loci, genes, and allelic variation associated with horticulturally important traits that have been the target of domestication and selection in watermelon, including fruit quality traits. The diploid nature and relatively small genome size of watermelon ($\sim$425 Mbp) (Arumuganathan and Earle 1991) meant that new genomic tools were quickly adopted, and currently, two high-quality draft genomes are publicly available (Guo et al. 2013, 2019; Wu et al. 2019) at the Cucurbit Genomics Database (CuGenDB; http://cucurbitgenomics.org/) (Zheng et al. 2018). CuGenDB also includes transcriptome and comparative genomics tools (e.g., SyntenyViewer) for watermelon (Zheng et al. 2018). Genetic mapping, QTL-Seq, and comparative genomics have been used extensively to identify important watermelon fruit quality genes and loci (Sandlin et al. 2012; Ren et al. 2014; Pan et al. 2020; Branham et al. 2017; Wang et al. 2019; Dou et al. 2018b; Park et al. 2016) that drive consumer preference and thus breeding efforts for this important horticultural crop.

5.2 Fruit Bitterness

The bitter compound found in Cucurbit crops, cucurbitacin, is an undesirable trait in edible watermelon. Understanding the genetic regulation of cucurbitacin is a crucial aspect of watermelon breeding programs. Several studies investigated cultivated watermelon and its crop wild relatives (CWRs) to understand the correlation between bitterness and cucurbitacin. Different types of cucurbitacin, cucurbitacin B (CuB), cucurbitacin C (CuC), cucurbitacin E (CuE), and cucurbitacin E-2-O-glucoside (CuE-Glu) are found in *Citrullus* fruits, leaves, roots, and stems (Zhou et al. 2016; Kim et al. 2018). CuE, also known as elaterinide, is the main bitter compound in *Citrullus* fruit (Matsuo et al. 1999; Chambliss et al. 1968).

The dominant *Bi* gene controls cucurbitacin synthesis responsible for bitterness in *Citrullus* fruits (Chambliss et al. 1968; Navot et al. 1990), while the recessive *su* (suppressor of bitterness) gene controls the presence or absence of bitterness in watermelon fruit (Chambliss et al. 1968; Robinson et al. 1976). The *Bi* gene has been shown to be an oxidosqualene cyclase (*OSC*; *Cla007080*) gene on chromosome 6 of the watermelon genome (Table 5.1) (Davidovich-Rikanati et al. 2015; Zhou et al. 2016). *OSC* genes are responsible for the first step in cucurbitacin biosynthesis (Shibuya et al. 2004). Accumulation of cucurbitacin is not uniform throughout the plant, and the *Bl* (*bitter_leaf*), *Br* (*bitter_root*), and *Bt* (*bitter_fruit*) genes are associated with accumulation in leaves, roots, and fruits, respectively (Shang et al. 2014; Zhou et al. 2016). The *Bt* gene was identified as a basic helix–loop–helix transcription factor (*bHLH; Cla011508*) on chromosome 1 (Table 5.1) (Zhou et al. 2016; Li et al. 2018; Zhang et al. 2018). A single base-pair mutation (C382T) in the second exon leads to a truncated protein associated with non-bitter fruit (Zhou et al. 2016; Li et al. 2018).

5.3 Sugar Content and Composition

5.3.1 Soluble Solids Content (SSC)

Watermelon is known for its sweet, juicy fruit with a SSC (measured in degrees Brix) of 10–13. Fruit from other *Citrullus* species has much lower SSC with *C. colocynthis* having 1.6 Brix and *C. amarus* and *C. mucosospermus* 3.4 Brix

Table 5.1 Candidate genes controlling fruit quality traits in watermelon

Trait	Locus/QTL	Chr	Gene ID	Allelic variation	References
Bitterness (*Bi*)	*Qbi6*	6	*Oxidosqualene cyclase* (*OSC; Cla007080*[a])	Differential expression	Zhou et al. (2016), Chambliss et al. (1968), Robinson et al. (1976)
Bitter fruit (*Bt*)	*Qbt1*	1	*Basic helix–loop–helix transcription factor* (*bHLH; Cla011508*[a])	SNP (C382T) in the second exon leads to a truncated protein	Zhang et al. (2018), Li et al. (2018), Zhou et al. (2016)
Sugar content	*QBRX2.1*	2	*Tonoplast Sugar Transporter* gene (*ClTST2; Cla000264*[a])	SNP in promoter	Sandlin et al. (2012), Ren et al. (2018), Ren et al. (2014)
Sugar content		4	*Alkaline α-galactosidase gene* (*ClAGA2; Cla97C04G070460*[b])	Two SNPs in promoter	Ren et al. (2021), Guo et al. (2019)
Sugar content		1	*Sugars Will Eventually Be Exported Transporter 3* (*ClSWEET3; Cla97C01G000640*[b])	Differential expression	Ren et al. (2021)
Sucrose content	*Qsuc2.1*	2	*Vacuolar sugar transporter* (*ClVST1; Cla97C02G031010*[b])	SNP (C99A) leading to truncated protein	Ren et al. (2020), Ren et al. (2014)
Fruit shape	*Qfsi3.1*	3	*Sun gene family* (*ClSUN25-26-27a; Cla011257*[a])	Deletion and SNP	Dou et al. (2018a, b), Legendre et al. (2020), Pan et al. (2020)
Andromonoecy (*a*), pleiotropic effect on fruit shape	*Qand3*	3	*1-aminocyclopropane-1-carboxylate synthase* (*CitACS4; Cla011230*[a])	SNP (C1477G) leading to amino acid substitution (C364W)	Prothro et al. (2013), Boualem et al. (2008), Boualem et al. (2016), Boualem et al. (2009), Aguado et al. (2018), Manzano et al. (2016)
Flesh color/lycopene content	*Qlcyb4.1*	4	*Lycopene β-cyclase* (*LCYB, Cla005011*[a])	SNPs leading to amino acid substitutions (V226F and K435N)	Bang et al. (2007), Wang et al. (2019), Zhang et al. (2020)
Flesh color	*Qflc1.1*	1	*Phytoene synthase 1* (*PSY1; Cla009122*[a])	Two non-synonymous mutations (N133D and K148E) associated with orange fleshed	Branham et al. (2017), Fang et al. (2020)
Flesh firmness	*Qffi6.1*	6	*Xyloglucan endotransglucosylase/hydrolase* (*XTH; Cla018816*[a])	Differential expression	Juarez et al. (2013), Gao (2018), Gao et al. (2016), Anees et al. (2021)

(continued)

Table 5.1 (continued)

Trait	Locus/QTL	Chr	Gene ID	Allelic variation	References
Rind pattern *Dominant golden color*	*Dgo*	4	*Magnesium-chelatase subunit H* (*Cla97C04G068530*[b])		Yang et al. (2015), Park et al. (2016), Guo et al. (2019)
Rind pattern *Depth*	*D*	8	*2-phytyl-1,4-beta-naphthoquinone methyltransferase* (*Cla97C08G161570*[b]),	Non-synonymous SNP in the fifth exon	Yang et al. (2015), Park et al. (2016), Li et al. (2019), Guo et al. (2019)
Rind pattern *Stripe*	*S*	6	WD40 repeat (*Cla97C06G126710*[a])	Non-synonymous SNP in the fifth exon	Yang et al. (2015), Park et al. (2016), Guo et al. (2019), Wu et al. (2019)
Rind pattern		9	*Arabidopsis pseudo-response regulator2-like* (*ClAPRR2; ClCG09G012330*[c])	SNP causing alternative splice junction leading to a 16 bp insertion in the mRNA	Oren et al. (2019)
Rind toughness (*E*)	*Qrto10.1*	10	*ethylene-responsive transcription factor 4* (*ClERF4; Cla97C10G187120*[b])	11 bp deletion	Liao et al. (2020)

[a]GeneID based on 97103 v1 genome (Guo et al. 2013)
[b]GeneID based on 97103 v2 genome (Guo et al. 2019)
[c]GeneID based on Charleston Gray genome (Wu et al. 2019)

(Guo et al. 2019). A number of QTLs have been identified for SSC (Fall et al. 2019; Liu et al. 2015; Ren et al. 2014; Sandlin et al. 2012), indicating the complex inheritance of this trait. Sandlin et al. (2012) identified four QTLs associated with brix, *QBRX1* (syn. *brix7.1*; PVE ~ 8%), *QBRX2.1* (syn. *brix 9.1*; PVE ~ 21%), *QBRX2.2* (syn. *brix9.2*; PVE ~ 12%), and *Qbrx7* (syn. *brix 8.1*; PVE ~ 7%) on chromosomes 1, 2, and 7, respectively. *Qbrx1, Qbrx2.2,* and *Qbrx7* were identified in a *C. lanatus* (Klondike Black Seeded × New Hampshire Midget) background, while *QBRX2.1* was identified in the *C. lanatus* × *C. mucosospermus* background. In a *C. lanatus* × *C. amarus* population (97103 × PI 296341-FR), Ren et al. (2014) also identified *QBRX2.2* (PVE ~ 28%), as well as additional QTLs on chromosomes 2 (*Qbrix2.3*; syn. *Qbrix2.1*; PVE ~ 22%), 6 (*Qbrix6*; PVE ~ 8%), and 8 (*Qbrix8*; PVE ~ 14%). *Qbrix2.3* overlaps with major QTLs for fructose (*Qfruc2.1*; PVE ~ 23%) and sucrose (*Qsuc2.1*; PVE ~ 10%) in the same population (Ren et al. 2014). The effect of *QBRIX2.1* has been suggested to be due to changes in expression of the *Tonoplast Sugar Transporter* gene (*ClTST2; Cla000264*) associated with a single SNP in the promoter region of the gene (Table 5.1) (Ren et al. 2018). Interestingly, overexpression of *ClTST2* is also associated with development of red flesh (Ren et al. 2018). This locus was found to be a target of selection during watermelon domestication and improvement, while *Qbrix6* was found to be under selection during domestication (Guo et al. 2019). A region on chromosome 4 containing an alkaline α-galactosidase gene (*ClAGA2; Cla97C04G070460*) was found to be associated with selection for sugar accumulation during domestication (Table 5.1) (Guo et al. 2019). *ClAGA2* plays a crucial role in raffinose hydrolysis into sucrose and galactose, and differential expression of this gene affects sugar content of watermelon fruit (Ren et al. 2021). The differential expression is associated with two SNPs in the *ClAGA2* promotor that interacts with the *nuclear transcription factor Y subunit C* transcription factor (Ren et al. 2021). Together with *Sugars Will Eventually Be Exported Transporter 3* (Table 5.1) (*ClSWEET3*; *Cla97C01G000640*) on chromosome 1 that plays a role in sugar storage in vacuoles, *ClTST2* and *ClAGA2* are key genes for sugar accumulation in watermelon (Ren et al. 2021).

5.3.2 Sugar Composition

The sugar composition of fruit contributes to fruit sweetness and plays an important role in taste and consumer preference (Brueckner et al. 2007; Evans 2008; Kyriacou and Rouphael 2018). The main sugars in mature watermelon fruit are sucrose, fructose, and glucose, with considerable variation in ratios of these sugars among genotypes (Pardo et al. 1997; Yoo et al. 2012; Yativ et al. 2010).

Ren et al. (2014) identified five QTLs associated with fructose content, one for glucose content and three for fructose content in a *C. lanatus* × *C. amarus* population (97103 × PI 296341-FR). The QTL for fructose content was on chromosomes 2 (*Qfru2.1*, PVE ~ 23%; *Qfru2.2*, PVE ~ 25%; *Qfru2.3*, PVE ~ 28%), 6 (*Qfru6*, PVE ~ 21%), and 8 (*Qfru8*; PVE ~ 15%). The single QTL for glucose content was identified on chromosome 6 (*Qglu6*, PVE ~ 23%), and for sucrose content, two were identified on chromosome 2 (*Qsuc2.1*, PVE ~ 1-%; *Qsuc2.2*; PVE ~ 14%) and one on chromosome 5 (*Qsuc5*, PVE ~ 11%). *Qfru2.1* and *Qsuc2.1* colocalized with *Qbrx2.3*, while *Qfru2.2* and *Qsuc2.2* colocalized with *Qbrx2.2* and *Qfru2.3* with *Qbrx2.1*. The glucose (*Qclu6*) and fructose (*Qfru6*) content QTLs on chromosome 6 are also colocalized, but in a different region than *Qbrix6*. A SNP which leads to the truncation of *Vacuolar sugar transporter* (*ClVST1*) underlies *Qsuc2.1* and is associated with sucrose accumulation in the interspecific 97103 × PI 296341-FR population (Table 5.1) (Ren et al. 2021).

In the *C. lanatus* background, QTL for fruit glucose content (*Qglu5.1*, PVE ~ 13%; *Qglu5.2*, PVE ~ 10%; *Qglu5.3*, PVE ~ 12%; *Qglu8.1*, PVE ~ 17%; *Qglu8.2*, PVE ~ 15%), fructose content (*Qfru5.1*, PVE ~ 18%; *Qfru5.2*, PVE ~ 18%), and sucrose content (*Qsur5.2*, PVE ~ 25%; *Qsur8.1*, PVE ~ 9%) were identified in overlapping loci on chromosomes 5 and 8 (Fall et al. 2019). A QTL for fructose, glucose, and sucrose content of the edge portion of the flesh was identified on chromosomes 10 (*Qfru10.1* syn. *FCE10.1*, PVE ~ 11.8%) and 1 (*Qglu1* syn. *SCE 1.1*, PVE ~ 7% and *Qsuc1* syn. *GCE*, PVE ~ 6.5%) (Cheng et al. 2016). The difference in the QTLs identified in the interspecific populations compared to the elite *C. lanatus* background suggests that different regions were under selection during domestication and improvement.

5.4 Fruit Shape and Size

Fruit shape and size are important traits both for consumer preference and transportation logistics. There are often regional preferences for specific shapes of sizes, and recent trends are toward smaller, blocky fruit (Fig. 5.1) (Wehner 2008). Like other cucurbits, fruit shape is largely determined pre-anthesis, leading to a high correlation between the shape of the ovary and fruit shape (McKay 1936; Pan et al. 2020; Weetman 1937; Legendre et al. 2020). The *O* gene has been described as controlling fruit shape with *OO*, *Oo*, and *oo* leading to elongate, oval, and spherical, fruit, respectively (Poole and Grimball 1945; Tanaka et al. 1995; Weetman 1937).

More than 50 QTLs associated with fruit shape and size have been identified in watermelon (Pan et al. 2020). Due to the relationship between fruit shape (FSI), fruit length (FL), and fruit diameter (FD), where FSI = FL/FD, QTLs associated with these traits are often colocalized. A stable major QTL for FSI (*Qfsi3.1*) in a broad range of genetic backgrounds is located on chromosome 3 of the watermelon genome (Dou et al. 2018b; Liu et al. 2014, 2016; Sandlin et al. 2012). This is the presumptive location of the *O* gene and *Cla011257*, a *SUN* family gene (*ClSUN25-26-27a*), was identified as the candidate gene for *Qfsi3.1* (Table 5.1) (Dou et al.

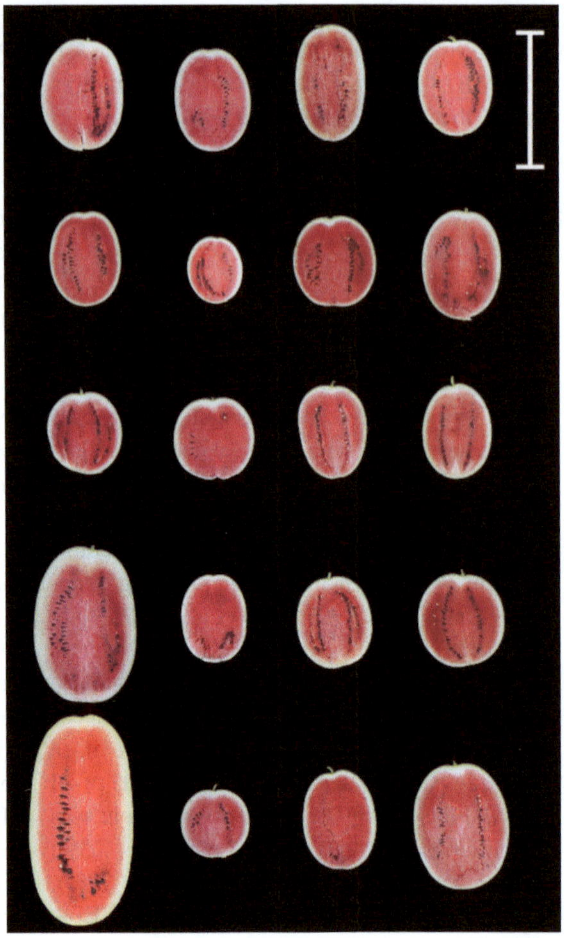

Fig. 5.1 Phenotypic variation in fruit shape, size, and rind thickness in *C. lanatus* (image by Reeve Legendre)

2018b; Kim et al. 2015; Legendre et al. 2020; Pan et al. 2020). The wild-type allele (cultivar 97103) is associated with round fruit, while a SNP and a 159 bp deletion in the third exon of the gene are associated with blocky and elongated fruit, respectively (Dou et al. 2018b; Kim et al. 2015; Legendre et al. 2020; Pan et al. 2020). It has long been known that the *SUN* gene, a member of the IQ Domain (IQD) family, influences tomato fruit shape by controlling fruit elongation early in fruit development, after pollination and fertilization (van der Knaap and Tanksley 2001). In tomato, most of the effect on fruit shape facilitated by this gene is detectable during the early stages of fruit development, though the cell patterns are established pre-anthesis (Wu et al. 2011; Xiao et al. 2008). In contrast to watermelon, where mutations in the coding regions of *ClSUN25-26-27a* are associated with shape variation, overexpression of *SUN*, due to gene duplication, is responsible for more elongated fruit in tomato (Xiao et al. 2008).

Sandlin et al. (2012) also identified a major QTL associated with FD (and fruit weight) on chromosome 2 (*Qfd2.2; Qfwt2.2*) in populations derived from *C. lanatus*, *C. mucosospermus*, and *C. amarus*. Other QTLs that have been identified in multiple genetic backgrounds are a QTL for FL on chromosome 4 (*Qfl4.1*) and QTL for FW and FD on chromosomes 2 (*Qfw2.1* and *Qfd2.1*), 8 (*Qfw8.1* and *Qfd8.1*) (Cheng et al. 2016; Liu et al. 2016; Lu et al. 2016; Pan et al. 2020; Reddy et al. 2015; Ren et al. 2014; Sandlin et al. 2012). The genes underlying these QTL are not currently known.

Sex expression is known to have a pleiotropic effect on fruit shape in watermelon (Loy 2006; Poole and Grimball 1945; Rosa 1928). Pistillate flowers have a more elongated fruit shape than hermaphroditic flowers. A major QTL associated with sex expression, and the location of the *a* locus was mapped on chromosome 3 (*Qand3*) and an *ACS* (*CitACS4*; *1-aminocyclopropane-1-carboxylate synthase*) gene (*Cla011230*) was identified as the candidate gene (Prothro et al. 2013). *ACS* genes are known to play a critical role in sex expression in cucurbits (Boualem et al. 2008, 2016, 2009). A missense mutation in *CitACS4* was shown to be associated with andromonoecy and fruit shape (Aguado et al. 2018; Manzano et al. 2016) (Table 5.1).

5.5 Flesh Color

A large variety of fruit flesh colors have been observed in watermelon (Fig. 5.2). Numerous genes and interactions controlling flesh color have been described, including an epistatic interaction between *Wf* (white flesh) and *B* (yellow flesh; *b*: red flesh) (Henderson 1992; Robinson et al. 1976; Shimotsuma 1963). Flesh colors, other than white (*Wf*), are recessive to *C* (canary yellow) and salmon yellow (y) is recessive to coral red (Y^{Crl}), while Y^{Crl} is dominant over orange flesh (y^{O}) (Henderson et al. 1998; Henderson 1989; Poole 1944; Porter 1937). However an inhibitor to C (*i-C*) has been described with epistatic interaction leading to red flesh in the presence of the *C* allele (Poole 1944; Henderson et al. 1998). Scarlet red flesh (Y^{Scr}) was also found to be dominant over coral red (Y^{Crl}) flesh (Gusmini and Wehner 2006).

The different watermelon flesh colors are due to accumulation of different carotenoids (Fang et al. 2020; Perkins-Veazie et al. 2006; Subburaj et al. 2019; Sun et al. 2018; Zhao et al. 2013). Lycopene is the dominant carotenoid in red fleshed watermelon, while in orange-fleshed fruit, the combination of β-carotene, Pro-lycopene, phytoene, and ξ-carotene seems to be the most

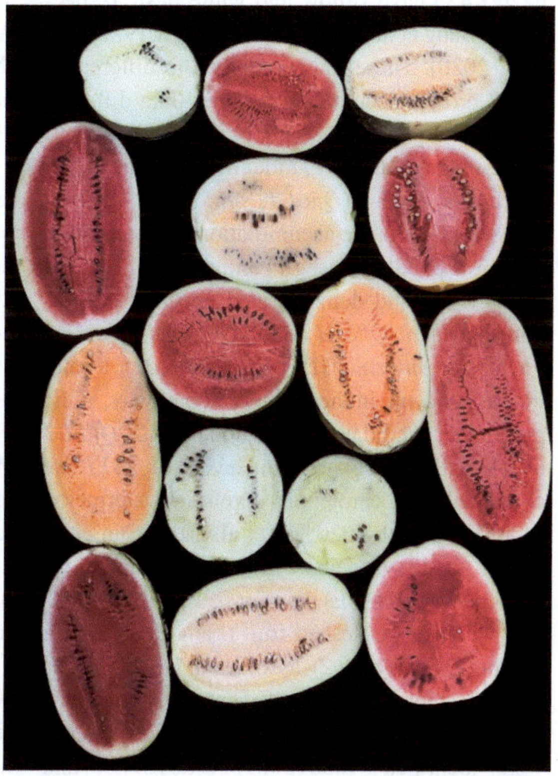

Fig. 5.2 Phenotypic variation in fruit flesh color in *C. lanatus*

important carotenoids contributing to flesh color (Fang et al. 2020; Subburaj et al. 2019; Tadmor et al. 2005). In yellow fleshed fruit, violaxanthin, phytoene, β-carotene, and lutein are among the main carotenoids accumulating in the flesh (Fang et al. 2020; Subburaj et al. 2019).

A major QTL associate with flesh color in a population developed from a cross between LSW-177 (red flesh) and Cream of Saskatchewan (COS, pale yellow flesh) was identified on chromosome 4 (*Qflc4.1*; PVE ~ 32) (Liu et al. 2015). *Qflc4.1* colocalized with a QTL (*Qlcyb4.1*) associated with 92% of the PVE for lycopene content in this population and the location of the *lycopene β-cyclase* (*LCYB*, *Cla005011*) gene in the watermelon genome (Table 5.1) (Liu et al. 2015; Nan et al. 2016; Subburaj et al. 2019; Wang et al. 2019). *LCYB* converts lycopene to β-carotene via γ-carotene (Giuliano et al. 2008). In watermelon germplasm, there are two SNPs (G676T and G1305C) leading to amino acid substitutions (V226F and K435N) within the single exon of *LCYB*. Genotypes with V226/K435 have red flesh, while genotypes with F226/N435 and F226/K435 have yellow and white flesh, respectively (Bang et al. 2007; Wang et al. 2019; Zhang et al. 2020). These mutations lead to differential *LCYB* protein accumulation rather than differential RNA expression (Zhang et al. 2020).

Branham et al. (2017) identified a major QTL on chromosome 1 (*Qflc1.1*, syn. *qFC.1*) associated with β-carotene accumulation in watermelon fruit in a population segregating for yellow and orange flesh and demonstrated epistatic interaction between *Qflc1.1 and Qlcyb4.1* (*LCYB*). *Phytoene synthase 1* (*PSY1*; *Cla009122*), a rate-limiting enzyme in the carotenoid pathway, is the presumed candidate gene underlying *Qflc1.1* (Table 5.1). The expression levels of *PSY1* are correlated with carotenoid levels in watermelon flesh (Fang et al. 2020; Lv et al. 2015; Sun et al. 2018), and two non-synonymous mutations (N133D and K148E) in *Cla009122* are associated with orange-fleshed fruit (Fang et al. 2020).

The QTL (*Qflc6.2*) associated with the Y^{Scr} gene was mapped to a 40 Kb region of chromosome 6 and three potential candidate genes were identified (*Cla018769*, *Cla018770*, and *Cla018771*) (Li et al. 2020). Additional minor QTLs associated with watermelon flesh color have been identified on chromosomes 3 (*Qflc*3.1), 6 (*Qflc6.1*), and 11 (*Qflc 11.1*) (Fall et al. 2019; Liu et al. 2015; Subburaj et al. 2019).

5.6 Rind Pattern

Watermelon fruit can have solid or striped rind (patterned rind) (Fig. 5.3). Rind appearance is an important trait for consumer preference, with many consumers preferring fruit with striped rinds over those with solid rinds (Lou and Wehner 2016). The stripes can vary in width and color, with the darker color usually considered the stripe and the lighter color the background. The edges of the stripes can range from diffuse to sharp with some stripes being very distinct, while others can be inconspicuous, marbled (UPOV 2013) or intermitted (Gusmini and Wehner 2006). Watermelon fruit rind patterns can also appear mottled and have reticulated netting (veins) or appear as a ground spot (Lou and Wehner 2016; UPOV 2013; Weetman 1937). Additional unique patterns of yellow spots (Moon and Stars) or yellow rind (Royal Golden) have also been described (Barham 1956; Poole 1944) (Fig. 5.3).

The genetic control of watermelon rind patterns has been studied since the 1930s and the qualitative inheritance of many patterns has been described (Barham 1956; Gusmini and Wehner 2006; Kumar and Wehner 2011; Lou and Wehner 2016; Poole 1944; Porter 1937; Weetman 1937). In this model, solid dark green rind (*G*) is dominant over striped (g^s) and gray rind (*g-1* and *g-2*) (Kumar and Wehner 2011; Porter 1937; Weetman 1937). A dominant allele at either locus will produce dark green rind and intermittent stripes (*ins*) are recessive to stripes (Gusmini and Wehner 2006). The Moon and Stars pattern of large and small yellow spots on the fruit and foliage is dominant (*Sp*) (Poole 1944; Rhodes 1986), as is the yellow-colored ground spot (*Yb*) (Gusmini and Wehner 2006), while the yellow rind from Royal Golden is recessive (*go*) (Barham 1956; Robinson et al. 1976).

Fig. 5.3 Phenotypic variation in rind color and pattern in *Citrullus*

Recently, Yang et al. (2015) developed an alternative model based on three loci, *Dgo* (*Dominant golden color*), *D* (*Depth*), and *S* (*Stripe*), rather than three alleles at the same locus. According to this model, *Dgo* controls the color of the background (yellow > green) and is a different locus than the recessive *go* (green > yellow) described earlier (Barham 1956; Robinson et al. 1976). The *D* locus controls the depth of rind color, with deep (dark) color being dominant over standard color, while the *S* locus controls the presence or absence of foreground stripes (presence > absence). Park et al. (2016) mapped the *Dgo, D,* and *S* loci on chromosomes 4, 8, and 6, respectively, and designed markers for selection of the loci. The location of these loci was confirmed in GWAS studies, and *Cla97C04G068530* (*Magnesium-chelatase subunit H*), *Cla97C08G161570* (*2-phytyl-1,4-beta-naphthoquinone methyltransferase*), and *Cla97C06G126710* (*WD40 repeat*) have been suggested as candidate genes for *Dgo, D,* and *S,* respectively (Table 5.1) (Guo et al. 2019; Wu et al. 2019). Dou et al. (2018a) also identified the location of dominant yellow rind color on chromosome 4, but identified a gene of unknown function (*Cla002755*) as a candidate gene. Li et al. (2019) demonstrated that *ClCG08G017810* (*Cla97C08G161570* homolog in Charleston Gray genome) was expressed higher in dark rind and developed a CAPS marker for the non-synonymous SNP in the fifth exon of the gene associated with the phenotype. *Citrullus lanatus Stripe Pattern (ClSP)* was also mapped on chromosome 6, with some overlap with the previously mapped *S* locus, but at the present time it is not known whether they represent the same locus (Yue et al. 2021).

Oren et al. (2019) identified a QTL on chromosome 9 associated with dark (dominant) vs. light rind color. A SNP in *ClCG09G012330* (*ClAPRR2*; *Arabidopsis pseudo-response regulator2-like*) causing an alternative splice junction leading to a 16 bp insertion in the mRNA of the dark rind parent is associated with the phenotype (Table 5.1). A homolog of *ClAPRR2* (*Melo3C003375; CmAPRR2*) is also responsible for darkness of rind color in melon. Interestingly, *CmAPRR2* also plays a role in pigment accumulation (carotenoids) in melon flesh (Oren et al. 2019).

The distinctive small and large yellow spots on the rind (and leaves) of the heirloom cultivar Moon and Stars were determined to be controlled by two complimentary loci on chromosomes 1 (*Sp1*) and 8 (*Sp2*) (Liu et al. 2021). This result is in contrast to the findings of Rhodes (1986) that a single gene controls the trait, possibly due to different genetic backgrounds used in the two studies (Liu et al. 2021).

5.7 Flesh Firmness

Flesh firmness is an important component of watermelon flesh quality and sensory perception. The lower amount of juice in hard fleshed fruit can negatively affect taste, while fruit with soft flesh can have reduced post-harvest storage (Harker et al. 2003; Kyriacou and Rouphael 2018; Risse et al. 1990). Flesh firmness is regulated by metabolic processes that include synthesis and degradation of cell wall components (Gao et al. 2020). Flesh pulp softening during ripening is a complex process and is mainly due to changes in cell structure, specifically pectin, cellulose, hemicellulose, and protopectin (Brummell and Harpster 2001; Sun et al. 2020; Anees et al. 2021).

A major QTL (*Qffi6.1*) for flesh firmness from *C. amarus* is located on chromosome 6 of the watermelon genome (Gao 2018; Gao et al. 2016; Juarez et al. 2013). *Cla018816*, a *xyloglucan endotransglucosylase/hydrolase* (*XTH*) gene that is differentially expressed between firm and soft fleshed near-isogenic lines (Anees et al. 2021), is the presumptive candidate gene for *Qffi6.1*. *XTH* genes are involved in cell wall modification during fruit maturation (Matsui et al. 2005; Muñoz-Bertomeu et al. 2013). Transcriptome analysis also revealed pivotal roles for another *XTH* gene (*Cla006648*), cellulose synthase (*Cla012351*), galactosyltransferase (*Cla006648*), pectinesterase gene (*Cla004251*), ethylene response element transcription factor 1 (*Cla004120*) and ethylene response element transcription factor 2a (*Cla007092*) in watermelon flesh firmness (Anees et al. 2021). While the role of the ethylene-responsive element in watermelon fruit flesh is not clear, its involvement in the regulation of other cell wall genes, mainly related to pectin, is suggestive (Anees et al. 2021).

Sun et al. (2020) identified QTLs for flesh firmness from *C. mucosospermus* on chromosomes 2 (*Qffi2.1*; PVE ~ 14) and 8 (*Qffi8.1*; PVE ~ 24). Based on differential transcription during fruit ripening of firm and soft fleshed genotypes, a DUF579 family member (*Cla016033*) and MADS-box transcription factor (*Cla012507*) were identified as the most likely candidate genes for *Qffi2.1* and *Qffi8.1*, respectively. An 11-bp (GGCAATCCAAA) insertion at the promoter region of *Cla016033* in the *C. mucosospermus* parent is suggested as the causal mutation associated with firm flesh (Sun et al. 2020).

5.8 Rind Thickness and Toughness

Internal rind characteristics, including rind thickness (Fig. 5.1) and toughness, are important for consumer preference and shipping ability. Thick rinds allow watermelon fruit to survive the rigors of shipping, allowing growers to service a global market. However, this desire for thick rinds is kept in check by consumer preference, which demands the rind make up a small portion of the overall fruit (Wehner 2008). This, combined with the trend in consumer preference toward smaller, icebox style watermelons, means that breeders must balance shippability with internal fruit quality from a consumer perspective. A significant positive correlation (r = 0.38–0.71) was observed between rind thickness and fruit size traits and a major QTL on chromosome 2 (*Qrth2*) associated with rind thickness colocalized with QTL for FD (*Qfd2.2*) and FW in two different populations (Sandlin et al. 2012; Yang et al. 2021). However, large fruit with thin rind and small fruit with thick rind are not uncommon and additional minor QTLs for rind thickness have been identified on chromosomes 5, 6, and 8 (McGregor 2017; Sandlin et al. 2012).

Early studies found rind thickness to be independent of rind toughness (*E*), which is dominant over explosive rind (*e*) (Poole 1944; Porter 1937). Fruits with explosive rind ruptures or cracks when cut and make the fruit unsuitable for shipping (Wehner 2012). Explosive rind has been extensively used in breeding of non-harvestable pollinizers, making the small non-harvested watermelons easier to crush (Wehner 2012). Rind toughness is usually described in terms of rind hardness and/or cracking resistance and can be measured by several methods, including cutting with a knife, the "thumb" test, or using a texture analyzer (Yang et al. 2021; Wehner 2008; Liao et al. 2020). A major QTL associated with rind toughness is located on chromosome 10 of the watermelon genome (*Qrto10.1*) (Liao et al. 2020). An 11 bp deletion in the *ethylene-responsive transcription factor 4* (*ClERF4*) gene was suggested as the causal mutation underlying *Qrto10.1* (Table 5.1). Yang et al. (2021) also identified QTLs associated with rind toughness and rind hardness on chromosomes 9 (Qrto9; PVE = 5.4) and 10 (*Qrto10.2*; PVE = 49.1); however, the location of *Qrto10.2* does not overlap with *Qrto10.1*. In the latter study, a positive correlation was observed between rind thickness and rind hardness and rind toughness.

5.9 Fruit Quality Traits and Domestication

The differences in fruit bitterness, size, flesh color, and sweetness between the CWRs of watermelon and cultivated watermelon demonstrate the remarkable influence of selection pressure, domestication, and breeding efforts. Edible watermelon suffers from a severe lack of genetic diversity compared to its CWRs due to a genetic bottleneck during domestication (Guo et al. 2013, 2019; Levi et al. 2001). Even though this is true for most domesticated crops, it is especially acute in watermelon (Hardigan et al. 2017). Genes involved in fruit quality traits are often targets of selection during domestication or diversification of crops (Doebley et al. 2006).

C. lanatus genotypes are fixed for the non-bitter alleles at the *Qbt1* gene (*Cla011508*), while *C. amarus* and *C. colocynths* accessions are fixed for the bitter alleles, and *C. mucosospermus* alleles at the locus are segregating, indicating that selection happened early in domestication (Guo et al. 2019; Renner et al. 2021). Fruit sugar content also appears to have been selected early in domestication as indicated by the reduction in nucleotide diversity in *C. lanatus* and *C. mucosospermus* in the *ClAGA2* and *ClSWEET3* genes associated with fruit sweetness (Guo et al. 2019; Ren et al. 2021). The region containing *QBRX2.1* (*ClTST2*) seemed to have be selected during both domestication and improvement, while *Qbrix6* was selected during domestication (Guo et al. 2019). *Qfru10.1,* identified in the *C. lanatus* background, seemed to have been selected during improvement (Cheng et al. 2016; Guo et al. 2019).

The vibrant flesh colors sweet watermelon are known for, is absent in pale fleshed *C. mucosospermus* and *C. amarus*. Variation associated with flesh color in *PSY1* (*Cla009122*) and *LCYB* (*Cla005011*) was selected during domestication (Guo et al. 2019). The white flesh and absence of the V226F mutation in *LCYB* in *C. lanatus* subsp. *cordophanus* indicate that this trait was a target of selection early in domestication (Renner et al. 2021). Increased size of the edible part of fruit is a common target of selection during domestication (Doebley et al. 2006). *Qfsi3.1*and *Qfwt2.1* associated with fruit shape and size were targets of selection during both domestication and improvement (Ren et al. 2014).

5.10 Conclusion

Advances in genomics have led to the discovery of genes and allelic variation associated with important watermelon fruit quality phenotypes that have been the target of research for nearly a hundred years. As our understanding of the molecular mechanisms underlying these traits improves, more efficient and targeted selection will advance breeding efficiency for traits of importance to watermelon consumers.

References

Aguado E, García A, Manzano S et al (2018) The sex-determining gene *CitACS4* is a pleiotropic regulator of flower and fruit development in watermelon (*Citrullus lanatus*). Plant Reprod 31(4):411–426. https://doi.org/10.1007/s00497-018-0346-1

Anees M, Gao L, Umer MJ et al (2021) Identification of key gene networks associated with cell wall components leading to flesh firmness in watermelon. Front Plant Sci 12(1164):630243. https://doi.org/10.3389/fpls.2021.630243

Arumuganathan K, Earle E (1991) Nuclear DNA content of some important plant species. Plant Mol Biol Rep 9:208–218

Bailey LH (1930) Three discussions in cucurbitaceae. Gentes Herb 2:175–186

Bang H, Kim S, Leskovar D, King S (2007) Development of a codominant CAPS marker for allelic selection between canary yellow and red watermelon based on SNP in *lycopene β-cyclase* (*LCYB*) gene. Mol Breed 20 (1):63–72. https://doi.org/10.1007/s11032-006-9076-4

Barham WS (1956) A study of the Royal Golden watermelon with emphasis on the inheritance of the chlorotic condition characteristic of this variety. Proc Amer Soc Hort Sci 67:487–489

Boualem A, Fergany M, Fernandez R et al (2008) A conserved mutation in an ethylene biosynthesis enzyme leads to andromonoecy in melons. Science 321(5890):836–838. https://doi.org/10.1126/science.1159023

Boualem A, Lemhemdi A, Sari MA et al (2016) The andromonoecious sex determination gene predates the separation of *Cucumis* and *Citrullus* genera. PLoS ONE 11:e0155444

Boualem A, Troadec C, Kovalski I et al (2009) A conserved ethylene biosynthesis enzyme leads to andromonoecy in two *Cucumis* species. PLoS ONE 4(7):e6144

Branham S, Vexler L, Meir A et al (2017) Genetic mapping of a major codominant QTL associated with β-carotene accumulation in watermelon. Mol Breed 37 (12):146. https://doi.org/10.1007/s11032-017-0747-0

Brueckner B, Schonhof I, Schroedter R et al (2007) Improved flavour acceptability of cherry tomatoes. Target group: children. Food Qual Prefer 18(1):152–160. https://doi.org/10.1016/j.foodqual.2005.09.011

Brummell DA, Harpster MH (2001) Cell wall metabolism in fruit softening and quality and its manipulation in transgenic plants. Plant Mol Biol 47:311–340

Chambliss OL, Erickson HT, Jones CM (1968) Genetic control of bitterness in watermelon fruits. Proc Amer Soc Hort Sci 93:539–546

Cheng Y, Luan F, Wang X et al (2016) Construction of a genetic linkage map of watermelon (*Citrullus lanatus*) using CAPS and SSR markers and QTL analysis for fruit quality traits. Sci Hortic 202:25–31. https://doi.org/10.1016/j.scienta.2016.01.004

Chomicki G, Renner SS (2015) Watermelon origin solved with molecular phylogenetics including Linnaean material: another example of museomics. New Phytol 205(2):526–532. https://doi.org/10.1111/nph.13163

Davidovich-Rikanati R, Shalev L, Baranes N et al (2015) Recombinant yeast as a functional tool for understanding bitterness and cucurbitacin biosynthesis in watermelon (*Citrullus* spp.). Yeast 32(1):103–114

Doebley JF, Gaut BS, Smith BD (2006) The molecular genetics of crop domestication. Cell 127(7):1309–1321. https://doi.org/10.1016/j.cell.2006.12.006

Dou J, Lu X, Ali A et al (2018a) Genetic mapping reveals a marker for yellow skin in watermelon (*Citrullus lanatus* L.). PLOS ONE 13(9):e0200617. https://doi.org/10.1371/journal.pone.0200617

Dou J, Zhao S, Lu X et al (2018b) Genetic mapping reveals a candidate gene (*ClFS1*) for fruit shape in watermelon (*Citrullus lanatus* L.). Theor Appl Genet 131(4):947–958. https://doi.org/10.1007/s00122-018-3050-5

Evans CB (2008) Consumer preferences for watermelons: a conjoint analysis. Auburn University

Fall LA, Perkins-Veazie P, Ma G et al (2019) QTLs associated with flesh quality traits in an elite × elite watermelon population. Euphytica 215(2):30. https://doi.org/10.1007/s10681-019-2356-y

Fang X, Liu S, Gao P et al (2020) Expression of *ClPAP* and *ClPSY1* in watermelon correlates with chromoplast differentiation, carotenoid accumulation, and flesh color formation. Sci Hortic 270:109437. https://doi.org/10.1016/j.scienta.2020.109437

FAOSTAT (2020) Food and agriculture organization of the united nations, crop production. http://www.fao.org/faostat/en/#data/QC. December 2020

Gao L (2018) Transcriptome analysis and fine mapping of major genes controlling flesh firmness and sour flesh in watermelon. Chinese Academy of Agricultural Sciences, Beijing

Gao L, Zhao SJ, Lu XQ et al (2016) Linkage analysis on flesh firmness of watermelon [*Citrullus lanatus* (Thunb.) Mansf.] by SSR molecular marker. J Plant Genet Resour 17:866–870

Gao Y, Guo Y, Su Z et al (2020) Transcriptome analysis of genes related to fruit texture in watermelon. Sci Hortic 262:109075. https://doi.org/10.1016/j.scienta.2019.109075

Giuliano G, Tavazza R, Diretto G et al (2008) Metabolic engineering of carotenoid biosynthesis in plants. Trends Biotechnol 26(3):139–145. https://doi.org/10.1016/j.tibtech.2007.12.003

Guo S, Zhang J, Sun H et al (2013) The draft genome of watermelon (*Citrullus lanatus*) and resequencing of 20 diverse accessions. Nat Genet 45:51–58

Guo S, Zhao S, Sun H et al (2019) Resequencing of 414 cultivated and wild watermelon accessions identifies selection for fruit quality traits. Nat Genet 51(11):1616–1623. https://doi.org/10.1038/s41588-019-0518-4

Gusmini G, Wehner TC (2006) Qualitative inheritance of rind pattern and flesh color in watermelon. J Heredity 97:177–185

Gusmini G, Wehner TC, Jarret RL (2004) Inheritance of egusi seed type in watermelon. J Hered 95(3):268–270. https://doi.org/10.1093/jhered/esh031

Hardigan MA, Laimbeer FPE, Newton L et al (2017) Genome diversity of tuber-bearing *Solanum* uncovers complex evolutionary history and targets of domestication in the cultivated potato. Proc Natl Acad Sci 114 (46):E9999–E10008. https://doi.org/10.1073/pnas.1714380114

Harker FR, Gunson FA, Jaeger SR (2003) The case for fruit quality: an interpretive review of consumer

attitudes, and preferences for apples. Postharvest Biol Technol 28:333–347
Henderson WR (1989) Inheritance of orange flesh color in watermelon. Cucurbit Genet Coop Rpt 12:59–63
Henderson WR (1992) Corrigenda to the 1991 Watermelon gene list (CGC 14:129–137). Cucurbit Genet Coop Rpt 15:110
Henderson WR, Scott GH, Wehner TC (1998) Interaction of flesh color genes in watermelon. J Heredity 89:50–53
Juarez B, King JJ, Bachlava E et al (2013) Methods and compositions for watermelon firmness, Patent Application Publication. Patent No: WO2013033611A1, Pub.No.: US 2013/0055466A1
Kim KH, Hwang JH, Han DY et al (2015) Major quantitative trait loci and putative candidate genes for powdery mildew resistance and fruitrelated traits revealed by an intraspecific genetic map for watermelon (*Citrullus lanatus* var. *lanatus*). PLoS ONE 10: e0145665
Kim Y-C, Choi D, Zhang C et al (2018) Profiling cucurbitacins from diverse watermelons (*Citrullus* spp.). Hortic Environ Biotech 59(4):557–566
Kumar R, Wehner TC (2011) Discovery of second gene for solid dark green versus light green rind pattern in watermelon. J Heredity 102(4):489–493
Kyriacou MC, Rouphael Y (2018) Towards a new definition of quality for fresh fruits and vegetables. Sci Hortic 234:463–469. https://doi.org/10.1016/j.scienta.2017.09.046
Legendre R, Kuzy J, McGregor C (2020) Markers for selection of three alleles of *ClSUN25-26-27a* (*Cla011257*) associated with fruit shape in watermelon. Mol Breed 40(2):19. https://doi.org/10.1007/s11032-020-1104-2
Levi A, Thomas CE, Wehner TC et al (2001) Low genetic diversity indicates the need to broaden the genetic base of cultivated watermelon. HortScience 36 (6):1096–1101
Li B, Lu X, Dou J et al (2018) Construction of a high-density genetic map and mapping of fruit traits in watermelon (*Citrullus lanatus* L.) based on whole-genome resequencing. Int J Mol Sci 19:3268
Li B, Zhao S, Dou J et al (2019) Genetic mapping and development of molecular markers for a candidate gene locus controlling rind color in watermelon. Theor Appl Genet 132(10):2741–2753. https://doi.org/10.1007/s00122-019-03384-3
Li N, Shang J, Wang J et al (2020) Discovery of the genomic region and candidate genes of the *Scarlet Red Flesh Color* (*Yscr*) locus in watermelon (*Citrullus Lanatus* L.). Front Plant Sci 11(116). https://doi.org/10.3389/fpls.2020.00116
Liao N, Hu Z, Li Yet al (2020) Ethylene-responsive factor 4 is associated with the desirable rind hardness trait conferring cracking resistance in fresh fruits of watermelon. Plant Biotechnol J 18(4):1066–1077. https://doi.org/10.1111/pbi.13276
Liu CQ, Gao P, Luan FS (2014) A linkage map for watermelon and QTL analysis for fruit related traits. Sci Agric Sin 47:2814–2829
Liu D, Sun D, Liang J et al (2021) Characterization and bulk segregant analysis of 'moon and star' appearance in watermelon. Sci Hortic 285:110140. doi:https://doi.org/10.1016/j.scienta.2021.110140
Liu S, Gao P, Wang X et al (2015) Mapping of quantitative trait loci for lycopene content and fruit traits in *Citrullus lanatus*. Euphytica 202(3):411–426. https://doi.org/10.1007/s10681-014-1308-9
Liu S, Gao P, Zhu QL et al (2016) Development of cleaved amplified polymorphic sequence markers and a CAPS-based genetic linkage map in watermelon (*Citrullus lanatus* [Thunb.] Matsum. and Nakai) constructed using wholegenome re-sequencing data. Breed Sci 66:244–259
Lou L, Wehner TC (2016) Qualitative inheritance of external fruit traits in watermelon. HortScience 51 (5):487
Loy BJ (2006) Fruit size in melon in monoecious and andromonoecious isolines. Cucurbit Genet Coop Rpt 2005–2006(28–29):12–13
Lu BY, Zhou HW, Chen X et al (2016) QTL analysis of fruit traits in watermelon. J Fruit Sci 33:1206–1218
Lv P, Li N, Liu H et al (2015) Changes in carotenoid profiles and in the expression pattern of the genes in carotenoid metabolisms during fruit development and ripening in four watermelon cultivars. Food Chem 174:52–59. https://doi.org/10.1016/j.foodchem.2014.11.022
Manzano S, Aguado E, Martínez C et al (2016) The ethylene biosynthesis gene *CitACS4* regulates monoecy/andromonoecy in watermelon (*Citrullus lanatus*). PLOS ONE 11(5):e0154362. https://doi.org/10.1371/journal.pone.0154362
Matsui A, Yokoyama R, Seki M et al (2005) *AtXTH27* plays an essential role in cell wall modification during the development of tracheary elements. Plant J 42:525–534
Matsuo K, DeMilo AB, Schroder RFW et al (1999) Rapid high-performance liquid chromatography method to quantitate elaterinide in juice and reconstituted residues from a bitter mutant of Hawkesbury watermelon. J Agric Food Chem 47:2755–2759
McGregor CE (2017) Genetic control of rind thickness in watermelon. HortScience 52(9):S59
McKay JW (1936) Factor interaction in *Citrullus:* seed-coat color, fruit shape and markings show evidence of Mendelian inheritance in watermelon crosses J Hered 27:110–112
Muñoz-Bertomeu J, Miedes E, Lorences EP (2013) Expression of *xyloglucan endotransglucosylase/hydrolase* (*XTH*) genes and *XET* activity in ethylene treated apple and tomato fruits. J Plant Physiol 170(13):1194–1201. https://doi.org/10.1016/j.jplph.2013.03.015
Nan W, Shi L, Peng G et al (2016) Developmental changes in gene expression drive accumulation of lycopene and β-Carotene in watermelon. J Amer Soc Hort Sci 141 (5):434–443. https://doi.org/10.21273/JASHS03690-16
Navot N, Sarfatti M, Zamir D (1990) Linkage relationships of genes affecting bitterness and flash color in watermelon. J Heredity 81:162–165

Oren E, Tzuri G, Vexler L et al (2019) The multi-allelic *APRR2* gene is associated with fruit pigment accumulation in melon and watermelon. J Exp Bot 70 (15):3781–3794. https://doi.org/10.1093/jxb/erz182

Pan Y, Wang Y, McGregor C et al (2020) Genetic architecture of fruit size and shape variation in cucurbits: a comparative perspective. Theor Appl Genet 133(1):1–21. https://doi.org/10.1007/s00122-019-03481-3

Pardo JE, Gomez R, Tardaguila J et al (1997) Quality evaluation of watermelon varieties (*Citrullus vulgaris* S.). J Food Quality (20):547–557

Park S, Kim K-T, Kang S-C et al (2016) Rapid and practical molecular marker development for rind traits in watermelon. Hortic Environ Biotechnol 57:385–391

Perkins-Veazie P, Collins JK, Davis AR et al (2006) Carotenoid content of 50 watermelon cultivars. J Agric Food Chem 54:2593–2597

Poole CF (1944) Genetics of cultivated cucurbits. J Hered 35:122–128

Poole CF, Grimball PC (1945) Interaction of sex, shape, and weight genes in watermelon. J Agr Res 71:533–552

Porter DR (1937) Inheritance of certain fruit and seed characters in watermelons. Hilgardia 10:489–509

Porter DR, Bission CS, Allinger HW (1940) Factors affecting the total soluble solids, reducing sugars, and sucrose in watermelons. Hilgardia 13:31–66

Prothro J, Abdel-Haleem H, Bachlava E al (2013) Quantitative trait loci associated with sex expression in an inter-subspecific watermelon population. J Am Soc Hort Sci 138(2):125–130

Reddy UK, Abburi L, Abburi VL et al (2015) A genome-wide scan of selective sweeps and association mapping of fruit traits using microsatellite markers in watermelon. J Hered 106(2):166–176. https://doi.org/10.1093/jhered/esu077

Ren Y, Guo S, Zhang J et al (2018) A tonoplast sugar transporter underlies a sugar accumulation QTL in watermelon. Plant Physiol 176(1):836–850. https://doi.org/10.1104/pp.17.01290

Ren Y, Li M, Guo S et al (2021) Evolutionary gain of oligosaccharide hydrolysis and sugar transport enhanced carbohydrate partitioning in sweet watermelon fruits. The Plant Cell koab055. https://doi.org/10.1093/plcell/koab055

Ren Y, McGregor C, Zhang Y et al (2014) An integrated genetic map based on four mapping populations and quantitative trait loci associated with economically important traits in watermelon (*Citrullus lanatus*). BMC Plant Biol 14(1):33. https://doi.org/10.1186/1471-2229-14-33

Renner SS, Wu S, Pérez-Escobar OA et al (2021) A chromosome-level genome of a Kordofan melon illuminates the origin of domesticated watermelons. Proc Nat Acad Sci 118(23):e2101486118. https://doi.org/10.1073/pnas.2101486118

Rhodes B (1986) Gene affecting foliage color in watermelon. J Hered 77:134–135

Risse LA, Brecht JK, Sargent SA et al (1990) Storage characteristics of small watermelon cultivars. J Am Soc Hortic Sci 115:440–443

Robinson RW, Munger HM, Whitaker TW et al (1976) Genes of the cucurbitaceae. HortScience 11:554–568

Rosa JT (1928) The inheritance of flower types in *Cucumis* and *Citrullus*. Hilgardia 3:233–250

Sandlin KC, Prothro JM, Heesacker AF et al (2012) Comparative mapping in watermelon [*Citrullus lanatus* (Thunb.) Matsum. et Nakai]. Theor Appl Genet 125:1603–1618

Shang Y, Ma Y, Zhou Y et al (2014) Biosynthesis, regulation, and domestication of bitterness in cucumber. Science 346(6213):1084–1088. https://doi.org/10.1126/science.1259215

Shibuya M, Adachi S, Ebizuka Y (2004) Cucurbitadienol synthase, the first committed enzyme for cucurbitacin biosynthesis, is a distinct enzyme from cycloartenol synthase for phytosterol biosynthesis. Tetrahedron Lett 60:6995–7003

Shimotsuma M (1963) Cytogenetical studies in the genus *Citrullus*. VII. Inheritance of several characters in watermelons. Jap J Breeding 13:235–240

Subburaj S, Lee K, Jeon Y et al (2019) Whole genome resequencing of watermelons to identify single nucleotide polymorphisms related to flesh color and lycopene content. PLOS ONE 14(10):e0223441. https://doi.org/10.1371/journal.pone.0223441

Sun L, Zhang Y, Cui H et al (2020) Linkage mapping and comparative transcriptome analysis of firmness in watermelon (*Citrullus lanatus*). Front Plant Sci 11 (831). https://doi.org/10.3389/fpls.2020.00831

Sun T, Yuan H, Cao H et al (2018) Carotenoid metabolism in plants: the role of plastids. Molecular Plant 11(1):58–74. https://doi.org/10.1016/j.molp.2017.09.010

Tadmor Y, King S, Levi A et al (2005) Comparative fruit colouration in watermelon and tomato. Food Res Int 38(8):837–841. https://doi.org/10.1016/j.foodres.2004.07.011

Tanaka T, Wimol S, Mizutani T (1995) Inheritance of fruit shape and seed size of watermelon. J Japan Soc Hort Sci 64:543–548

UPOV (2013) Guidelines for the conduct of tests for distinctness, uniformity and stability. International union for the protection of new varieties—Watermelon TG/142/5

van der Knaap E, Tanksley SD (2001) Identification and characterization of a novel locus controlling early fruit development in tomato. Theor Appl Genet 103 (2):353–358. https://doi.org/10.1007/s001220100623

Wang C, Qiao A, Fang X et al (2019) Fine mapping of lycopene content and flesh color related gene and development of molecular marker–sssisted selection for flesh color in watermelon (*Citrullus lanatus*). Front Plant Sci 10(1240). https://doi.org/10.3389/fpls.2019.01240

Weetman LM (1937) Inheritance and correlation of shape, size and color in the watermelon *Citrullus vulgaris* Schrad. Iowa Agric Exp Stat Ann Bull 228:224–256

Wehner T (2008) Watermelon. In: Prohens J, Nuez F (eds) Vegetables I: asteraceae, brassicaceae, chenopodicaceae, and cucurbitaceae, vol 2. Springer. New York, pp 381–418

Wehner T (2012) 2012 Gene list for watermelon. The cucurbit genetics cooperative. Accessed 9 June 2021

Wu S, Wang X, Reddy U et al (2019) Genome of 'Charleston Gray', the principal American watermelon cultivar, and genetic characterization of 1,365 accessions in the U.S. National Plant Germplasm System watermelon collection. Plant Biotechnol J 17:2246–2258. https://doi.org/10.1111/pbi.13136

Wu S, Xiao H, Cabrera A et al (2011) *SUN* regulates vegetative and reproductive organ shape by changing cell division patterns. Plant Physiol 157:1175–1186

Xiao H, Jiang N, Schaffner E et al (2008) A retrotransposon-mediated gene duplication underlies morphological variation of tomato fruit. Science 319:1527–1530

Yang H, Park S, Park Y-H et al (2015) Linkage analysis of the three loci determining rind color and stripe pattern in watermelon. Korean J Hortic Sci Technol 33:559–565

Yang T, Amanullah S, Pan J et al (2021) Identification of putative genetic regions for watermelon rind hardness and related traits by BSA-seq and QTL mapping. Euphytica 217:19. https://doi.org/10.1007/s10681-020-02758-9

Yativ M, Harary I, Wolf S (2010) Sucrose accumulation in watermelon fruits: genetic variation and biochemical analysis. J Plant Physiol 167(8):589–596. https://doi.org/10.1016/j.jplph.2009.11.009

Yoo KS, Bang H, Lee EJet al(2012) Variation of carotenoid, sugar, and ascorbic acid concentrations in watermelon genotypes and genetic analysis. Hortic Environ Biotechnol 53(6):552–560. https://doi.org/10.1007/s13580-012-0014-6

Yue Z, Ma R, Cheng D et al (2021) Candidate gene analysis of watermelon stripe pattern locus *ClSP* ongoing recombination suppression. Theor Appl Genet. https://doi.org/10.1007/s00122-021-03891-2

Zhang J, Sun H, Guo S et al (2020) Decreased protein abundance *of Lycopene β-Cyclase* contributes to red flesh in domesticated watermelon. Plant Physiol 183(3): 1171–1183. doi:https://doi.org/10.1104/pp.19.01409

Zhang XP, Jiang Y (1990) Edible seed watermelons (*Citrullus lanatus* (Thunb.) Matsum. & Nakai) in northwest China. Cucurbit Genet Coop Rpt 13:40–42

Zhang Z, Zhang Y, Sun L et al (2018) Construction of a genetic map for *Citrullus lanatus* based on CAPS markers and mapping of three qualitative traits. Scientia Hort 233:532–538. https://doi.org/10.1016/j.scienta.2017.10.029

Zhao W, Lv P, Gu H (2013) Studies on carotenoids in watermelon flesh. Agric Sci 04:13–20

Zheng Y, Wu S, Bai Y et al (2018) Cucurbit Genomics Database (CuGenDB): a central portal for comparative and functional genomics of cucurbit crops. Nucleic Acids Res 47(D1):D1128–D1136. https://doi.org/10.1093/nar/gky944

Zhou Y, Ma Y, Zeng J et al (2016) Convergence and divergence of bitterness biosynthesis and regulation in *Cucurbitaceae*. Nat Plants 2:16183–16183. https://doi.org/10.1038/nplants.2016.183

Challenges of Traditional Breeding in Watermelon

6

Harshawardhan Choudhary, K. Padmanabha, Gograj Singh Jat, and Tusar Kanti Behera

6.1 Background

Plant domestication is a long-term process that results in complex phenotypes being genetically modified in response to human selection. The response of civilisation to climate change has resulted in significant changes in fauna and flora. The Domestication of plants is one of the most significant examples of human impact on ecological niche evolution.

Domestication is a reciprocal process in which plant species develop to rely on people for survival while also providing humans with a variety of benefits. Domestication has played a significant role in human population growth and spatial expansion during the evolution of civilisation. The importance of domesticated plants in agriculture cannot be overstated. Agriculture is responsible for the complex, technologically advanced societies and enormous human populations that exist today. Agriculture allowed humans to settle in a single location, paving the way for the development of civilised societies, which in turn led to the survival of human communities through the cultivation of domesticated plants. Hunters and gatherers, early hominids subsisted on naturally available flora, fruits, nuts, carrion and game. To achieve the same calories of food energy, hunting and gathering require more energy than farming operations. Hunters and gatherers are compelled to enter agriculture due to external pressures such as climate change and population expansion (MacDonald 2003). The favourable temperature conditions, along with adequate rainfall, resulted in major changes in the ecology, resulting in vegetation exuberance and plant community diversification. Which initiates the process of wild species domesticating crop plants for the benefit of mankind, sowing the first seeds of plant breeding.

Domestication of crops, particularly cereals, leads to the formation of large, permanent settlements. High-yielding domesticates, fertilisation, and in some circumstances, irrigation aided modern agriculture. Pesticides, herbicides and genetically modified crops have all enhanced yields, but with significant environmental consequences. Wheat (*Triticum aestivum*) is a cereal crop grown in Europe and Asia, rice (*Oryza sativa*) is grown in Southeast Asia, and maize (*Zea mays*) is grown in America.

Domesticated species are those that have been shaped by human selection to the point that they rely on human for survival. Previously, breeders chose traits that improved a species food suitability. Cultivar's ability to survive in the wild was reduced as an unintended consequence.

H. Choudhary (✉) · K. Padmanabha · G. S. Jat · T. K. Behera
Division of Vegetable Science, ICAR-Indian Agricultural Research Institute, Pusa Campus, New Delhi, India
e-mail: harshahit2001@yahoo.co.in

S. Kr. Dutta et al. (eds.), *The Watermelon Genome*, Compendium of Plant Genomes,
https://doi.org/10.1007/978-3-031-34716-0_7

6.1.1 Characteristics of Domesticates

Increased size of edible parts (e.g. fruits or tubers), increased palatability, decreased armament, lack of dormancy, decreased toxicity, shorter blooming, synchronised phenology, non-shattering fruits (in *Poaceae*) and thinner seed coats have all been selected for in different taxa.

Domesticated plants typically have larger fruits or other edible tissues than their wild counterparts. Given the presence of defence alleochemical substances, wild relatives of domesticates are frequently unpleasant. Cucurbitacin, a bitter compound found in wild cucurbit species, is abundant. Herbivory and plant pathogens were prevented in large part by chemicals that gave off an unpleasant taste. Cultivars have fewer mechanical defences. Some wild species are well known for their spininess. One of the reasons that domesticated species rely on humans for existence is the lack of these defences. In the domestication of cereal crops like wheat, non-shattering infructescences, polyploidy and free-threshing cultivars are important events.

Nikolai Vavilov, a Russian geneticist, was a dedicated field collector who travelled extensively in quest of domestication loci. His 1940 Russian publication (the theory of cultivated plant origins) was translated into English in 1951. Vavilov claimed that the centres of a crop's greatest diversity represented its origin. He initially proposed six centres later added two additional ones: China, India and Indochina, Central Asia, Near East, Mediterranean, Ethiopia, Mesoamerica and north-eastern South America. Vavilov's places of origin, according to Jack Harlan, were centres of diversity and long-standing agricultural activity, which may or might not be hubs of crop evolution or domestication.

6.2 Domestication of Watermelon

6.2.1 General Characteristics of Watermelon

The watermelon, *Citrullus lanate* (Thunb.) Matsum. & Nakai, is one of the most refreshing food items on hot summer days. It is grown mainly in the warmer regions of the world (Wehner 2008). The pinnatifid shape of leaf laminae is a distinguished feature of watermelon. Also, the flowers are borne singly with a female or hermaphrodite flower appearing at every seventh or eighth leaf axil and male flowers occupying the remainder. The ovaries and primordial fruits are hairy, becoming smooth and lustrous as they grow. The exocarp is green and exhibits striping, and the stripes have acute, jagged edges. Immediately underneath the green exterior is the thick, watery, pale green or white rind which is the mesocarp. This encloses the endocarp or fruit flesh within which are distributed the seeds. The seed coats, depending on accession or cultivar, can be brown, black, beige, yellow or other colours.

6.2.2 Classification of Citrullus

Other cultivated members of the genus include the citron, egusi, and colocynth watermelons, in addition to the dessert watermelon. The defining feature of colocynth is the small size of the seeds, which are about 8 mm long. Seed coatings are absent in Egusis. Young, developing citron and dessert watermelons can be distinguished by the hairiness of the former.

The watermelon species, *Citrullus ecirrhosus* Cogn., *Citrullus rehmii* De Winter and *Citrullus naudinianus* (Sond.) Hooker f., are not cultivated whereas *Citrullus colocynthis* Schrad. is sparingly cultivated. The citron, egusi and dessert watermelon have been largely treated as subspecies, botanical varieties or cultivar groups of *C. lanatus*, but Chomicki and Renner (2015), based on genomic analysis, proposed that each be designated as a separate species Indeed, barriers to crossing among these three, although rather weak, have been documented (Levi et al. 2003; Gusmini et al. 2004; Paris 2017), which would justify treating them as separate species. The citron watermelon is thus *C. amarus* Schrad., the egusi *C. mucosospermus* (Fursa) Fursa and the desert watermelon is *C. lanatus* even though it is in fact less lanate than the citron watermelon.

6.2.3 Historical Evidences

Ten sites in Egypt contain watermelon remains dating back to predynastic (5000 years ago) and dynastic eras (De Vartavan and Asensi Amoros 1997) A fruit with seeds, from ca. 1500 BCE, was recovered from the foundation of a temple in Sudan (Van Zeist 1983). Germer (1988) listed remains, including > 4500 years old seeds and > 4000 years old fruits belonging to *C. lanatus* var. *colocynthoides*. Roman-era seeds have also been found in Egypt (Cox and van der Veen 2008).

Egypt has preserved depictions of a variety of ancient flora and fruits. Watermelon is represented by a spherical, striped fruit attached to a short section of vine with two strongly split leaves in one illustration (Keimer 1924). A third depicts a giant, oblong striped fruit on a platter (Feliks 2005) and another depicts nine enormous, round, striped fruits in a basket (Feliks 2005). (Manniche 1989).

Watermelon is mentioned in the Hebrew Bible (*Numbers* 11:5), in the name of *avattihim.* The Hebrew word, *avattihim* is derived from an ancient Egyptian language root (Manniche 1989). Greek, Latin and Hebrew literature of Roman times, collectively, are rich in references to watermelons although the descriptions of the fruits are meagre. The *pepon* was described by Greek-writing physicians as cold and wet. Quintus Gargilius Martialis, writing in Latin (ca. 260 CE), mentioned that *pepone* were good in eating quality after the removal of the rind and pits and Anthimus (ca. 516 CE) wrote that they were to be eaten when fully ripened. In Hebrew-language codices of Jewish Law, the *avattihim* (watermelons), in most instances, are considered together with other edible cucurbits.

6.2.4 Origin of Watermelon

Despite the fact that the Citrullus genus is of African origin (Whitaker and Davis 1962; Zohary et al. 2012), there is still debate on where the watermelon originated in Africa. It's mostly due to phenotypic overlap between citron, egusi, and dessert watermelons and colocynths, as well as the fact that wild forms and primitive fruits of *Citrullus*, independent of taxon, possess hard, bitter, or bland, poor coloured flesh. Above all, many hypotheses for sweet desert watermelon origin have emerged as a result of taxonomic uncertainty and the weakening of crossability barriers among *Citrullus* species.

One of the most popular theories for the origin of watermelon is that it is descended from the northern African colocynth. However, hybrids of these two species show chromosome abnormalities and decreased fertility (Whitaker and Davis 1962; Wehner 2008). Colocynths are more distant from dessert watermelons than citron watermelons, according to molecular genetics and genomics findings (Levi et al. 2013; Chomicki and Renner 2015).

Second, the dessert watermelon is descended from the southern African citron watermelon. However, the discovery of 4000-year-old watermelon artefacts in Egypt is incongruent, as farming was not yet established in southern Africa at the time (Zohary et al. 2012). According to recent research, there have been significant changes at genomic level between the two species (Guo et al. 2013; Chomicki and Renner 2015). Furthermore, it has been suggested that the egusi watermelon of western Africa is considered the 'recent progenitor of present-day farmed watermelon' (Guo et al. 2013).

A fourth theory is that the dessert watermelon originated in north-eastern Africa and was originally domesticated there. As a result, wild, frequently hard, bitter or insipid watermelons with the scientific variety name colocynthoides or subspecies name cordophanus are living representatives of the sweet dessert watermelon's ancestor. Watermelons grow wild and 'semi-cultivated' in semi-arid and dry Sudan (Andrews 1950), with the highest populations probably occurring in the Nile Valley (Mariod et al. 2009). They've been spotted in the desert, where locals take advantage of them for water and food (Abdel-Magid 1989). Their range extends into Egypt to the north (Muschler 1912).

Citrullus lanatus used to have an obvious wild or domesticated dimorphism due to continuous selection for improved horticultural features, according to the newest sequencing and genomics technology (Abbo et al. 2014). The following evidence suggests that the dessert watermelon originated in north-eastern Africa, and that watermelons were cultivated for water and food there around 4000 years ago.

Watermelon has evolved from wild watermelons having small fruits containing hard, pale-coloured, bitter- or bland-tasting flesh to present-day watermelon with a greater number of fruits containing crisp, sweet and red flesh with a thin peel, due to domestication and contemporary breeding. The common sweet dessert watermelon (*C. lanatus*) has a restricted genetic base, believed that it evolved through a sequence of selections in a particular ancestral population.

6.3 Transformation of Watermelon from Bitter to Non-bitter

The present-day sweet dessert watermelon is the result of several years of cultivation and selection for desirable qualities in watermelon that are acceptable to humans. New forms of table watermelon have arisen due to extensive cultivation and continuous selection procedures resulting in the development of present-day farm cultivars that exhibit little relation to the ancient African variants. Watermelon is one of the most popular fresh fruits in the world, thanks to its sweetness and appealing flesh colour and flavour. *Citrullus lanatus* var *citroides* and *Citrullus colocynthis* have hard white, green or yellow flesh and are generally unsweet or bitter (2–3 Brix) (Liu et al. 2013) Watermelon cultivars have seen a significant increase in fruit quality (10–12 Brix) as a result of many years of domestication and breeding selection.

Watermelon has evolved from a late maturing vine with little fruit with hard, white flesh and bland or bitter flavour to an early maturing, more compact plant with large fruit with edible sweet flesh due to domestication and formal plant breeding.

Another Cucurbitaceae crop, watermelon (*Citrullus lanatus*), was domesticated around 4000 years ago. The researchers created a genome-variation map and inferred ancestral relationships and gene flow between *C. lanatus* and *Citrullus mucosospermus* based on enhanced genome assembly and resequencing of 414 watermelon accessions (Guo et al. 2019). Despite the fact that *C. mucosospermus* and *C. lanatus* are close relatives, the fruit flesh qualities of *C. mucosospermus* and *C. lanatus* were not selected by humans. The ancestor *C. mucosospermus* was presumably domesticated separately for seed eating, according to existing scientific evidence. *C. lanatus*, on the other hand, was cultivated for its fruit flesh (Guo et al. 2019). In all *C. lanatus*, a non-bitterness allele of the ClBt gene was discovered, which is homologous to cucumber bitterness genes.

The alkaline-galactosidase gene (ClAGA2) and a sugar-transporter gene, ClTST2, were found to be co-located in the fruit flesh sweetness quantitative trait locus. Watermelon with red flesh has a strong link to a single amino acid swap in the LCYB gene, which is thought to be involved in lycopene production. Watermelon's bitterness was removed due to convergent evolution (Guo et al. 2019). The domestication syndrome's genetic architecture has been revealed to be quite simple. As a result, regions of the genome linked to qualities like bitterness in watermelon can be quickly mapped.

Cucurbitacins are a type of bitter, highly oxygenated tetracyclic triterpenes produced mostly by the Cucurbitaceae plant family. In the cucurbitaceous family, there are 12 different types of cucurbitacins. The biosynthesis module for cucurbitacin C (CuC), which consists of nine genes and four pathway enzymes, has been identified. Cucurbitacins also have a diverse pharmacological profile, including anti-inflammatory, purgative, and anti-tumour properties. The loss of bitterness traits associated with the convergent domestication of wild watermelon may be due to a syntenic gene cluster of transcription factors that regulates the tissue-specific production of cucurbitacins.

The bitter chemical Cucurbitacin E (CuE) is found in abundance in watermelon. Understanding the evolutionary and genetic origins of this feature could provide significant information for developing low-cucurbitacin watermelon.

Domesticated from their exceedingly bitter ancestors, non-bitter watermelon cultivars are now used to produce melons for human consumption. In watermelon, expression of the CuE biosynthetic genes is controlled by two tissue-specific basic helix–loop–helix (bHLH) transcription factors (TFs) in the leaves (Bl, bitter leaf) and fruit (Bt, bitter fruit). Fruit bitterness is efficiently removed by mutations in the Bt promoter, and this trait of non-bitterness has been selected and fixed during the domestication process. Independent mutations in watermelon syntenic transcription factor genes may result in significant reductions in fruit bitterness, leading to the domestication of bitter wild watermelon.

Ten CuE biosynthetic enzymes, including an oxidosqualene cyclase (OSC, the Mendelian gene Bi), eight cytochromes P450 (CYPs) and acyltransferase (ACT) are expressed in the leaves of cultivated watermelon and the fruit of the wild ancestor, five of which are clustered (Bi, three CYPs and an ACT) on chromosome 6 (Bi cluster).

Gene ClBr (Br, bitter root) is expressed in the CuE biosynthetic pathway in the roots of watermelon, which is compatible with the bitterness distribution in these plants, according to the expression patterns of candidate genes and the cucurbitacin content in different tissues of seedlings and fruit. As a result, these transcription factors (TFs) are likely to be root-specific regulators of cucurbitacin production and likely play a role in watermelon's chemical defence against underground herbivores.

The potential regulators of fruit bitterness in watermelon were found by comparing the expression patterns of transcription factors (TF) and the bitterness content in the fruits of wild and cultivated cultivars. ClBt expression was found to be higher in bitter fruit of wild lines, but lower in cultivated lines, suggesting that these TF may govern fruit bitterness in watermelon.

These potential root- and fruit-specific regulators can directly stimulate transcription of the CuE biosynthesis enzymes' expressed genes. CuE biosynthesis was induced by temporary expression of transcription factors ClBt or ClBr in watermelon cotyledons. These two transcription factors (ClBt and ClBr) are thought to be the tissue-specific cucurbitacin regulators in watermelon. Domestication of wild watermelon was likely conferred by mutations within the conserved fruit-specific regulators, as disruption of bitterness regulation in the fruit of wild cucumber led to the domesticated non-bitter cultivar. ClBt is found in the domestication sweep zone, implying that it may play a role in crop domestication.

By comparing the sequences of ClBt from bitter wild lines with the genes from non-bitter cultivars, genetic variations related with the bitter fruit phenotype in watermelon were discovered. In cultivated watermelon, a single base pair mutation in the second exon of ClBt causes premature protein translation. The nucleotide at position 382 is T in non-bitter cultivars and C in bitter wild lines, and this single nucleotide polymorphism relates to the bitter fruit phenotype. ClBt that has been mutated encodes a truncated protein that is expected to be about half the length of the wild-type protein (127 vs. 249 amino acids), making it non-functional. The shortened ClBt gene did not demonstrate regulatory ability, as expected, which could explain the lack of CuE synthesis in farmed watermelon fruit. Watermelon has convergent domestication due to changes in the homologues of fruit bitterness regulators.

Human needs have resulted in convergent phenotypic development (e.g., increased fruit size and improved fruit taste) during the domestication of the watermelon crop. Although the molecular pathways behind these convergent phenotype variations are currently poorly understood, a rising number of cases supports the hypothesis that causal mutations in orthologous genes may be responsible for convergent changes in critical features in crop mitigation. Mutations in fruit bitterness regulators caused the domestication of wild watermelon to lose its bitterness. During the domestication of wild watermelon, breeding has resulted in selection for the loss of function of these direct fruit-specific regulators (ClBt).

Modern commercial watermelons have a high sugar content that is commensurate with soluble solids content (with a Brix value of 10–14°). By a chemical mechanism that is still unclear, large amounts of carbohydrates, primarily sucrose, accumulate in the cell vacuoles of the fruit flesh. A large sugar content QTL was discovered on chromosome 2 with the Tonoplast Sugar Transport gene, which is important for sucrose, fructose and glucose uptake into the vacuoles of the dessert watermelon (Zhang et al. 2017). Dessert watermelon (*C. lanatus*) landraces with low sugar content and Brix values could be a good source of bioactive substances with health benefits such as citrulline, ascorbic acid, potassium, flavonoids and carotenoids including alpha and beta carotene and lycopene.

6.4 Genetic Resources and Genetic Bottlenecks in Watermelon

Around 1100 and 800 AD, watermelon was introduced to China and India, respectively (Zhao 2015). Citrullus species can be found in the wild in southern and central Africa, C. colocynthis also grows wild in India. For the genus, India and China could be considered secondary centres of diversification. Watermelon cultivation is supposed to have started in ancient Egypt and India and expanded over the Mediterranean region, the Near East and Asia. Since 1629, the crop has been cultivated in the USA. The majority of modern watermelon cultivars grown in India and China are crosses of native, older cultivars and cultivars imported from Japan and the USA. The 'edible-seed watermelon' is a unique watermelon germplasm developed and commercially grown (on at least a half million acres annually) in China for seed consumption. Watermelon cultivars with edible seeds have large seeds (1000 seed weight > 250 g). With small thin leaves, slender vines, and a high number of branches, edible-seed watermelons may thrive on marginal land and are drought tolerant. They are, however, extremely vulnerable to common watermelon diseases.

Many cultivars have been developed over the world, and many of them have a genetic background in common (Levi et al. 2001a, b). The 'founder effect,' in which cultivars are generated from a small number of accessions, could explain the restricted genetic basis among them (Nimmakayala et al. 2014; Mayr 1954).

Seedless watermelons (triploid hybrid cultivars) have been the most widely grown commercially around the globe since the early twenty-first century. Seeded (diploid) watermelons, on the other hand, are still grown and consumed all over the world, primarily in Asia, Africa, the Middle East and South America. The fruit flesh of 'Crimson Sweet' and 'All Sweet' is of excellent quality, with few seeds. These cultivars, developed by Charles V. Hall and introduced in 1966 and 1972, are possibly the most important diploid cultivars in terms of current world production, and they are the parents of many seedless cultivars today.

The global demand for seedless varieties continues to rise. A seedless watermelon is a triploid hybrid that is produced by combining a diploid variety with a tetraploid line as the female parent (Kihara 1951). This is the world's first open gate for production triploid hybrids.

6.4.1 Genetic Diversity of Watermelon

Watermelons are a type of fruit that is grown in temperate and tropical climates around the world and is used as a source of water and sustenance for both animals and people (Wehner 2008; Paris 2015). *Citrullus* has its origins in Africa, where wild, feral, and landrace populations can be found throughout the continent's vast arid and semi-arid regions. *Citrullus* species can also be found in the Middle East's wild and desert areas, as well as central and southern Asia, and Asia Minor (Anatolia). Many of the plant materials in remote places are poorly defined, and more research is needed to adequately collect, classify and evaluate them (Levi et al. 2013; Nimmakayala et al. 2010; Reddy et al. 2013).

Citrullus lanatus has expanded around the world due to its sweet flesh, and it is now one of the most widely consumed vegetable fruit crops. Watermelon species include the egusi (*Citrullus mucosospermus* (Fursa) Fursa), citron (*Citrullus amarus* Schrad.) and colocynth (*Citrullus colocynthis* (L.) Schrad.). *Citrullus mucosospermus* is a plant endemic to western Africa's sub-Saharan region that is grown for its oily seeds (Jarret et al. 1997; Dahl Jensen et al. 2011). *Citrullus lanatus* and *Citrullus mucosospermus* have similar genome sequences and may usually hybridise to create extremely fertile offspring (Levi et al. 2011b). *Citrullus amarus* is a plant native to southern Africa that is grown for its hard fruit flesh, which is frequently fried or pickled (Bush 1978). *Citrullus amarus* crosses easily with *Citrullus lanatus* and *Citrullus mucosospermus.* When compared to *Citrullus lanatus* or *Citrullus mucosospermus,* however, the genomic sequences of *Citrullus amarus* show significant differences. For most genomic regions, the genetic populations resulting from these crosses with *Citrullus amarus* yield skewed (non-Mendelian) segregation ratios (Levi et al. 2002).

Citrullus colocynthis, popularly known as bitter apple, is grown for its medicinal virtues as well as the oil extracted from its seeds (Hussain et al. 2014). *Citrullus colocynthis* is found in northern Africa's deserts and semi-arid regions, as well as southwestern and central Asia's deserts and semi-arid regions, from the Mediterranean islands eastward to Afghanistan, Pakistan and India (Jeffrey 1967; Burkill 1985; Dane and Liu 2007; Paris 2015). The desert annual *Citrullus rehmii* (De Winter 1990; Jarret and Newman 2000) and the desert perennials *Citrullus ecirrhosus* (Cogn.) and *Citrullus naudinianus* Sond. are also related (Meeuse 1962). Several species (e.g. *Citrullus amarus, Citrullus mucosospermus and Citrullus lanatus*) appear to be natural admixtures, according to genetic and genomic evidence (Levi et al. 2013; Reddy et al. 2014a, b), blurring the distinction between them.

The nomadic peoples of Africa, the Middle East and Asia used watermelon as a source of water and nutrition in the deserts of Africa, the Middle East and Asia (Paris 2015). Many years of domestication and selection for optimum fruit quality by early agrarians and subsequent breeders resulted in a narrow genetic base among watermelon cultivars (Levi et al. 2001a, b). Low levels of polymorphism linked with the chromosomal regions flanking chosen alleles imply long-term selection for optimal fruit quality (Nimmakayala et al. 2010, 2014; Reddy et al. 2014a, b). The loss of genes linked with resistance and the narrow genetic basis among watermelon cultivars may have led to the susceptibility of today's cultivated watermelon to a wide range of diseases and pests (Levi et al. 2001a).

The recent sequencing and assembly of the elite Chinese watermelon accession 97103 and the American 'Charleston Gray' (Levi et al. 2011a) confirmed the presence of a small number of single nucleotide polymorphisms (SNPs) (approximately 1 SNP per 1300 bp) between these two morphologically distinct watermelon types. The cultivar 'Charleston Gray,' introduced in 1954, was the first to be resistant to both *Fusarium* wilt and anthracnose. 'Charleston Gray' has been employed in a variety of breeding projects and can be found in the pedigrees of many watermelon cultivars around the world (Wehner and Barrett 1996), including the popular 'Crimson Sweet' and 'All sweet.'

The limited genetic base of dessert watermelon (*Citrullus lanatus*) cultivars makes improving disease resistance a constant problem for researchers and breeders. The main goals of most watermelon breeders are high output, high fruit quality and early maturity (Gusmini and Wehner 2005a). Because of the little genetic diversity among watermelon cultivars, vigour (heterosis) in F1 hybrid watermelon lines is likely to be low. Diallel experiments using several watermelon lines indicated inconsistencies in estimates of heterosis across experiments (Gusmini and Wehner 2005a).

Primitive landraces and related materials could be a good source of genes for improving genetic diversity and hybrid vigour in seedless watermelon cultivars, especially in diploid and triploid types. Recent developments in molecular biology should be put to good use in examining genetic diversity, population structure and identifying gene loci that contribute to heterosis as well as those that confer resistance to biotic and abiotic stresses. These methods make it easier to include and use a variety of germplasm into watermelon breeding projects (Reddy et al. 2014a, b; Ren et al. 2014).

Disease resistance genes or alleles are prominent in *Citrullus mucosospermus* PIs. When compared to *Citrullus lanatus* 'Sugar Baby' and 'Mickylee,' the fruits of disease-resistant egusi watermelons (*Citrullus mucosospermus*) were resistant to *Phytophtora capsici* infection at all stages of growth (Kousik et al. 2014). Several PIs of *Citrullus lanatus* and *Citrullus mucosospermus* have been chosen and distributed as germplasm lines for watermelon breeding (Gillaspie and Wright 1993; Kousik et al. 2014).

Citrullus lanatus, *Citrullus mucosospermus* and *Citrullus amarus* genebank accessions have proved a valuable source of resistance to powdery mildew (PM), a serious disease of watermelon (Davis et al. 2007).

Various Citrullus PIs have been discovered as sources of resistance to race 1 or 2 of the powdery mildew pathogens, and the mode of resistance inheritance has been determined in numerous watermelon PIs (Tetteh et al. 2010; Ben-Naim and Cohen 2015). *Citrullus lanatus* and *Citrullus mucosospermus* germplasm lines with various disease resistance (Powdery mildew and Phytophthora fruit rot) have been identified and created by scientists for use in watermelon breeding efforts. Squash vein yellowing virus (SqVYV) causes watermelon vine decline, and some *Citrullus colocynthis* and *Citrullus lanatus* PIs could be a source of resistance (Kousik et al. 2009).

Various *Citrullus amarus* PIs are all resistant to root-knot nematodes (Thies and Levi 2003), *Fusarium* wilt race 2 (Wechter et al. 2012), gummy stem blight (Gusmini et al. 2005), anthracnose (races 1, 2 or 3) (Boyhan et al. 1994), powdery mildew (Davis et al. 2007) and poty viruses (Strange et al. 2002). *Citrullus amarus* accessions have resistance genes or alleles that do not exist in *Citrullus lanatus* cultivars or are not expressed (Levi et al. 2013). According to recent research (Levi et al. 2013), *Citrullus amarus* could be useful as a source of germplasm for the development of rootstocks for grafted watermelon cultivars because it produces large root systems and significantly higher yields than other cucurbit rootstocks in root-knot nematode-infested fields. Some *Citrullus amarus* accessions are also good sources of widespread mite tolerance (Kousik et al. 2007).

Citrullus colocynthis is another key species with the potential to contribute to the improvement of dessert watermelons. This species thrives in a variety of desert environments and has a large genetic variation. *Citrullus colocynthis* PIs are a good source of genes for improving biotic and abiotic stress resistance in cultivated watermelons, such as drought resistance, whitefly resistance (Coffey et al. 2015) and potyvirus resistance (Levi et al. 2016). *Citrullus ecirrhosus* and *Citrullus rehmii*, both of which lack tendrils, are desert-adapted species indigenous to southern Africa (Chomicki and Renner 2015). These species have a tap root (*Citrullus rehmii*) or a water-saving caudex (*Citrullus rehmii*) (*Citrullus ecirrhosus*). The gemsbok cucumber, *Citrullus naudinianus*, is the most physically unusual *Citrullus* species, with a massive subterranean storage root and fruits that resemble some *Cucumis* spp. This genus member is widespread in Sub-Saharan Africa and has long been regarded as a vital source of food and water for humans and animals (Chomicki and Renner 2015).

Interspecific crosses between *Citrullus* species are feasible to varying degrees (Robinson and Decker-Walters 1997), and they are frequently successful but result in low fruit and seed set and/or pollen viability (Sain and Joshi 2003). *Citrullus amarus, Citrullus colocynthis, Citrullus mucosopermus* and possibly other spp. are a valuable source of genes for use in breeding programmes aimed at increasing dessert watermelon yields through more efficient root systems that can withstand drought and extreme temperatures, in addition to their potential for disease and pest resistance (McGregor 2012). As previously stated, the root systems of *Citrullus rehmii, Citrullus ecirrhosus and Citrullus naudinianus* are all well adapted to desert environments.

According to a study of drought resistance utilising citron watermelon, *Citrullus amarus*, plants obtained in the wild in Botswana, the root system functioned to store water and extended deeper into the soil in response to dryness (Yoshimura et al. 2008). Furthermore, during drought conditions, root growth is modest in watermelon cultivars, whereas strong root growth occurs in *C. amarus* accessions (Yoshimura et al. 2008). As a result, several *Citrullus* spp. could be effective in improving watermelon cultivars with abiotic stress resistance, such as heat or cold tolerance, drought tolerance and/or water usage efficiency (Zhang et al. 2011).

Watermelon breeding initiatives must conserve genetically and morphologically varied germplasm in order to succeed now and in future. Plant material used to identify sources of resistance to diseases and pests is primarily obtained from gene banks. Given the growing human population, there is a greater demand for high-yielding, disease-resistant watermelon crops. As a result, efforts should be focused on collecting, maintaining and evaluating *Citrullus* germplasm.

Many resistance-containing PIs have been refined and developed into improved germplasm or breeding lines that have been effective in genetic studies and initiatives aimed at incorporating or improving resistance in the watermelon crop. This collection has been extensively used to investigate trait inheritance as well as *Citrullus* taxonomy and evolution (Branham et al. 2016).

In recent years, germplasm collections have been exploited to generate robust rootstocks with nematode and soil-borne disease resistance, which are desirable features for grafted watermelons (Thies et al. 2015).

Watermelon genetic resources must be collected and maintained at all levels of institutions. In conventional breeding procedures, the success of any breeding effort is largely determined by the breeder's genetic resource reservoir. Approximately 400 watermelon accessions were gathered in different institutions. Many of these watermelon accessions, including open-pollinated and F1 hybrid cultivars, have a similar genetic history, according to several genetic analyses.

6.4.2 Importance of Genetic Resources for the Development of Improved Varieties in Watermelon Through Traditional Breeding

Modern dessert watermelon cultivars have a limited genetic background and are sensitive to a wide range of diseases and pests, especially in tropical and subtropical climates with heavy rainfall and humidity. The relevance of identifying and introducing novel sources of disease and pest resistance genes into the watermelon is highlighted by recent consumer awareness of the safe usage of fruits and vegetables with lower pesticidal residual toxicity and environmental concerns. Many studies have found that cultivated watermelons have low levels of DNA polymorphism (Nimmakayala et al. 2010, 2014; Reddy et al. 2014a, b; Saha et al. 2022) although *Citrullus* sub-species have substantial genetic variation. As a result, there is a need to use the abundance of genes present within the *Citrullus* germplasm that has been gathered (and that is yet to be collected) to achieve this desired goal. Regardless of whether traditional or advanced molecular breeding was used, the results were the same. Improving genetic diversity and host plant resistance in watermelon cultivars is a top priority for maintaining and improving existing production levels. To create watermelon varieties with specified objectives, breeders and researchers around the world must access a large number of genetic resources collected at gene banks. Still, collecting wild watermelon germplasm and conserving and utilising genetic resources to continue generating robust watermelon breeding lines and cultivars is critical.

6.4.3 Sex Expression in Watermelon

Andromonoecism is the most common sex form in melon, resulting in distinct perfect (hermaphroditic) and staminate flowers in a single plant. Monoecism, which produces separate pistillate and staminate flowers on one plant, is the most common sex form in cultivated cucumber and watermelon. Trimonoecy, which produces all three types of flowers in one plant (staminate, pistillate and hermaphroditic), has been recorded in watermelon on rare occasions (Rosa 1928). Gynoecious watermelon, unlike cucumber and melon, was not discovered until 2007, when Jiang and Lin discovered a spontaneous gynoecious mutant in China. Still, the genetic link between andromonoecism and gynoecism in watermelon has not been clearly documented. A greater knowledge of sex form inheritance in watermelon would aid in the development of gynoecious cultivars and, eventually, the discovery of the molecular process underlying watermelon sex determination.

Sex form inheritance has been demonstrated in major Cucurbitaceae crops such as cucumber and melon. According to Rosa (1928), andromonoecism is a recessive trait, whereas monoecism is a dominant trait. Furthermore, significant progress was achieved in uncovering the cucumber sex determination process during the 1960s. In cucumber, the sex form is controlled by three genes: the female (F) gene, the male (m) gene and the androecious (a) gene (Kubicki 1969a, b, c, d). Furthermore, the inheritance of melon sex form has been thoroughly documented (Poole and Grimball 1939; Kenigsbuch and Cohen 1987, 1990). These researchers discovered that three genes, *g* (gynoecious), *a* (andromonoecious), and *m* (male), govern the melon's sex form. Trimonoecy and gynomonoecy are controlled by the *m* gene.

Watermelon's sex determination method is unknown, unlike cucumber and melon. Monoecy is dominant over andromonoecy, and the andromonoecious sex form has been known for a long time (Rosa 1928). Porter (1937) has also backed up this conclusion based on his studies. Poole and Grimball (1944) also established a link between fruit shape and hermaphroditic flowers, as well as supporting the idea that andromonoecism is a recessive trait. However, little progress has been made in the study of sex form inheritance in watermelon until recently. One of the difficulties in researching sex inheritance in watermelon is that, in addition to genetic factors, environmental factors and hormone levels also influence sex expression (Salman-Minkov et al. 2008; Manzano et al. 2014).

Ji et al. (2015) used crosses of gynoecious lines with monoecious or andromonoecious lines, as well as crosses of andromonoecious lines with monoecious lines, to investigate the genetic control of sex forms in watermelon. This would make it easier to establish elite gynoecious lines that can be used to inexpensively produce F_1 hybrids, revolutionising watermelon hybrid seed manufacturing.

The inheritance of a plant's sex form and sex expression by individual flowers in watermelon was studied using five crossing groups. SL3H or AKKZW, the andromonoecious parent, was crossed with either monoecious XHB or gynoecious XHBGM. The gynoecious XHBGM was also crossed with its progenitor, the monoecious XHB. The flowering patterns of each cross's F_1, F_2, BC_1P_1 and BC_1P_2 generations were studied in the field (from March to July) and in the greenhouse (Autumn from June to October). In order to assess their sex forms and the percentage of pistillate flowers in different seasons, the types of flowers within the first 30 nodes of the parents and their F_1 plants were recorded. To investigate the seasonal impact on sex expression in watermelon, the u test was utilised in the significant differences analysis of the percentage of pistillate flowers or hermaphrodite flowers.

Ji et al. (2015) investigated the type of flowers in the first 30 nodes of both parents and F_1 plants in all crossing groups in the spring and autumn to determine the sex form and percentage of pistillate flowers for each plant to determine the environmental influence on sex forms and sex expression of individual flowers in watermelon. In both the spring and autumn, the u test was employed to describe differences in the percentage of pistillate flowers. SL3H and AKKZW produced staminate and hermaphroditic flowers on single plants in both seasons, whereas XHB produced staminate and pistillate flowers on single plants in both seasons. In the spring, XHBGM produced pistillate flowers on all nodes in the spring, however in the autumn, while the majority of the plants produced predominantly pistillate flowers, a small number of 'abnormal pistillate flowers' with aberrant ovaries and sterile stamens were produced by a few plants. Furthermore, in both seasons, F_1 plants in all crossing groups were monoecious. This result reveals that monoecism is dominant over andromonoecism and gynoecism, and that environmental influences have no effect on a plant's sex shape.

The u test was performed by Ji et al. (2015) to determine whether there were significant differences in the percentage of pistillate flowers among individuals. The andromonoecious SL3H had 17.3% hermaphroditic flowers in the spring and 11.4% in the fall, whereas the other andromonoecious cultivar AKKZW had 18.5% in the spring and 12.8% in the fall. The pistillate flower percentage of the monoecious cultivar XHB was 23.0% in the spring and 19.1% in the autumn. The findings of the u test revealed that there is a considerable seasonal effect on individual flower sex expression in various cultivars. Furthermore, in the F_1 plants generated from the cross of SL3H × XHB, the percentage of pistillate flower was 21.6% in the spring and 16.0% in the autumn, but in the F_1 plants obtained from the cross of AKKZW × XHB, it was 21.8% in the spring and 16.0% in the autumn. The percentage of pistillate flower in F_1 plants of the XHBGM × XHB cross was 23.8% in the spring and 17.1% in the autumn, whereas it was 21.1% in the spring and 16.4% in the autumn in F_1 plants of the XHBGM × SL3H cross, and 22.0% in the spring and 15.6% in the autumn in F_1 plants of the cross XHBGM × AKKZW. The u test revealed a significant seasonal (environmental) effect on the percentage of pistillate flowers in F_1 plants derived from various crosses. In this study, the critical stage for flower initiation for the spring–summer experiment occurred in the spring when the daylight was short and temperatures were low, whereas the critical stage for flower initiation for the summer–autumn experiment occurred in the summer when the daylight was long and temperatures were high. Therefore, results showed that short days and low temperatures inhibited the development of staminate flowers and increases the number of pistillate flowers per plant while long days and high temperatures promoted the development of staminate flowers (Manzano et al. 2014).

Furthermore, environmental conditions determined where the first pistillate/hermaphroditic flower appeared on the node. The initial hermaphroditic flower was produced by andromonoecious parent SL3H around the 13th node in the spring–summer experiment, with successive hermaphroditic flowers produced every four to five nodes away from the preceding one. SL3H, on the other hand, produced the first hermaphroditic flower around the 15th node in the summer–autumn experiment, with following hermaphroditic flowers generated every six to seven nodes apart from the preceding one. The initial hermaphroditic flower was formed by AKKZW, an andromonoecious parent, around the 10th node, with successive hermaphroditic flowers produced every four to five nodes away from the previous one. However, in the summer–autumn trial, AKKZW produced the first hermaphroditic flower around the 13th node, with successive hermaphroditic flowers appearing every five to six nodes. The first pistillate flower was generated by the monoecious parent, XHB, around the 5th node, with additional pistillate blooms produced every four to five nodes away from the previous one. XHB generated the initial pistillate flower around the 7th node in the summer–autumn experiment, with successive pistillate blooms generated every five to six nodes away from the preceding one. These findings suggested that the first pistillate or hermaphroditic flower developed at a lower node location in the spring than in the summer. Low temperatures and short days can encourage the formation of pistillate/hermaphroditic flowers. Furthermore, XHBGM, a gynoecious mutant developed from XHB, produced pistillate flowers at all 30 nodes in the spring, but the percentage of pistillate flowers within the first 30 nodes fell to 95.4% in the autumn due to the formation of the 'abnormal female flowers.' These findings show that environmental influences have little effect on the sex form of individual watermelon plants (Ji et al. 2015).

The segregation ratio in the F_2 population of the cross of andromonoecious (SL3H or AKKZW) × monoecious (XHB) is 9 monoecious: 3 trimonoecious: 4 andromonoecious, and in BC_1P_1 (F_1 × andromonoecious parent) is 1 monoecious: 1 trimonoecious: 2 andromonoecious. This is owing to the fact that in these populations, two recessive genes determine the sex forms, one of which is epistatic to the other. In an earlier work (Rosa 1924), the recessive control of andromonoecy was reported, with the recessive gene denoted as *a*. Another recessive gene, which we call trimonoecious (*tm*), is believed to control the trimonoecious sex form, and the *a* locus is epistatic to *tm* (Ji et al. 2015).

In the F_2 population of the gynoecious (XHBGM) × monoecious (XHB), the segregation ratio is 3 monoecious: 1 gynoecious, but in the BC_1P_1 (F_1 × gynoecious parent), the segregation ratio is 1 monoecious: 1 gynoecious. These findings suggest that gynoecism is controlled by a single recessive gene called *gy*.

In a gynoecious × andromonoecious cross F_2 population, the segregation ratio is 27 monoecious: 12 andromonoecious: 9 gynoecious: 9 trimonoecious: 4 hermaphroditic: 3 gynomonoecious. The BC_1P_1 population (F_1 × gynoecious) has a segregation ratio of 1 monoecious: 1 gynoecious, whereas the BC_1P_2 population (F_1 × andromonoecious) has a segregation ratio of 1 monoecious: 1 trimonoecious: 2 andromonoecious. The findings revealed that three recessive genes, denoted by the letters *a*, *gy*, and *tm*, govern the sex forms of watermelon.

Finally, three recessive genes control sex forms in watermelon, with a being recessive epistatic to tm, and the two interact to define monoecy, trimonoecy, and andromonoecy in watermelon. In monoecious, trimonoecious or andromonoecious plants, the Gy genotype governs the formation of staminate blooms. The gygy genotype transforms staminate flowers into pistillate flowers, which results in gynoecy, gynomonoecy or hermaphrodites. Based on our results, we propose genotype for each of the six sex forms as follows: monoecious (AGyTm), trimonoecious (AGytmtm); andromonoecious (aaGyTm or aaGytmtm); gynoecious (AgygyTm); gynomonoecious (Agygytmtm); and hermaphrodite (aagygyTm or aagygytmtm) (Ji et al. 2015).

6.5 Breeding Methods and Cultivar Development in Watermelon

The process of plant domestication and breeding has resulted in a decrease in genetic diversity in watermelon. Yield, fruit size and shape, flesh colour and texture, sweetness, earliness, dwarf habit, and disease resistance have all been used as breeding targets for watermelon. In the private seed industry, watermelon breeding programmes are primarily focused on the development and production of F_1 hybrid cultivars.

The worldwide seed market now requires a wide range of features. Tolerance to biotic and abiotic stresses such as novel disease pathogens, drought, high temperatures, chilling injuries, and customers' need for health-enhancing features are among them. As a result, modern watermelon breeding requires an awareness of the population structure of adapted and wild-type genotypes, as well as the production of novel traits through varied cross-combinations.

Modern breeding strategies have emphasised the introduction of new genetic variety from neglected germplasm accessions, particularly for disease resistance.

Watermelon fruit comes in a variety of sizes, shapes and rind patterns, including ice box, small, medium, large or giant; round, oval, blocky or elongate; grey, narrow stripe, medium stripe, wide stripe, solid or dark solid; white, yellow, orange, or red flesh; and seeded or seedless flesh. Red flesh, blocky shape, and large size (8–11 kg) cultivars, such as All sweet, are the most popular seeded cultivars commercially. Red flesh, oval shape, and medium size (5–8 kg) are common cultivars for seedless watermelons, such as the cultivar Tri-X-313.

Plant breeders have released disease-resistant cultivars, dwarf vines, larger fruit, greater sugar content, higher lycopene content, seedlessness and novel flesh colours including fiery red, dark orange and canary yellow during the course of the year. Watermelons have become more popular as a result of the development of seedless triploid hybrids.

Diverse germplasm collection and evaluation are important criteria for watermelon breeding. Priorities for Citrullus germplasm collection include India, particularly the Indo-Gangetic plains and areas in the country's northwest; Africa, particularly the south and southwest (Kalahari Region); former Soviet Union and Iran's southern areas; and tropical Africa has to be provided.

6.5.1 Important Cultivars

To become familiar with the diversity of germplasm, watermelon breeders must analyse a sample of cultivars. Horticultural characteristics were improved over time in cultivars that were produced throughout time. The breeding effort begins by intercrossing the best cultivars now available or by crossing the best cultivars with accessions that have one or more important features not seen in the elite cultivars. Watermelon breeders will need seeds from the best cultivars, a set of cultivars developed at various times in the past, a collection of accessions from germplasm repositories, and lines with beneficial or interesting gene mutations in the early stages of breeding.

The majority of commercial cultivars were hybrids, with only a few open-pollinated cultivars. Popular diploid (seeded) open-pollinated cultivars ('All sweet,' 'Black Diamond,' 'Cal sweet,' 'Crimson Sweet,' 'Jubilee II' and 'Legacy') were grown at first to meet local demand. Hybrids have a larger range of adaption. More over half of the cultivars named were of the 'All sweet' kind, which is typically considered to be of good quality. 'Sangria' and 'Royal Sweet,' 'Fiesta,' and 'Mardi Gras' and 'Regency' were the most popular diploid (seeded) cultivars. Almost half of the triploid (seedless) cultivars were of the 'Tri-X-313' variety. 'Tri-X-313,' 'Summer Sweet 5244,' 'Millionaire,' 'Genesis' and 'Tri-X-Shadow' were the most popular triploid cultivars.

In order to develop better cultivars for a specific location, seeds of cultivars, breeding lines, populations, plant introduction accessions and gene mutants that exhibit the desired features at a high degree will be required by the watermelon breeder. The breeder must determine whether the characteristic is quantitatively inherited (fruit yield, earliness, size and sweetness) or qualitatively inherited (dwarfness, anthracnose resistance, flesh colour). Adapted accession with the best genetic background is preferable than wild type.

From 1880 to 1900, systematic cultivar development initiatives were implemented. 'Cuban Queen,' developed and marketed by Burpee in 1881, 'Round Light Icing,' (1885), 'Kolb Gem,' developed by Reuben Kolb of Alabama in 1885 and marketed by D. M. Ferry, 'Florida Favourite,' selected from the cross of 'Pierson' × 'Georgia Rattlesnake,' by Girardeau in Monticello, Florida in 1887, and 'Tom Watson,' created by Alexander Seed Co. in Augusta, Georgia in 1906, and 'Stone Mountain,' developed by Hastings Co. in Atlanta, Georgia in, 1924, are two important cultivars created for transportation.

Among the notable cultivars developed in the second half of the previous century are *Fusarium* wilt resistance is strong in 'Charleston Gray' (USDA, Charleston, 1954), 'Crimson Sweet' (Kansas State University, 1963), 'Calhoun Gray' (Louisiana State University, 1965), and 'Dixielee' (1979), 'Jubilee' (1963), and 'Smokylee' (1971) (all from the University of Florida). Dark red flesh is found in 'Dixielee' (University of Florida, 1979) and 'Sangria' F_1 (Syngenta—Rogers Brand, 1985). High yields can be found in 'Millionaire' F_1, 3x (Harris Moran, 1992) and 'Royal Jubilee' F_1 (Seminis). High soluble solids are found in 'Crimson Sweet' (Kansas State University, 1963) and 'Sugarlee' (University of Florida, 1981). Dwarf vines characterise 'Kengarden' (University of Kentucky, 1975). Seedless 'Tri-X-313' F_1 3x (Syngenta—American Seedless, 1962). Icebox size varieties include 'Minilee' (University of Florida, 1986), 'Mickylee' (University of Florida, 1986), 'New Hampshire Midget' (University of New Hampshire, 1951), 'Sugar Baby' (M. Hardin, Oklahoma, 1955) and

'Tiger Baby' (Seminis). The flesh of 'Yellow Doll' (Seminis, 1977) is canary yellow.

6.5.2 Traditional Breeding Methods Followed in Watermelon

One of the most crucial parts of a breeding programme is the selection of parental materials. Breeders will choose the parental lines for crossing based on the needs and requirements of the customers. Recurrent selection and pedigree selection are required to produce a base population with general adaptation and inbred lines with desirable features. Backcross breeding can also be employed to correct defects in high-performing ruling varieties.

6.5.2.1 Recurrent Selection

Recurrent selection refers to the repeated selection and massing of selected plants. Because watermelon plants are large and have a long generation time of more than five months, single-plant selection with few generations in each cycle and some plants per family will be effective in genetic improvement. Inter-crossing of 20–25 elite cultivars by hand for two or more generations and utilising bees in a separate block for 2 or more generations before initiating a selection procedure for various quantitative traits can result in a population with a broad genetic base. For a set of highly heritable traits, simple recurrent selection could be applied. Reciprocal recurrent selection can be used to improve the combining ability of two populations at the same time. To improve quantitative traits like yield, recurrent selection could be applied. Inbred lines would be developed using the improved population.

6.5.2.2 Pedigree Selection

Pedigree selection is the most commonly used method of breeding in watermelon. This method employs the selection of two highly diverse and adapted parental lines. The goal of pedigree breeding is to create new lines that combine the traits of both parents. The F_1 generation is developed by crossing the two selected parents, one as female and one as male which is further self- or sib-pollinated to generate a segregating population. For the development of the F_3 generation, the F_2 is self- or sib-pollinated while selection is done for traits with high heritability and then choosing the top plants in each of the best F_3 families. Selection of qualitative character was mostly done in the F_3 generation. Selection for quantitative characters in the F_4 generation will be based on family-row performance. As the families reach six generations of self-pollination, they become more uniform and can then be used as inbred lines.

6.5.2.3 Single-Seed-Descent Method

Single-seed-descent method is a modification of pedigree method where inbred lines are developed rapidly by self- or sib-pollination and selection is performed at later generations. This strategy is used to improve quantitative characteristics such as yield. In watermelon, qualitative traits that can be selected in early generations by following traditional pedigree breeding in order to minimise undesired features in early generations.

6.5.2.4 Backcross Breeding

Backcross breeding is a technique for transferring one qualitative (very heritable) trait into a superior inbred. The recurrent parent is the superior inbred line. In this method, six generations of selection and then backcrossing to the recurrent parent is done to recover the recurrent parents genotypes (except for the addition of new traits) and undesirable traits of non-recurrent (donor) parents are eliminated. There are two backcrossing procedures used depending on the genes of interest, either recessive or dominant.

For the traits controlled by recessive genes, recurrent parent is crossed with the donor parent, and then resulting F_1 is backcrossed to the recurrent parent and also self-pollinated to develop segregating F_2 generations for the traits of interest. Individuals with specific characters are backcrossed to the recurrent parent resulting in the development of BC_1 generation. At this stage, BC_1 generation is tested for the traits of interest, and individuals possessing it are self-pollinated once again to develop a segregating population for superior selection and backcrossing to the recurrent parent. This process is repeated until the BC_6 generation and superior individuals are self-pollinated and selected for desired traits resulting in the development of the improved inbred line. This inbred line will be identical to the original but only with one new trait which will not require more testing in trials.

For the traits controlled by dominant genes, the recurrent parent is crossed with the donor parent, and then resulting F_1 is backcrossed to the recurrent parent. BC_1 generation is tested for the traits of interest, and individuals possessing it are backcrossed to the recurrent parent. This process is repeated until the BC_6 generation and superior individuals are self-pollinated and selected for desired traits using progeny test for homozygous expression of the trait.

6.5.2.5 Hybrid Development

The two monoecious inbreds were crossed to produce hybrids. After the development inbreds, they have been crossed in all possible combinations. It may contain too many combinations to fully analyse. For example, 25 inbreds could produce $(25 \times 24)/2 = 300$ different hybrids, without including reciprocals. So, it is necessary to make hybrids only from pairs of inbreds having complementing traits of each other. Experimental hybrids were tested in phases, with fewer hybrids being examined in later stages and more work being put into each hybrid. Trials were undertaken in two replications in each of two locations in the first year. The best-performing hybrids were examined in 8–12 locations in the second year. The hybrids were sent for regional trails in a greater area in the third year. Data should be obtained from least 10 of the 50 trials. Based on these trails, the best one or two hybrids released in the fourth year. Usually, in watermelon hybrids were not much advantage of over open-pollinated cultivars for most traits, but hybrids are more uniform. The major advantage is the ability to develop seedless triploids by making crosses between a tetraploid inbred line as female and diploid inbred line as male. In rare circumstances, hybrids outperform their best parent, resulting in heterosis. Watermelon heterosis was determined to be around 10%. Watermelon hybrids are usually created by making crosses between a large seeded female inbred line with small seeded male inbred line.

6.6 Seedless Watermelon Production

Seedless triploid watermelons are produced by crossing a tetraploid ($2n = 4x = 44$ chromosomes) inbred line as female parent with a diploid ($2n = 2x = 22$) inbred line as a male parent. The outcome is a triploid ($2n = 3x = 33$) hybrid. Triploid plants have three sets of chromosomes that can't be divided uniformly during meiosis and the presence of non-functional female and male gametes. Due to the fact that triploid hybrid is female sterile, upon selfing it will produce seedless fruits. Furthermore, the triploid lacks viable pollen grains; hence, a diploid watermelon line/variety must be grown in the vicinity of the production field to provide sufficient pollen for pollination for fruit development. In the field, one-third of the plants should be diploid and two-third triploids. The seeded diploid fruit should be easily distinguishable from the seedless triploid fruit for harvesting and sale. The tetraploid lines employed in triploid seed production have grey rinds, making it easier to distinguish self-pollinated from cross-pollinated progeny when crossed with a diploid line with striped rind.

Fruit set and enlargement in triploid seedless watermelon production are dependent on growth regulators from pollen grains and embryos in developing seeds within the fruit. Due to insufficient pollination, triploid watermelon fruits are triangular in shape and of inferior quality. Inadequate pollination has been linked to an increase in the occurrence of hollow heart syndrome. Watermelon flowers produced by triploids do not produce enough viable pollen to trigger fruit formation and development. As a result, pollen from a diploid seeded watermelon variety must be used. Because this variety will account for one-quarter to one-half of all watermelons produced in the field, it is critical to utilise a marketable diploid polleniser variety. To avoid confusion during harvest, the seeded polleniser fruit's rind pattern and/or form should be immediately distinguishable from that of the triploid fruit. When female blossoms on triploid plants are open and ready for pollination, it's critical that pollen from the diploid polleniser variety is available. To achieve fruit set, pollen from the male flower must be transferred to a female flower on that plant or another plant in the field. Pollen is transferred by a variety of naturally occurring insects, but the honeybee is the most successful.

6.6.1 Constraints in Triploid Watermelon Production

There are many constraints in triploid watermelon production including additional time for tetraploid production, extra selection against sterility and fruit abnormalities in tetraploid lines, selection of parents for low incidence of hard seed coat in hybrids, a reduction in seed size per unit area, reduces seed vigour for farmers, and the requirements for the diploid polliniser to use up to one-third of the farmers production field.

Another issue with triploid hybrids is the presence of chaffy seed coat (white or coloured) in the fruits. Fruits with enormous visible seed coats which are undesirable to consumers are generated under certain environmental circumstances. During trialling, triploid fruit should be examined for seed coat issues. Before triploid development, selections for some traits need to be done in the parental lines. If the parents have large seeds, the hybrids will have big seed coverings. At least three genes, *l*, *s*, and *ts*, are involved in controlling seed size. Tetraploid lines with small or tomato-sized seeds may be useful in resolving the issue. Aside from genetic influences, certain unknown environmental factors appear to enhance the quantity of hard seed coats in triploid hybrids that perform poorly.

It is recommended that the diploid polleniser variety be planted in the field's outside row and then every third row. The polleniser variety has been planted every third plant in each row as an alternative, although this complicates triploid fruit harvesting because mixed diploid and triploid fruit must be separated. This complicates planting since blank spaces must be left where the diploid should be. It is difficult to keep the three triploids to one diploid rotation going.

If the diploid type is planted too early, it will set fruit and stop generating male blossoms, whereas the triploid type will continue to produce many female blossoms. The triploid will be ready to set fruit if planted too late, but there will be insufficient pollen to provide fruit set. Bottleneck fruits of long-fruited watermelon types occur from poor or inadequate pollination. Poorly pollinated fruits in round-fruited cultivars might be flat-sided or deformed. Currently, one-quarter to one-half of the field must be planted to a diploid seeded variety in the creation of triploid seedless hybrids. As a result, a more efficient polleniser could result in a higher yield of seedless watermelon per acre, allowing more of the field to be planted to the triploid type.

Because no energy is spent in seed generation, seedless triploid hybrids should produce larger yields than diploid hybrids. In practice, however, this may not be the case. The availability of viable pollen to trigger fruit set limits fruit yield in triploids. Pollination issues are the cause of poor fruit growth. If the fruit is to develop entirely and without curvature, all three lobes of the stigma must be fully pollinated. In the case of triploid hybrids, up to one-third of the field must be planted to a diploid polleniser to ensure proper fruit development in the male sterile triploids.

6.6.2 Remedies to Overcome the Poor Triploid Watermelon Production

Planting a less space-consuming diploid pollinator (short vine diploid pollinator) in a triploid watermelon production field minimises the overall consumption inputs by diploid plants in the field. This strategy permits the triploid hybrid to produce better yields in part because the triploid plants have more space, water, sunlight and nutrients available to them, as well as because there are more triploid plants per field.

The employment of a short vine diploid pollinator also facilitates harvesting of the triploid fruits produced, as the diploid fruits remain in or near to the row where they were planted, making hand or automated harvesting easier. When diploid short vine pollinator seed and triploid seed are put in each row, a greater number of triploid transplants can be planted per field than with traditional methods. Because the short vine diploid plant used in this approach is small, it does not require additional row space, resulting in a 33–50% increase in triploid plants per acre. The total triploid fruit output per acre increases significantly as a result of the additional triploid plants.

Planting short-vined diploid pollinisers yields a supply of good pollen without the drawbacks of longer-vined diploid pollinator types. The above technique of planting eliminates the requirement for the existing planting strategy, which involves alternate rows of diploid and triploid plants every two rows. The diploid pollinator is easy to spot on the short vine and avoids tripping on when harvesting. When harvesting triploid fruits, the diploid pollinator is close to the row and should be easy to avoid. When put in the same row as the triploid plants, the short vine diploid pollinator triggered triploid fruit set 10 days earlier than when diploid plants were put in separate alternate rows from the triploid plants. For triploid hybrid production, using a short vine pollinator resulted in a considerable increase in triploid fruit output.

6.6.3 Testing of Triploids

Triploid hybrids are evaluated in the same way that diploid cultivars are. Triploid hybrids can be superior or inferior than their diploid parents since triploids are not inherently superior to diploids. As a result, many different superior combinations of parents should be tested in triploid yield trials to see which ones produce the finest hybrids. Triploid hybrids with poor horticultural performance will be produced by diploid inbred parents with low horticultural performance.

The superior triploid hybrid seed is generally produced using hand or honeybees' pollination in proper isolation blocks. Pollinated flowers should be covered using butter paper bag after pollination is over to avoid sib or self-pollination and capped the day before to keep bees out. The date should be written on the blooms so that the fruit can be picked 35–50 days later. The tetraploid blooms are sib- or cross-pollinated 84% of the time when seed production occurs by bee pollination in isolation blocks, yielding 3x and 4x seeds (progeny). The tetraploid parent's fruit should have a grey rind pattern, and the diploid parent's fruit should have wide stripes, so the triploid hybrid will have striped fruit that can be recognised from the grey fruited tetraploids that emerge from self- or sib-pollination of the female parent.

6.6.4 Tetraploid Production

The use of triploid hybrids has allowed seedless fruit to be produced. H. Kihara developed the tetraploid method for seedless watermelon production. Breeders who want to make seedless triploid hybrids must create tetraploid inbred lines that can be used as a female parent in a cross with diploid male parent. The restricted number of tetraploid inbreds available is one of the key limiting factors in breeding seedless watermelons.

6.6.5 Steps of Tetraploid Production

(1) Choice of selection of diploids, (2) Tetraploid plants production, (3) Development of tetraploid line, (4) Production of hybrid and testing.

6.6.6 Choice of Selection of Diploids

When crossed with a diploid line with stripped rind, it will be easy to differentiate self-pollinated progeny (seeded fruit from female parental line) from cross-pollinated progeny (seeded fruit from male parental line) (which will be seeded fruit from triploid hybrid). Here grey fruit should be discarded by the farmers so they are not mislabelled as seedless watermelons.

6.6.7 Tetraploid Plants Production

Tetraploids can be produced on a regular basis in watermelon utilising tissue culture plants or the insecticide oryzalin. Colchicine (C22H25O6N), a toxic alkaloid used in the treatment of gout, is a widely used method for producing tetraploid watermelon. Colchicine suppresses chromosomal separation during anaphase by inhibiting spindle formation. At seedling stage, the shoot apex treatment was determined to be the most successful of all the colchicine application methods.

For speedy and uniform germination, the selected diploid line is planted in flats (8×16 cells is a typical size) in the greenhouse on heating pads that maintains the soil medium at 85 °F. For small-seed size cultivars, the colchicine solution is employed at a concentration of 0.1%, 0.15–0.2% for medium-seed size cultivars, and 0.2–0.5% for large-seed size cultivars. Colchicine is administered to the seedling developing point once a day in the morning and once a day in the evening for three days, using one drop on small- or medium-seed size cultivars and two drops on large-seed size cultivars. Because the treatment creates diploid, tetraploid or aneuploid plants, the tetraploids must be identified and selected at a later stage. When T_0 diploids are treated with colchicine, around 1% of the seedlings (referred to as T_1 generation tetraploids) become tetraploids. Tetraploids are produced in greater numbers by some diploid cultivars and breeding lines than by others.

Tetraploids can be easily identified by counting the chromosomes of cells under a microscope or comparing stem, leaf, flower and pollen size to diploid controls. Counting the amount of chloroplasts in stomatal guard cells using a leaf peel under a microscope is a popular method. Tetraploid guard cells have roughly 10–14 chloroplasts (20–28 total on both sides of the stomata), whereas diploid guard cells have just 5–6 (10–12 total). The method can be used to check the ploidy level of a large number of plants in the seedling stage before transferring them to the main section of the greenhouse or field nursery to allow self-pollination. Identifying tetraploid seedlings using their phenotype in flats before transplanting, the chloroplast number in the stomatal guard cells of true leaves in seedling flats and greenhouse pots, and the appearance of the fruit and seeds at harvest after self-pollination in the greenhouse are the most common methods used. Tetraploid leaves are thicker, grow slower and have shorter stems than diploids.

6.6.8 Development of Tetraploid Line

In the T_0 generation (plants from colchicine-treated diploids), tetraploid plants are picked (using methods such as leaf guard cell chloroplast number) from the greenhouse flats where they were treated with colchicine. The T_1 generation must then be planted in flats to ensure that the plants in the following generation are tetraploids, and the selections must be transplanted to greenhouse pots for self-pollination. Seeds from those selections (T_2) can then be multiplied in bigger plantings, such as field isolation blocks, to yield enough seeds per tetraploid line for triploid hybrid development.

Over generations of self- or sib-pollination, the fertility and seed yield of tetraploid lines would rise, owing to the elimination of plants with chromosome anomalies, resulting in a tetraploid line with balanced chromosome number and regular creation of 11 quadrivalents. Tetraploid lines produce 50–100 seeds per fruit in early generations, and sometimes as few as 0–5 seeds, compared to 200–800 seeds for diploids. Poor seed germination is another issue with early generation tetraploids, making it difficult to establish homogeneous field plantings. Before sufficient seeds of tetraploid lines can be produced for commercial production of triploid hybrids, it may take up to ten years of self-pollination. When compared to the original lines, advanced generations of tetraploid lines usually have better fertility, seed output and germination rate. Before initiating commercial production of a triploid hybrid cultivar, several businesses require more than 100 pounds of seed from a tetraploid inbred. Each pound of triploid seeds necessitates the creation of around 110 tetraploid plants.

6.6.9 Evaluation of Tetraploids

The examination of tetraploids as parents of triploid hybrids (typically T_3 generation or later). Tetraploids should be tested directly for rind pattern, high seed yield and other features such as male sterility in hybrid seed production to save time and effort. However, after establishing controlled crossings with diploid male parents, the key test for tetraploids is as female parents in triploid hybrid seed generation. To provide adequate pollen for fruit set in triploid hybrids, yield experiments are conducted using two rows of triploid plots alternating with one row of diploid plots. For the market type and production area of interest, useful tetraploid inbreds should produce triploid hybrids with outstanding yield and quality.

6.6.10 Production of Hybrid and Testing

Triploid hybrids are evaluated in the same way that diploid cultivars are. Triploid hybrids can be better or worse than their diploid parental lines since triploids are not inherently superior to diploids. In triploid yield experiments, several different combinations of parental lines should be tested to see which ones produce the best hybrids. In general, triploid hybrids with low horticultural performance will be produced by diploid inbred parents with low horticultural performance.

Another issue with triploid hybrids is the presence of empty seed coats (coloured or white) in the fruit. Fruit with enormous visible seed coverings that are undesirable to consumers is generated under certain environmental circumstances. Seed coat concerns in triploid fruit should be investigated. Before triploid production, some selection needs be done on the parents. If the parents have large seeds, the hybrids will have big seed coverings. At least three genes, *l*, *s* and *ts*, are involved in controlling seed size. Tetraploid lines with small or tomato-sized seeds may be useful in resolving the issue. Aside from genetic influences, certain unknown environmental factors appear to enhance the quantity of hard seed coats in triploid hybrids that perform poorly.

6.7 Breeding for Lycopene Content in Watermelon

A diet rich in fruits and vegetables has been found to contain a variety of bioactive components, including phytochemicals that target life-threatening diseases. Lycopene is a well-studied antioxidant that may be found in watermelon in this context. Watermelon is one of the few sources of cis-isomeric lycopene that is easily available. It is useful in lowering the risk of cancer, cardiovascular disease, diabetes and macular degeneration. Watermelon is an excellent source of natural antioxidants, particularly lycopene, ascorbic acid and citruline. These active compounds defend against long-term health issues such as cancer recurrence and cardiovascular disease (Zhang and Hamauzu 2004; Omoni and Aluko 2005; Fenko et al. 2009). Fruits and vegetables containing lycopene have an unique red colour (Mutanen and Pajari 2004). The presence of a significant amount of lycopene in watermelon has prompted farmers and growers to create high red flesh variants in recent decades.

6.7.1 Nutrition Profile of Watermelon

A 100 g watermelon gives 30 kcal, according to the nutritional profile. It contains about 92% water and 7.55% carbohydrates, including 6.2% sugars and 0.4% dietary fibre. It's high in carotenoid, vitamin C, citrulline, carotenoids and flavonoids, and it's low in fat and cholesterol, thus it's a low-calorie fruit (Leskovar et al. 2004; Bruton et al. 2009). Watermelon also contains a lot of β-carotene, which is an antioxidant and a precursor to vitamin A. Aside from lycopene, it contains B vitamins, particularly B1 and B6, as well as minerals including potassium and magnesium (Huh et al. 2008). Watermelon has phenolic levels that are comparable to those seen in other fruits (Kaur and Kapoor 2001; Jaskani et al. 2005).

6.7.1.1 Watermelon: A Potential Source of Lycopene

Previously, only tomato and tomato products were thought to be viable sources of lycopene, but there is now evidence that watermelon contains a significant quantity of cis-configured lycopene. As a result, consumers are increasingly turning to watermelon and related products to address their health concerns. However, the amount of lycopene in tomatoes varies depending on the species and growing conditions (Fish and Davis 2003). Lycopene is found in crystalline form in cells and ranges from 2.30 to 7.20 mg/100 g fresh weight bases (Huh et al. 2008; Charoensiri et al. 2009; Artés-Hernández et al. 2010). Furthermore, the lycopene level of red-fleshed watermelon is over 40% more than that of tomato, with 4.81 and 3.03 mg/100 g, respectively. Yellow-orange and yellow-coloured-fleshed watermelon, on the other hand, have a lower lycopene concentration, with 3.68 and 2.51 mg/100 g, respectively (Jaskani 2005; Choudhary et al. 2009). Because the protein-carotenoid combination in tomatoes is broken down, lycopene is accessible in greater quantities following heat treatment. Watermelon lycopene, on the other hand, is immediately available to the human body after eating (Edwards et al. 2003; Perkins-Veazie and Collins 2004; Jaskani et al. 2006; Saftner et al. 2007).

Watermelon's higher lycopene to carotene ratio, 1:12, results in impressive antioxidant capacity (Mort et al. 2008). Foods high in lycopene concentration are referred to as functional foods because of this special attribute (Shi and Maguer 2000; Collins et al. 2005).

6.7.1.2 Synthesis Route of Lycopene

In the biosynthesis of lycopene, a complex process begins when chlorophyll degrades to generate white-coloured leucoplast, which then yields specific red colour pigmented organelles, such as chromoplast (Bowen et al. 2002). Lycopene is a carotenoid created as a by-product of the synthesis of xanthophylls such as cryptoxanthin, zeaxanthin and leutin. Tetraprenoids, which are made up of 40-C isoprenoids (5-C isoprene unit), are the building blocks of carotenoids. Isopentenyl diphosphate (IPP) is added step by step to dimethylallyl diphosphate (DMAPP) to produce geranylgeranyl diphosphate, a 20-C precursor (GGPP). Desaturation of GGPP produces 11 conjugated double bonds, which are found in nature as lycopene. From here, cyclic conversion occurs, turning it to α- and β-carotene, which, when oxidised, generate xanthophylls (Ishida and Bartley 2005). Lycopene crystals are present in the shape of tiny globules hanging throughout the fruit and are a bright red colour (Chandrika et al. 2009). Because of its lipophilic nature, lycopene is found in the thylakoid membrane as a protein lycopene complex at the cellular level. The presence of all-trans lycopene within the fruit, which is changed from cis-configured lycopene by the activity of the carotenoid isomerase enzyme, is well-documented. However, if this enzyme is not present in watermelon, keep it in its cis-form (Akhtar et al. 1999; Bangalore et al. 2008). High-lycopene intake is one effective treatment for free radical scavenging, which can damage DNA and proteins (Mortensen et al. 1997). Lycopene can help in cancer prevention, immune boosting and cardiovascular protection (Zou and Feng 2015).

Watermelon flesh colour is an essential commodity trait for consumers, and differing carotene compositions result in a wide spectrum of watermelon flesh colours. White, pale yellow, canary yellow, orange, pink, red and scarlet are the most common flesh colours for watermelon. Furthermore, certain accessions of the Citrullus amarus subspecies are bright green. In white-fleshed watermelon, only trace quantities of phytofluene were found (Zhao et al. 2008). The pigments in yellow-fleshed watermelons were discovered to be a mixture of β-xanthophyll derivatives derived from zeaxanthin, but the composition varied between accessions. The three main pigments were neoxanthin, violaxanthin and neochrome (1.66 and 0.29 g/g for canary yellow and light yellow, respectively) (Bang et al. 2010), while Liu (2.01–2.82 g/g) found all-trans-violaxanthin, 9-cis-violaxanthin and luteoxanthin in yellow-fleshed watermelon (Liu et al. 2012). Watermelons with orange flesh contained substantially more ζ-carotene, prolycopene or β-carotene than watermelons with other colours (Tadmor et al. 2005). Lycopene, trace amounts of phytoene, prolycopene and xanthophyll were the main pigments found in pink, red and scarlet watermelon germplasm resources (Perkins-Veazie et al. 2006), while ζ-carotene, zeaxanthin, violaxanthin, and other carotenoids were barely detectable in mature red fruit (Perkins-Veazie et al. 2006; Grassi et al. 2013).

It has previously been reported that the red, yellow and white flesh colours of watermelon are inherited. Except for white, which was called Wf, canary yellow (C) dominates most other flesh colours (c), such as red, pink, orange and pale yellow. The Wf gene is epistatic to the yellow flesh trait. Henderson et al. (1998) discovered the allele i-C (inhibition of C and c), which was epistatic to C and produced red flesh even when C was present; however, Bang et al. (2010), who investigated the genetic basis of red and canary yellow flesh colours in watermelon, could not confirm these findings. Bang et al. also discovered the py gene, which causes pale yellow flesh (2010).

6.7.1.3 Phenotypic Segregation Analysis of Flesh Colour and Lycopene Content

According to Wang et al., in a population of LSW-177 (red flesh) × COS (pale yellow), the flesh colour of the F_1 generation was canary yellow, near to pale yellow, showing pale yellow with an incomplete dominance over red. The segregating population had five categories of flesh colour: red (87 plants), pale yellow (48 plants), canary yellow (173 plants), and two irregular colour patterns in the heart and placental tissues of the fruit: red mixed with pale and canary yellow or red in mixed patterns (18 and 26 plants, respectively). Because the flesh colour of the majority of the mixed pale and canary yellow fruits was > 50% canary or pale yellow by cross-sectional area, the two mixed-colour plants could be categorised as canary yellow and pale yellow, respectively. According to these classification criteria, in a population of LSW-177 (red flesh) × COS (pale yellow), 199 (173 + 26), 66 (48 + 18) and 87 plants were judged to have canary yellow, pale yellow and red flesh colour, respectively, fitting a genetic segregation ratio of 9:3:4 (2 = 0.02 and 1.12, $P = 0.99$ and 0.57 for the years 2013 and 2014, respectively), indicating that flesh colour was affected by two major genes. Visual observation could also classify canary yellow and pale yellow into red and non-red groups. By statistical analysis, the segregation of these two groups resulted in a ratio that did not deviate substantially from a 3:1 ratio (non-red group: red group = 265:87, P 0.05). Based on the genetic background of population LSW-177 (red flesh) × COS, these findings suggested that a single main recessive gene determined red and non-red colour in watermelon (pale yellow).

LSW-177 has a high-lycopene concentration in ripe fruit, with an average value of 41.72 2.82 g/g, substantially greater than the COS and F_1 generation's 0.24 0.03 and 0.42 0.05 g/g. When the lycopene content and flesh colour data were compared, we discovered that when the lycopene content was greater than 13.57 g/g, the plants had a red flesh colour. At this threshold value, the population of LSW-177 (red flesh) COS (pale yellow) could be separated into two groups (high- and low-lycopene groups). The genetic ratio of the high-lycopene (87 plants) and low-lycopene (265 plants) groups in the F_2 progeny satisfactorily match a 1:3 (2 = 0.015, $P = 0.902$) ratio, suggesting that lycopene accumulation is influenced by one main gene. They then compared these findings to data from 2013 (Liu et al. 2015), when they obtained lycopene content data in an F_2 generation (234 plants) using the identical parental materials. In both years, the F_2 generations revealed the same genetic ratio for flesh colour and similarly diverging lycopene content trends.

The flesh colour of the F_1 generation in a population of LSW-177 (red flesh) × PI 186490 (white flesh) was canary yellow mixed with white. In the F_2 generation, four main flesh colours emerged: red combined with white, red mixed with white and yellow, canary yellow mixed with white, and white. The majority of the plants in the red and white mixed groups accumulated lycopene only around the seed region, with only two plants having totally red flesh colour. The flesh colour of two groups was separated: one with red or red combined with white skin, and the other with yellow, white, and mixed yellow and white flesh. These two groups had 97 and 262 individuals, respectively, fitting a genetic segregation ratio of 1:3 (2 = 0.780, $P = 0.377$).

The F_1 plants were backcrossed to garden female to obtain a BC_1 population, and the F_1 plants had canary yellow mixed with white flesh in the garden female (red flesh) × PI 186490 (white flesh) cross. The BC_1P_1 generation segregated approximately seven main flesh colours (red, red mixed with white, orange, orange mixed with white, canary yellow, canary yellow mixed with white, and mixed canary yellow, orange and red). Two groups were also divided, one with red and red mixed with white flesh, and the remainder fruits with different flesh colours, according to the categorisation norms of the LSW-177 (red flesh) × PI 186490 (white flesh) population. The genetic segregation between these two groups was 116:106, which corresponded to a 1:1 ratio ($P = 0.502$). Given that the white flesh trait has incomplete dominance over red flesh in watermelon, the F1 plants of population LSW-177 (red flesh) × PI 186490 (white flesh) and population garden female (red flesh) PI 186490 (white flesh) did not exhibit white flesh colour equivalent to that of PI 186490 (Wang et al. 2019).

6.7.2 Phytochemicals and Antioxidants in Watermelon

Consumers have been more health conscious in recent years as a result of the rising prevalence of noncommunicable diseases such as hypertension, diabetes, cancer and cardiovascular disease, among others, and there has been an increase in demand for high-quality fruits. As a result, research into the antioxidant composition of watermelon is becoming increasingly significant, aiding watermelon breeding for the development of high-quality, nutritious fruits. As a result, it's critical to characterise various watermelon genotypes for such compounds in order to determine their nutritional value, which varies based on cultivars and genotypes, sample location, and fruit ripening phases (Tlili et al. 2011). However, because information on phytochemicals and antioxidants in watermelon cultivars grown in India is limited, it is necessary to determine phytochemicals and antioxidants in selected genotypes of watermelon grown in India's hot arid climate in order to identify promising genotypes rich in antioxidants.

Choudhary et al. (2015) studied phytochemicals and antioxidants in ten different red-fleshed genotypes of watermelon, including eight released varieties from different Indian research institutes (Sugar Baby, Durgapur Lal, Charleston Grey, Asahi Yamato, Arka Manik, AHW 19, AHW 65, and Thar Manak), one advance breeding line (AHW/BR 16), and one indigenous collection (IC 582909). Total phenols differed significantly across the genotypes studied, ranging from 16.77 to 21.41 mg/g DW base, according to the findings. Asahi Yamato had the greatest total phenols (21.41 mg/g DW basis), followed by AHW/BR 16 (20.67 mg/g DW basis) and Sugar Baby (20.61 mg/g DW basis, respectively). Total phenolics in watermelon fruits have been observed to range from 13.05 to 18.08 mg gallic acid equivalent/100 g fresh weight basis (Nagal et al. 2012). Similarly, significant differences in total flavoniods (55.60–100.93 mg/100 g DW basis) were found among the genotypes. The cultivar Asahi Yamato had the greatest total flavonoids content (100.93 mg/100 g DW base), which was substantially greater than all genotypes and nearly twice as much as AHW 19 (55.60 mg/100 g DW).

Flavonoids are one of the largest contributors to vegetable antioxidant activity, and it is well known that flavonoids exhibit antioxidant activity and have a significant impact on human health by scavenging or chelating free radicals (Ebrahimzadeh et al. 2008). The differences in tannin concentration between watermelon genotypes were statistically significant, ranging from 35.07 to 60.83 mg/100 g DW basis, with the cultivar Durgapura Lal having the greatest tannin concentration. Several factors, including genotype, sample area and meteorological circumstances, could impact these differences in phenolic composition (Tlili et al. 2011; Nagal et al. 2012).

Different genotypes of watermelon had substantial differences in total carotenoids, lycopene and antioxidant activity. Total carotenoid content ranged from 4.90 to 8.06 mg/100 g FW, with Asahi Yamato (8.06 mg/100 g FW) having the highest total carotenoid content, followed by AHW/BR 16 and Sugar Baby (6.90 and 6.65 mg/100 g FW, respectively). Various researchers have previously reported similar findings on carotenoids in watermelon (Perkins-Veazie 2007; Zhao et al. 2013).

Lycopene levels in red-fleshed watermelon genotypes ranged from 3.74 to 6.80 mg/100 g FW, indicating a twofold variance. All other genotypes were found to be significantly inferior to the cultivars Asahi Yamato (6.80 mg/100 g FW) and AHW/BR 16 (6.01 mg/100 g FW). Choo and Sin (2012), on the other hand, found a low level of lycopene (0.95 mg/100 g) in red-fleshed watermelon. This was due to the fact that the lycopene content in red-fleshed watermelons varied depending on genotype and environmental factors (Perkins-Veazie et al. 2001). According to Kang et al. (2010), lycopene accounted for the majority of total carotenoids (84–97%). This study's total carotenoids and lycopene values are very similar to cultivars cultivated in different parts of the world (Nagal et al. 2012). The lycopene concentration of main watermelon cultivars has previously been reported as 4.26 mg/100 g (Tadmor et al. 2005), 3.3–12.0 mg/100 g (Perkins-Veazie 2007) and 3.46–8.0 mg/100 g (Nagal et al. 2012), both of which were greater than tomato (Tadmor et al. 2005).

The chemical diversity of phenolic antioxidants compounds renders it difficult to separate and quantify individual antioxidants from the plant matrix. As a result, because total antioxidant activity is an integrated characteristic of all antioxidants present in a complex sample, it is more useful for evaluating health benefits (Apak et al. 2007). According to the CUPRAC experiment, the average antioxidant activity of several watermelon genotypes ranged from 40.13 to 84.05 mol Trolox equivalent (TE)/100 g FW. The cultivar Asahi Yamato demonstrated statistically significant antioxidant activity (84.05 mol TE/100 g FW) across all genotypes, followed by AHW/BR 16 and Sugar Baby (72.98 and 66.79 mol TE/100 g FW, respectively). These genotypes' strong antioxidant activity can be attributed to the presence of rich phenols, flavonoids or lycopene, as well as other reducing agents that may diminish the oxidised state of antioxidant molecules. In watermelon, Nagal et al. (2012) and Choo and Sin (2012) made similar observations. The phenolic compounds are the most abundant antioxidants in vegetables, and they are effective at scavenging free radicals. A large range of phenolic compounds, in addition to vitamin C, have shown high in vitro antioxidant activity (Pantelidis et al. 2007). Since lycopene is the most abundant carotenoids in watermelon, accounting for 84–97% of total carotenoids (Kang et al. 2010), it is hypothesised that the total antioxidant content of fruits will vary under similar settings.

Total phenols ($r = 0.921$), total flavonoids ($r = 0.966$), total carotenoids ($r = 0.979$) and lycopene ($r = 0.992$) all had a strong positive connection with antioxidant activity evaluated by the CUPRAC assay ($P = 0.01$). However, there was no link found between antioxidant activity and tannin levels. Total phenols, total flavonoids, total carotenoids and lycopene levels are the key contributions to antioxidant activity in watermelon, according to this study. Nagal et al. (2012) found a strong positive relationship between antioxidant activity and phenolic concentration in watermelon.

6.7.3 Breeding for Quality Traits in Watermelon

Varied market requirements exist for different groups of shippers and consumers, so fruit size is a significant issue in a breeding programme. Mini (under 4.0 kg), icebox (4.0–5.5 kg), small (5.5–

8.0 kg), medium (8.0–11 kg), large (11–14.5 kg) and giant (> 14.5 kg) are the general categories. Fruit size is inherited in a polygenic manner, involving an estimated 25 genes. Shippers in the USA deal with certain weight categories, such as seeded at 8.0–11 kg and seedless at 6.5–8.0 kg.

The adaptation of the watermelon's specific environment resulted in small- or medium-fruited varieties. Early maturing cultivars were produced from Japanese and Russian varieties. Early cultivars generated from the aforesaid sources include 'White Mountain' and 'New Hampshire Midget,' which have 1.0–2.0 kg fruit with a 65-day maturity. The early cultivar 'Petite Sweet' produces fruit weighing 2.0–4.5 kg. M. Hardin selected 'Sugar Baby,' a small-fruited cultivar popular in some parts of the world, in Oklahoma in 1956.

Fruit quality is critical in the production of watermelon. As a result, breeders have always concentrated on improving fruit shape, rind pattern, flesh colour, flesh texture, sweetness, seed colour and other qualitative features. The genetics of these features in cultivated watermelon cultivars has been studied in several ways (Guner and Wehner 2003) Fruit shape is an important external quality attribute that exhibits a variety of phenotypic states, stressing its importance in breeding for customer preference or choices. Watermelon fruit shape is controlled by an incompletely dominant gene, according to several studies, resulting in elongate (OO), oval (Oo) and spherical (oo) fruits (Guner and Wehner 2004; Dou et al. 2018). Lou and Wehner (2016) recently presented a new set of alleles for fruit shape at the ob locus, including ObE for elongate, which is dominant over ObR for round and ob for oblong fruit forms.

Fruits are classified as round, oval, blocky or elongate in general. When it comes to round versus elongate, one gene is implicated, with the F1 being intermediate (blocky). Fruit shape is sometimes linked to cotyledon shape at the seedling stage. Elongate cotyledons are found on plants with elongate fruit, while round cotyledons are found on plants with round fruit. Others, on the other hand, have determined that fruit shape selection at the seedling stage is useless. There was a higher sensitivity to the formation of gourd-neck or bottleneck fruit, which are culls, among old varieties with elongate-shaped fruit. Hollow heart was more common in old varieties with round fruit. To lessen the incidence of these abnormalities, some of the first hybrids were created by crossing elongate and round inbreds. Genetic tolerance to those defects has recently been integrated into new cultivars, reducing the importance of fruit shape.

External fruit quality features such as rind colour and pattern, in addition to fruit shape, influence consumer attention. Porter and Weetman (1937) discovered the g locus, which has three alleles: G for solid dark green, gs for striped and g for grey rind patterns. The genes g-1 and g-2 were discovered by Kumar and Wehner (2011). The grey rind pattern is produced by the g1g1/g2g2 genotype, while the dark green rind pattern is produced by the others (G1—/–G2-). The rind pattern, which might be grey, striped or solid, is the third important aspect of market type. Stripes on the rind might be narrow, medium or wide, with the dark green portions acting as the stripes. On a light green or medium green background, the striped pattern can be found. The colour of the solid rind might be bright or dark green. The dark green rind pattern takes precedence over the grey rind pattern. The striped rind pattern is dominant over the solid dark green rind pattern, and the striped rind pattern is dominant over the solid light green rind pattern. However, when the colour has been faded by the sun, the striped pattern can be visible on a solid dark green fruit. Over a light or medium green backdrop, the stripes might be placed.

In addition to the common rind patterns, the recessive gene f controls furrowed versus smooth rind. The majority of modern cultivars have a smooth rind. Golden rind, which is regulated by the recessive gene go, is another fascinating mutant. Because the change in fruit colour during fruit maturity is accompanied with chlorosis of the leaves, its utility as a fruit ripeness indicator is restricted. Furthermore, it does not appear to be a reliable sign of maturity and it may be detrimental to productivity, particularly if a multiple harvest method is used.

6.7.3.1 External Fruit Quality

On cultivars that will be delivered to market, rind durability is critical. The rind on large-fruited cultivars should be thick and tough, whereas the rind on small-fruited cultivars should be thin and rough. To maintain a balanced proportion for the best appearance, the thickness of the rind should be a modest fraction of the flesh diameter. Large-fruited cultivars benefit from a thicker rind since they require additional protection during postharvest handling and shipment. The rind might be stiff and firm, like in the case of 'Peacock,' or tough and soft, like in the case of 'Calhoun Gray.' For cultivars that will be delivered to market, brittle rind, such as in 'New Hampshire Midget,' is not desirable. Cutting a 1/16 to 1/8 in. × 3 in. piece of rind from a fruit and bending it into an arc can be used to test peel flexibility. The rind is flexible and tough if it bends into a tight arc. It is sensitive and explosive if it breaks early in the endeavour. The firmness of the rind can be determined by punching it with a spring-loaded punch. Punching through a tough rind takes more force than punching through a soft or fragile rind. Watermelon breeders, on the other hand, frequently adopt speedier procedures to determine rind firmness. One way is to drop the fruit from a specific height (for example, knee height) and check if it breaks open. The drop height would be determined by the soil type in the field. Another method is the 'thumb' test, in which the breeder presses on the rind of each fruit at a certain area. It has a delicate rind if it breaks when only a tiny amount of force is applied; otherwise, it should be resistant to shipping damage.

6.7.3.2 Characteristics of Internal Fruit Quality

Watermelon flesh colour is one of the internal quality features that predict appeal and acceptability, as well as potential health benefits. The carotenoid content and composition are responsible for the colour of the fruit flesh. Different genes responsible for white, salmon yellow, canary yellow, scarlet red, coral red and orange colours of watermelon flesh have been identified in several reports on genetic investigation of watermelon flesh colour (Gusmini and Wehner 2006). Canary yellow (C) dominates all other colours (c) except white (Wf), which is epistatic to yellow flesh (B), where genotype Wf_ B_ or Wf_ bb was white-fleshed, wfwf B_ was yellow-fleshed, and wfwf bb was red-fleshed. Red, orange and salmon-yellow flesh colours are produced by the y locus, which has three alleles (Y, yo and y, respectively). Y has the upper hand over yo and y, and yo has the upper hand over y. (Henderson et al. 1998). Canary yellow flesh (C) is dominant to pink (c), epistatic to coral red (Y), and influenced by an inhibitor of canary yellow (i C), where homozygous recessive iiC_ gentopye results in red flesh even when C is present (Henderson et al. 1998). A third yellow flesh gene (B) has been identified as being hypostatic to the white flesh gene (Wf) (Henderson 1992). Scarlet red flesh colour is monogenic and dominant over coral red, according to Gusmini and Wehner (2006). Despite several of these observations, the genetics of flesh colour remain a mystery, possibly due to the imprecise nature of flesh colour descriptions and the reappearance of abnormal/intermediate phenotypes in segregating populations. Gusmini and Wehner (2006) discovered phenotypes that were halfway between red and salmon yellow flesh. As a result, they proposed that distinct ratings of colour at different sections of the flesh be used to acquire more information on the genetics of flesh colour.

Another important feature in watermelon crop improvement is sweetness, which is quantified as TSS (per cent). TSS is said to be linked to the sweetness and flavour of watermelon flesh (Pardo et al. 1997). Several studies of TSS inheritance have produced a variety of results, including three incompletely dominant genes (El-Hafez et al. 1981); partial dominance (Brar and Nandpuri 1977); dominance gene effects and dominance by dominance epistasis (Sharma and Choudhury 1988); and additive gene effect (Lou and Wehner 2009), depending on backgrounds. Sugar concentration is chosen by breeders based on flavour and refractometer findings. Readings from a refractometer can be taken in the field using a handheld unit and provide information on the percentage of soluble solids (°Brix). These refer to the amount of sugar in the product, which should be at least 10%. Brix levels as high as 14% have been reported in newer cultivars. Some cultivars have more fructose, which has a sweeter flavour than sucrose. A refractometer does not quantify the difference in taste.

Watermelon flavour should be prioritised over sweetness in the selection process (sugar content). Freedom from bitterness, which is regulated by a single dominant gene, should be included in the flavour profile and can be introduced by crosses with *C. colocynthis* accessions. Another ingredient is the caramel flavour found in the 'Sugar Baby' fruit, which some taste testers dislike. The flavour is sometimes linked to the colour of the flesh, which is dark red. The origin of caramel flavour is unknown; however, it does respond to selection. Breeders should look for lines that have a mild (not bitter) taste, a high sugar content (°Brix), no caramel flavour, and a great 'watermelon' flavour. To give a comparison for the plant breeder, cultivars with outstanding taste should be included as checks in all selection blocks. 'Allsweet,' 'Crimson Sweet' and 'Sweet Princess' are examples of high-quality cultivars that are commonly used.

Internal quality is influenced by the texture of the flesh. The flesh of a watermelon might be mushy or solid, fibrous or crisp. Plant breeders should strive to create cultivars with firm, crisp flesh as their goal. The genes that control certain characteristics are unknown, yet they are inherited.

White, green, tan, brown, black and red are some of the coat colours seen on watermelon seeds. White seed colour is usually avoided since it indicates immaturity in the fruit and makes it harder to discern mature from immature seeds. White seeds, on the other hand, could be a good target for developing near-seedless cultivars with few, small and inconspicuous seeds. The fiery red or canary yellow flesh colour contrasts well with the black seed colour. With orange flesh, black, brown or tan seeds look great.

Tan (RR tt WW) and green (rr TT WW) seed coats were found to be more dominating than red (rr tt WW) (Poole 1944). Poole et al. (1941) presented a three-gene model with the genotypes RR TT WW for black seeds, which are dominant over all other colours, RR tt WW for tan, RR TT ww for clump, RR tt ww for white tan-tip, rr ttWW for red and rrttWW for white pink-tip. Furthermore, he proposed the d gene as a modifying factor for the black seed coat colour, which is only effective when combined with the RR TT WW genotype, resulting in RR TT WW DD being black and RR TT WW dd being dotted black (Poole et al. 1941).

For the confectionery (edible seeded) kind, seed size should be high, whereas for the standard (edible flesh) kind, seed size should be small or medium. The new seed size mutant just identified in watermelon is tomato seed type. The seed size is around half that of a tiny watermelon seed and is determined by a single recessive gene called *ts*. For the confectionary, the number of seeds should be high, but for the edible flesh type, it should be low to medium is considered good. Small-fruited cultivars should have fewer seeds so that the seeds do not appear to make up more than the usual amount of the fruit volume.

6.7.3.3 Genetics of Fruit Quality Traits in Watermelon

Maragal et al. (2019) undertook research to better understand the inheritance of essential exterior and internal fruit quality parameters in *Citrullus amarus* prebred lines. Two backcross inbred lines (BIL-53 and BIL-99 as female parents) and IIHR-140-152, an icebox-type elite inbred line as a common pollen parent, were used to create the experimental materials. BIL-53 and BIL-99 are the back cross inbred lines (BC1F6) derived from an intraspecific cross between IIHR-82 (*Citrullus amarus*) with an improved cultivar, Arka Manik (recurrent parent) which is a popular variety of India. BIL-53 has a creamish white leathery flesh with a TSS of 3–5%, red-coloured seeds, flat fruits with a medium green marbled rind pattern and yellowish white blotchy inter-stripe pattern, and yellowish white blotchy inter-stripe pattern. BIL-99 has a salmon yellow flesh colour, a TSS of 5–7%, tan/brown seeds, and spherical fruits with a grey-type rind (medium to light green narrow reticulations on a light green background). The male parent in both crossings was an ice box type inbred line (IIHR-140-152) with dark red-coloured flesh, TSS ranging from 10 to 13%, black-coloured seeds, and rectangular fruits with dark green solid stripes and greenish wavy inter-stripe pattern. In terms of rind stripe pattern, fruit shape, seed colour, TSS and flesh colour, BIL-53 is similar to its wild citron parent (IIHR-82). By crossing both BILs to IIHR-140-152 for genetic study, controlled pollinations were done to create two sets of families. Each family consists of six generations of P_1, P_2, F_1, F_2, BC_1 (backcross with female parent) and BC_2 (backcross with male parent).

The F_1 of the BIL-99 × IIHR-140-152 cross-produced oval fruits, which are in between the spherical and oblong fruit morphologies of the parents. Plants segregated 17:52:27 (spherical: oval: oblong) in the F_2 generation, according to a 1:2:1 ratio (2 = 2.75, $P = 0.25$). In BC_1 and BC_2 generations, the fruit shape segregated as 1:1 for spherical: oval and oblong: oval (both 2 = 0.02, $P = 0.88$), and (both 2 = 0.03, $P = 0.86$) correspondingly. As a result, fruit shape is inherited as a single gene with partial dominance, with the oblong shape outnumbering the spherical shape. However, no simple Mendelian ratio fits the segregation for flat fruit shape in the several progenies of the cross BIL-53 × IIHR-140-152 (Maragal et al. 2019).

All of the fruits in the F_1 generation of the cross BIL-53 (marbled stripes) × IIHR-140-152 (solid stripes) had solid stripes. The F_2 was 113:41 (solid: marbled type), with a 3:1 ratio of goodness of fit (2 = 0.22, $P = 0.63$). Stripe pattern segregated as 1:1 (2 = 0.55, $P = 0.45$) and 1:0 (2 = 0.00, $P = 1$) ratios in BC_1 and BC_2 generations, respectively. Thus, in BIL-53 × IIHR-140-152, a single gene controls the stripe pattern, with solid stripes predominating over marbled stripes.

All of the F_1 fruits in the cross of BIL-99 (grey stripes) × IIHR-140-152 (solid stripes) were solid striped. With a goodness of fit for 3:1 expectation of 2 = 0.57, $P = 0.45$, the F_2 segregated in the ratio 63:18 (solid stripes: grey). The segregation ratios for solid and grey stripes in BC_1 and BC_2 were 1:1 (2 = 0.02, $P = 0.88$) and 1:0 (2 = 0, $P = 1$), respectively, confirming that stripes are monogenic and that solid stripes are dominant over grey type stripes. The results of this experiment indicate that the stripe pattern is monogenic in both crosses, with solid-type dominating marbled and grey stripe patterns. These alleles were given the names gs for striped and g for grey rind patterns by Porter (1937) and Weetman (1937). For marbled rind pattern stripes, we offer gm.

The F_1 generation of the cross BIL-53 (blotchy type) × IIHR-140-152 (wavy type) showed a blotchy inter-stripe pattern. In accordance with a 3:1 ratio (2 = 1.95, $P = 0.16$), the F_2 generation was segregated as 123:31 (blotchy type: wavy type). BC_1 fruits all displayed a blotchy inter-stripe pattern, but BC_2 fruits was separated in a 1:1 ratio ($P = 0.72$) (Maragal et al. 2019).

Similarly, the grey type inters tripe pattern was displayed in the F_1 generation of BIL-99 (grey type) × IIHR-140-152 (wavy type). The F_2 generation was classified as 75:9 (grey type: wavy type) with a 15:1 ratio of goodness of fit ($\chi^2 = 2.86$, $P = 0.09$). Fruits in BC_1 exhibited a grey-type inter-stripe pattern throughout, but fruits in BC_2 were segregated in a 3:1 ratio (grey: wavy) ($\chi^2 = 0.08$, $P = 0.77$). The results of this experiment indicate that blotchy and grey inter-stripes are dominant over wavy inter-stripes. In the BIL-53 × IIHR-140-152 cross, monogenic inheritance was observed, but the BIL-99 × IIHR-140-152 hybrid was digenic with duplicate gene action. Stripe and inter-stripe patterns were shown to be segregated separately, showing that they represent two distinct traits.

Green colour is inherited as a dominant characteristic in the F_1 generation of BIL-53 (yellowish white) × IIHR-140-152 (green), since all of the fruits had green-coloured inter-stripes. Individuals from F_2 generations were found to be 109:45 (green: yellowish white), with a 3:1 ratio having the best fit ($\chi^2 = 1.46$, $P = 0.22$). In the BC_1 generation, the inter-stripe colour segregated 1:1 (green: yellowish white) ($\chi^2 = 1.48$, $P = 0.22$). All of the fruits in the BC_2 generation had green-coloured inter-stripes.

All of the F_1 fruits in the cross BIL-99 (light green) × IIHR-140-152 (green) had green inter-stripes. The F2 generation was 66:18 (green:

light green), which corresponded to the 3:1 expectation ($\chi^2 = 0.57$, $P = 0.45$). The 1:1 and 1:0 (green:light green) segregation ratios in BC_1 and BC_2 showed monogenic inheritance, with green dominating light green. Inter-stripe colour is a monogenic characteristic in both crossings, according to genetic study, and green colour is dominant over yellowish white and light green hues (Maragal et al. 2019).

All of the fruits in the F_1 generation of BIL-53 (red) × IIHR-140-152 (black) had black-coloured seeds, demonstrating that seed coat colour is a dominant feature. Individuals from the F_2 generation were split 110:40 (black: red), with a 3:1 ratio fitting best ($\chi^2 = 0.22$, $P = 0.63$). The seed coat colour segregated 1:1 (black: red) in the BC_1 generation ($\chi^2 = 3.06$, $P = 0.08$), but all seeds in the BC_2 generation were black.

Fruits exhibited black-coloured seeds in the F_1 generation of BIL-99 (tan) × IIHR-140-152 (black), showing that black colour seed coat is inherited as a dominant feature. Seed colour segregation in F_2 plants was 56:25 (black: tan), fitting a 3:1 ratio ($\chi^2 = 1.49$, $P = 0.22$). BC1 generation seeds segregated in a 1:1 ratio ($\chi^2 = 0.02$, $P = 0.88$), whereas BC2 generation fruits all had black pigmented seeds. The black seed colour dominates over the red and tan in these results, indicating single gene control. Segregation in tan × red seed coat colours in crosses needs to be proven to see if it's due to numerous alleles at the same locus or separate genes (Maragal et al. 2019).

All of the fruits in the F_1 generation of BIL-53 (white) × IIHR-140-152 (red) were yellow, showing inhibitory or partial dominance. The flesh colour around seeds in the F_2 generation was segregating for three colours, yellow, red and white, in a 9:3:4 ratio ($\chi^2 = 3.02$, $P = 0.22$). Flesh colour separated in the BC_1 and BC_2 generations in the ratios of 1:0:1 (yellow: red: white) ($\chi^2 = 0.02$, $P = 0.88$) and 1:1:0 (yellow: red: white) ($\chi^2 = 1.43$, $P = 0.23$). They conclude that the presence of two genes with recessive inhibitory gene action may be responsible for the emergence of yellow colour surrounding seeds in F_1, F_2, BC_1 and BC_2 generations. In homozygous recessive (aa) state, the inhibitory gene produces white and blocks the expression of other genes. As a result, aaY_ or aayy genotypes may control white colour, A yy genotypes control red colour and A Y_ genotypes control yellow colour (Maragal et al. 2019).

In the BIL-99 (salmon yellow) × IIHR-140-152 (red) hybrid, F_1 fruits showed the salmon yellow colour around the seeds, indicating that salmon yellow had prevailed over red. The F_2 segregated as 41:37 (salmon yellow: red) with a 9:7 ratio with a goodness of fit of ($\chi^2 = 0.43$, $P = 0.51$). The flesh colour around the seeds segregated in a 1:0 (red: salmon yellow) ($\chi^2 = 0.08$, $P = 0.77$) and 3:1 (salmon yellow: red) ($\chi^2 = 0.23$, $P = 0.63$) ratio in the BC1 and BC2 generations, respectively. This finding suggests that the colour of the flesh around the seeds may be controlled by two genes that function complementarily. We propose the genes Y1Y1Y2Y2 and Y1Y1Y2Y2 for salmon yellow flesh and red colour around the seeds, respectively. It also suggests that the set of genes that control colour around seeds may differ from those that control colour in other parts of the fruit (Maragal et al. 2019).

TSS showed continual fluctuation in both crosses, demonstrating polygenic control. To further understand gene function, the six generation mean analysis and scaling test were used. The significance of the scaling test revealed that the basic additive dominance model was insufficient and that non-allelic interaction occurred in both crosses. For both crossings, the estimations of several genetic components, such as mean (m), additive (d), dominance (h), additive × additive (i) additive × dominance (j) and dominance × dominance (l). The results of non-allelic interactions were interpreted using Hayman's (1958) method. The additive and additive × additive components were prominent in the cross BIL-53 × IIHR-140-152, whereas the dominant and dominant × dominant components were significant in BIL-99 × IIHR-140-152. In both crosses, opposite signals of h and l suggested a duplicate kind of epistasis (Maragal et al. 2019).

6.8 Breeding for Diseases Resistance in Watermelon

6.8.1 Fusarium Wilt of Watermelon

Watermelon *Fusarium wilt* (FW) was the second Fusarium wilt disease to be reported, following the first case in cotton (Martyn 2012). Fusarium wilt is a deadly disease of watermelon caused major damage in the early 1890s in the southern USA. Many modern watermelon cultivars have a small genetic base and are susceptible to *Fusarium wilt* (Levi et al. 2001a; Lambel et al. 2014). *Fusarium wilt* resistance for race 1 and race 2 is now relatively rare in commercial watermelon hybrids.

Many vascular wilts caused by *Fusarium oxysporum* formae speciales that are generally considered host specific affect the *Cucurbitaceae* family (Martyn 2012). Watermelon FW is a serious soil-borne disease caused by *Fusarium oxysporum* f. sp. *niveum* (FON). Watermelon crop yield has been severely harmed by FW, which has resulted in the yield loss of several commercial cultivars. Fusarium develops a septate mycelium that produces conidia and creates chlamydospores. Infection spreads through the root-tip region, wounds and the germinating seed's radicle. The fungus infects the root cortex, then establishes itself and spreads through the xylem elements. Other tissues aren't invaded until the plant is dead or close to death. The ideal soil temperature for the organism in watermelon plants is 27 °C. FON races 0, 1, 2 and 3 have been identified as pathogenic (Zhou et al. 2010). Earlier, FON race 1 was the most common prevailing race (Xu et al. 2000), but later FON race 2 has become more common in the previous two decades. Because of the various strains of the fungus and the lack of a perfect method for precisely discriminating between susceptible and resistant individuals in a mixed population, nothing is known about resistance inheritance. Orton (1911) proved that resistance could be transferred by hybridisation and selection by crossing the resistant Stock Citron with the susceptible Eden variety. Porter and Melhus (1932) revealed that wilt-resistant individuals can be found in commercial types, eliminating the need for inedible citron for breeding programmes. Braun (1942) attempted to uncover the physiological basis of wilt resistance by showing that acetic acid inhibits fungal growth and that a comparable substance was discovered in the stems of resistant Citron watermelon but not in the roots of either the resistant or susceptible varieties. Stock Citron and Iowa Belle are both resistant to watermelon wilting. Range of resistance would include, Summit and Calhoun Gray are highly resistant; Charleston Gray, Hawksbury and Garrisonian are moderately resistant while Florida Giant is susceptible. In places with a short growing season, fusarium-resistant cultivars are essential. In 1991, a highly resistant to race 2 inbred line (PI 296341-FR) was released for the first time (Martyn and Netzer 1991). The FON race 1 resistance is controlled by a dominant gene, but it is also influenced by modifying genes (s) Netzer and Weintall (1980). Resistance to FON race 2 was controlled by at least one recessive pair of genes, but in the citron-type PI 296341-FR, a dominant gene from a susceptible parent was epistatic over the recessive gene for resistance (Martyn 2012). As a result, it's been difficult to transfer the high level of resistance from wild citron-type PI 296341-FR into commercial cultivars for FON race 2. The development of SY630 (a pollinating parent for seedless watermelon production), which has stronger genetic relationship with commercial cultivars than PI 296341-FR, showed that transferring the high level of resistance from semi-wild SY630 into commercial cultivars is possible for FON race 1 and race 2.

The disease *Fusarium* wilt can be effectively controlled by breeding-resistant lines. Planting-resistant cultivars will improve watermelon quality and yield while also reducing the need for fungicides. Marker-assisted selection (MAS) can be a more successful technique of selecting quantitative trait loci (QTL) related with disease resistance and other economically significant features than standard phenotype-based selection. Although various *Fusarium* wilt-resistant

Table 6.1 Watermelon genotypes used to differentiate races of *Fusarium oxysporum* f. sp. *niveum*

Cultivar or genotype	Disease response to[a]		
	Race 0	Race 1	Race 2
Sugar Baby, Black Diamond	S	S	S
Charleston Gray	R	S	S
Calhoun Gray	R	R	S
PI 296341-FR[b]	R	R	R

[a] S = susceptible. R = resistant

[b] Extreme care should be taken when using this PI as a differential, as the resistance is not fixed and seed increases may contain susceptible individuals, resulting in false positives

watermelon germplasms have been identified, none of the commercially developed watermelon cultivars showed resistance for both FON race 1 and race 2. The transfer of FON-resistant loci to commercial cultivars is hampered by the paucity of molecular markers connected to FON resistance (Table 6.1).

6.8.2 Anthracnose

Watermelon anthracnose affects all of the crop's aboveground portions. Vine defoliation may occur in circumstances where infection is strong and lesions on the leaves are numerous, resulting in yield loss or inferior quality fruit. The fruit's marketability can also be harmed by direct infection. Lesions on leaves are often uneven and jagged in appearance. Larger, older leaf lesions' cores may fall off, giving the leaf a 'shot-hole' appearance. The lesions on the stem are light brown and spindle-shaped. Anthracnose lesions in watermelon fruit can be sunken and round, and they can be orange or salmon in colour. These lesions usually begin on the fruit's lower surface, where moisture accumulates. Anthracnose infection and spread are aided by warm, moist conditions. Infection requires moisture, and rain aids in the dispersal of fungal spores from plant to plant. When the plant canopy has developed sufficiently to create a favourable habitat for the fungus to invade, the symptoms might become severe.

The anthracnose pathogen has been identified in seven different races. Watermelon races 4, 5 and 6 are the most dangerous, but races 1 and 3 are the most significant. Many cultivars are resistant to races 1 and 3, but race 2 resistance will be required soon. The first source of anthracnose resistance was discovered in an accession, Africa 8, which was sent to D. V. Layton of the USDA by R. F. Wagner in Umtali, South Africa. From that source, Layton created anthracnose-resistant 'Congo,' 'Fairfax' and 'Charleston Gray.' Resistance was later discovered to be inherited through Ar-1, a single dominant gene. Races 1 and 3 are resistant to the gene, whereas race 2 is not. That source of resistance is present in 'Crimson Sweet' and many other modern cultivars. Resistance to Race 2 has been linked to a number of genes. Resistance to complex Colletotrichum species has been reported in PI 189225, PI 271775, PI 299379 and PI 271778. PI 203551, PI 270550, PI 326515, PI 271775, PI 271779 and PI 203551 are some of the other sources of anthracnose resistance identified in the literature. 'R 143' was reported to be resistant to race 2 of the pathogen. In a screening test comprising 76 plant introductions, PI 512385 demonstrated the highest resistance to race 2 of the disease.

6.8.3 Gummy Stem Blight

Gummy stem blight (GSB) is a destructive fungal disease that affects cucurbitaceous vegetable crop farming all over the world, resulting in substantial yield losses (Stewart et al. 2015). It has been found to infect at least 12 *Cucurbitaceae* genera and 23 species, including watermelon (*Citrullus lanatus*) (Keinath 2011).

Warm and humid climates, which are favourable to spore germination and disease development, exacerbate the prevalence of GSB (Robinson and Decker-Walters 1997). GSB was formerly assumed to be caused by a single pathogen, *Didymella bryoniae* (syn. *Stagonosporopsis cucurbitacearum*) (Aveskamp et al. 2010), but it has since been discovered that three *Stagonosporopsis* species are responsible for the disease: *S. cucurbitacearum* (syn. *D bryoniae*), *S. citrulli*, and *S. caricae* (Stewart et al. 2015). Crown blight, stem cankers and severe defoliation are all indications of gummy stem blight on watermelon plants, with symptoms appearing on the cotyledons, hypocotyls, leaves and fruit (Maynard and Hopkins 1999).

Cultural methods and fungicide application are now used to manage GSB in watermelon. However, because the three causal *Stagonosporopsis* species cannot be distinguished based on symptoms, new reports of differential fungicide resistance pose a substantial challenge to growers (Brewer et al. 2015). Furthermore, fungicide applications raise production costs significantly, and recurrent usage may have a harmful influence on the environment, especially if residues remain in the soil. The optimum option would be to use GSB-resistant cultivars, but there are currently no commercial watermelon cultivars with high levels of GSB genetic resistance.

Citrullus amarus, a wild relative of watermelon (*Citrullus lanatus*) (Renner et al. 2017), has been an important source of disease resistance alleles in watermelon breeding due to the small genetic base of cultivated watermelon following domestication (Levi et al. 2017). *Citrullus* germplasm sources with varying degrees of GSB host resistance have been identified (Gusmini et al. 2005). Resistance to GSB in C. PI 189225 (Sowell and Pointer 1962) was the first to characterise *amarus*, while PI 271778 was the second (Sowell 1975; Norton 1979). The release of AU-Producer, AU Jubilant, AU-Golden Producer and AU-Sweet Scarlet was the result of attempts to introgress resistance from these two sources into commercial cultivars (Norton et al. 1995). However, in commercial production areas, these cultivars did not prove to be resistant (Song et al. 2002). New sources of resistance that included accessions from both *Citrullus amarus* and *Citrullus lanatus* species were later described by Gusmini et al. (2005) and included PI 164248, PI 244019, PI 254744, PI 271771, PI 279461, PI 296332, PI 482379, PI 490383, PI 526233 and PI 482276. PI 482276 was found to be resistant to various isolates from all three *Stagonosporopsis* species.

Resistance to GSB was initially thought to be mediated by a single gene, *db*, in PI 189225. (Norton 1979). Later research on PI 189225, PI 482283 and PI 526233 discovered that multiple genes with small effects are likely to be responsible for this trait (Ren et al. 2019). On chromosome 8 of watermelon, a quantitative trait locus (QTL) underlying GSB resistance in PI 189225 was recently discovered (Ren et al. 2019). This QTL accounts for around 32% of the population's phenotypic diversity.

Gusmini et al. (2017) used three PI watermelon accessions (PI 189225, PI 482283 and PI 526233) to investigate the inheritance of GSB resistance. From four crosses of resistant PI accessions by susceptible cultivars, four families of six progenies (Pr, Ps, F_1, F_2, BC_1Pr and BC_1Ps) were generated. Each family was tested for GSB resistance in two seasons in North Carolina, in both field and greenhouse environments. In order to create homogeneous and robust epidemics, artificial inoculation was used. The influence of the Mendelian gene db, which confers resistance, was investigated. The partial failure of the findings to fit the single gene inheritance model showed that PI 482283 and PI 526233 resistance to GSB may be controlled by a more complex genetic mechanism.

6.8.4 Powdery Mildew

Watermelon Powdery Mildew (PM) is a fungal disease caused by *Podosphaera xanthii* (castagne) USA. N. Braun and N. Braun Shishkoff is the most important agent. A moderately high temperature of 25–30 °C and > 98% humidity are ideal conditions for pathogen sporulation. The entire plant, including petioles, stems, cotyledons and leaves, shows symptoms of mycelial and condial growth. Chlorotic leaves, a reduced canopy, and decreased fruit quality and yield are all symptoms of severely afflicted plants (Keinath and Dubose 2004). Watermelon PM has developed as a major disease concern, and two pathogenically different races, 1W and 2W, have been reported (Tetteh et al. 2010). Watermelon race confirmation was carried out using a collection of diverse melon cultivars (Davis et al. 2007). In the USA, several cultivars and wild-type species that are resistant to these races have been identified (Tetteh et al. 2010). The genetic inheritance of resistances, on the other hand, has not been thoroughly investigated (Davis et al. 2002). Due to a lack of resistant cultivars, the disease is mostly treated by applying fungicides repeatedly (6–7 times) throughout the growing season.

One twenty watermelon accessions were screened in an effort to generate resistant cultivars, and 23 cultivars, including IT207182, were identified as somewhat resistant to PM (Lee et al. 2010). Both PI254744 and 'Arka Manik,' an open-pollinated cultivar from India (Rai et al. 2008), were found to be highly resistant to PM. PM resistance was studied in F_2, $F_{2:3}$, and reciprocal backcross generations produced from the resistant cultivar 'Arka Manik' × (HS3355) (susceptible line). Resistance to PM in 'Arka Manik' is conditioned by a single incompletely dominant gene (Pm1.1), according to the segregation ratios (Kim et al. 2013).

6.8.5 Zucchini Yellow Mosaic Virus

Virus-induced plant diseases are a key limiting factor in commercial watermelon production around the world. Papaya ringspot virus-watermelon strain (PRSV-W), Zucchini yellow mosaic virus (ZYMV) and Watermelon mosaic virus are the most common viruses that infect watermelon (WMV). Several species of aphids transmit all three viruses intermittently, and mixed infections are common (Ali et al. 2012). Chemical control of vectors is not usually an effective means of disease control. Cross-protection with weak ZYMV isolates, as well as treatment with mineral oil sprays and light-reflective surfaces, had limited effectiveness and needed additional input expenditures. As a result, genetic resistance remains the most straightforward, effective and efficient method of minimising disease-related losses (Ali et al. 2012). Muskmelon yellow stunt virus (ZYMV) was initially identified in squash growing in northern Italy and France in 1981 (Lisa and Dellvalle 1981; Tiwari and Rao 2014). ZYMV is the most devastating virus in watermelon production globally, infecting all agriculturally important Cucurbitaceae species (e.g. *C. lanatus*, *Cucumis sativus*, *Cucumis melo* and *Cucurbita* spp.) (Nagendran et al. 2017). Several aphid species (e.g. *Aphis gossypii* Glover) transmit ZYMV in a nonpersistent way, and it is easily transmitted mechanically. The virus overwinters on wild species in locations where cucurbit crops are not grown continually. Although natural infection appears to be limited to *Cucurbitaceae* species, representatives of 11 dicotyledon families are regarded diagnostic hosts (Tiwari and Rao 2014). ZYMV belongs to the genus Potyvirus in the family Potyviridae, having flexuous particles measuring 750 nm in length and containing a single strand of RNA (Romay et al. 2014). There are at least 25 strains of ZYMV known (Desbiez and Lecoq 1997). The occurrence of Connecticut (CT) and Florida (FL) strains of ZYMV was documented by Provvidenti et al. (1984), with the FL strain being more widespread in the USA. In the 1990s, Provvidenti discovered a new ZYMV strain, Zucchini yellow mosaic virus-China strain, affecting cucurbit farms in Beijing, China (ZYMV-CH).

Plants infected with any of the ZYMV strains have a reduced photosynthetic capacity, stunted growth, malformed fruit and premature death (Guner and Wehner 2008). Yellow mosaic, stunted blistering, and laminar decrease on leaves are symptoms of severe ZYMV infection in cucurbit crops, and fruit remains small, forming knobby patches, significantly deformed and mottled (Nagendran et al. 2017).

Guner (2018) conducted an experiment in watermelon to investigate the inheritance of resistance to the Zucchini yellow mosaic virus. Resistance to ZYMV-FL was examined in the resistant watermelon accession PI 595203 (*Citrullus mucosospermus*). The inheritance of resistance to ZYMV-FL was studied using the F_1, F_2 and BC_1 progenies produced from the cross 'Calhoun Gray' (CHG) × PI 595203 and 'New Hampshire Midget' (NHM) × PI 595203. Seedlings were injected with a severe isolate of ZYMV-FL at the first true leaf stage and graded on a scale of 1–9 on the severity of viral symptoms for at least 6 weeks. The high level of resistance to ZYMV-FL in PI 595203 was revealed to be controlled by a single recessive gene (zym-FL).

If ZYMV resistance was polygenic in inheritance, resistance may be enhanced by crossing the most resistant accessions or crossing high-resistance accessions with moderate-resistance accessions. Backcrossing may be used to integrate ZYMV resistance into breeding lines from donor PIs if the resistance was monogenic in inheritance. Improved breeding lines could then be utilised to generate more resistant cultivars.

6.8.6 Watermelon Bud Necrosis Virus

Bud necrosis disease (BND), caused by watermelon bud necrosis orthotospovirus (WBNV), has emerged a major deterrent in watermelon cultivation in Indian subcontinent (Holkar et al. 2018). Watermelon bud necrosis orthotospovirus (WBNV), (Family: *Tospoviridae*, Bunyavirales) belonging to the watermelon silver mottle orthotospovirus (WSMoV) serogroup (Jain et al. 1998, 2014; Adams et al. 2017). It was first reported during 1991 infecting watermelon at Indian Institute of Horticultural Research (IIHR), Bangalore, India (Singh and Krishna Reddy 1996) and later found to affect several other cucurbits, such as cucumber, ridge gourd and muskmelon (Jain et al. 1998, 2007; Mandal et al. 2003; Kumar et al. 2010). WBNV is widely distributed and endemic in majority of the watermelon growing states of India (Mandal et al. 2012). The virus is reported to be transmitted by melon thrips, Thrips palmi Karny (Rajasekharam 2010; Rebijith et al. 2012, 2016). Several studies have reported up to 100% yield loss due to this disease (Singh and Krishna Reddy 1996; Jain et al. 1998, 2007; Kunkalikar et al. 2011). The field symptoms of WBNV appear as leaf chlorosis, silver mottling of leaves, shortened internodes, brittle and upright growth of younger shoots, necrosis on apical bud, stem, petiole and fruit stalk. Infected plants produce unmarketable, small, deformed fruits with uneven surface and necrotic or chlorotic rings, depending on the cultivar (Mandal et al. 2012; Nagesh et al. 2018). Currently, vector management (cultural, physical, chemical and biological) has been the focus of managing this disease, but with limited success (Krishna Kumar et al. 2006; Mandal et al. 2012; Anonymous 2008). Hence, host plant resistance has been suggested as the economically viable and environment-friendly option for managing this disease (Riley and Pappu 2004). In this connection, breeding efforts are underway at IARI, New Delhi and IIHR, Bengaluru, India, to deploy resistant varieties to WBNV. The germplasm accession from *Citrullus colosynthis* was identified as the source of resistance against WBNV, and it is being utilised in the development of WBNV-resistant varieties in the Indian conditions (Holkar et al. 2018). Jamatia et al. (2022) screened 93 watermelon genotypes comprising of core collection of watermelon germplasm from USDA and local collection of wild germplasm against WBNV and only three genotypes namely PI482334 (*Citrullus lanatus* var. *citroides*), PI219691 (*Citrullus lanatus* var. *lanatus*) and DWM 210 (*Citrullus colocynthis*) having OD 0.08, 0.07 and 0.06, respectively, were identified as highly resistant to bud necrosis disease. The identified genotypes maybe used as donors in resistance breeding programmes as well as resistance mapping followed by the identification of genes. Another source of resistance in IIHR-82, an accession belonging to *Citrullus lanatus* var. *citroides* was identified and understanding its genetics (Nagesh et al. 2018) from IIHR, Bangalore.

6.9 Improvement of Watermelon Through Reverse Breeding

Reverse breeding (RB) is a unique plant breeding strategy that aims to produce direct parental lines for any heterozygous plant, which is one of the most desired goals in plant breeding. Through controlled meiosis, RB creates perfectly complementing homozygous parental lines. By preventing meiotic crossing over, the approach reduces genetic recombination in the selected heterozygote. Male or female spores from such plants contain non-recombinant parental chromosomal combinations that can be grown in vitro to produce homozygous doubled haploid plants (DHs). Complementary parents can be chosen from these DHs and employed to recreate the heterozygote in perpetuity. Because traditional plant breeding makes it impossible to fix unknown heterozygous genotypes, RB has the potential to revolutionise future plant breeding.

By creating complementing homozygous lines, reverse breeding allows complex heterozygous genomes to be fixed (Dirks et al. 2003). Knocking down meiotic crossings and subsequent fixing of non-recombinant chromosomes in homozygous doubled haploid strains achieves this (DHs). The method not only allows for the fixation of uncharacterised germplasm, but it also gives breeders with a breeding tool that, when used on plants with known backgrounds, allows for the quick creation of chromosome substitutions that will help with individual chromosome breeding.

The suppression of crossover recombination in a selected plant is followed by the regeneration of DHs from spores bearing non-recombinant chromosomes in reverse breeding. The selection of a genotypically uncharacterised plant with a favourable combination of features from a segregating population (in this case, a segregating F_2). In this plant, crossing over is inhibited, and achiasmatic gametes are collected, cultured and used to create DHs. The DH lines can then be utilised to commercially recreate the elite heterozygote.

RB can also be used on plants with a known background in another application. RB can be used to develop chromosomal substitution lines if crossing over is eliminated in the F_1 hybrid generation rather than the F_2 generation. One or more chromosomes from one parent are present in the background of the other parent in these lines. Backcrossing the chromosome substitution lines to the parental lines results in populations that only segregate for the heterozygous chromosome(s). In theory, reverse breeding allows for the complete re-shuffling of chromosomes between two homozygous plants.

6.9.1 Doubled Haploids

Unfertilised ovules (gynogenesis) or microspore and anther cultures (androgenesis) can be used to produce doubled haploid plants as a result of achiasmatic meiosis, according to well-established protocols that have been devised for a range of plant species, including crops (Jain et al. 1996). The efficiency with which DH is formed from haploid spores varies by species (Forster et al. 2007). The non-recombinant parental chromosomes are a distinctive feature of DHs generated from spores formed during achiasmatic meiosis. The development of RB is restricted to crops where DH technology is widely used. This technology is well-established for the vast majority of crop species, and professional breeding companies commonly employ it in their breeding efforts (Maluszynski et al. 2003; Forster et al. 2007).

6.9.1.1 Regeneration of Doubled Haploid Plants in Watermelon

Several generations are necessary to make the breeding parents homozygous, traditional watermelon breeding through sexual hybridisation is slow and cannot fulfil the needs of today's market. Using homozygous double haploid plants as parents could cut down on the number of generations needed, allowing breeding programmes to go more quickly. To reduce the watermelon breeding cycle, double haploid plants might be employed as parents.

Gynogenesis, also known as plant regeneration from unfertilised female gametophytes, has been frequently employed in vitro ovule or ovary cultures to create haploid embryos and plants. Breeders and researchers can use in vitro biotechnological techniques to create haplotypes, which can be screened for superior appearance, flavour and disease resistance more rapidly and efficiently. Up to now, haploid breeding has been used on several species of Cucurbitaceae family (Min et al. 2016; Rakha et al. 2012), Cucurbita family (Kurtar et al. 2017; Wei et al. 2012), as well as other family.

The advancement of watermelon breeding and cultivation will be aided by the development of in vitro procedures for watermelon haploid generation, which is a noteworthy achievement in the field of biotechnology. Despite the fact that a number of species have been successfully obtained through haploid cultivation using biotechnology techniques, regenerated plants induction of watermelon from haploid still faces significant challenges. It has been reported that a number of factors influence induction, including maternal genotypes, physiology status donor plants, developmental stage of embryo sac, medium component and culture cohesion. In this context, Zhu et al. (2019) performed an experiment to obtain double haploid not only by callus pathway re-differentiation but also through direct embryoid. They talked about the following topics:

(1) Ovule expansion as a result of sterilisation, culture conditions and hormones.
(2) The role of culture media and donor genotype in embryoid induction.
(3) Ploidy and homozygozity in regeneration plants.

They standardised the best medium and culture conditions for watermelon double haploid breeding.

To test the effect of sterile period of sodium hypochlorite on the rate of ovule expansion, Zhu et al. (2019) sterile the ovary sections for 10 or 15 min. When the hypochlorite concentration is 10% and the sterile period is 10 min, the ratio of ovule enlargement is highest (62.12%), but when the sodium hypochlorite concentration is 10% and the sterile period is 15 min, the ratio of ovule enlargement is lowest (30.48%). This is because high concentrations of sodium hypochlorite sterilisation for a long time reduced the ratio of ovule enlargement, whereas low concentrations of sodium hypochlorite sterilisation for a long period increased the ratio of ovule enlargement. The results could be attributed to the fact that sodium hypochlorite has a dual effect on ovules; it not only acts as a disinfectant, but it also suppresses ovarian function.

The sections of ovary were cultured under varied dark times to better understand the function of dark culture on ovule expansion. The ratio of ovule enlargement rose considerably after 7 and 14 days of dark culture. There was no significant difference in the ratio of ovule enlargement at 7 and 14 days. The findings showed that dark culture for a specific period of time can enhance ovule expansion. They also discovered that dark culture for 14 days had the best effect on ovule enlargement, suggesting that dark culture could reduce callus development while still promoting ovule enlargement (Zhu et al. 2019).

Eight different differentiation mediums were employed to investigate the impact of differentiation media on embryoid induction. The media with the largest callus percentage was DM1 (NAA 0.1 mg/L and BA 0.5 mg/L), the media with the highest browning mortality was DM7 (BA 4 mg/L, GA3 5 mg/L and Adenine 30 mg/L), and the medium with the greatest embryo rate was DM5 (NAA 0.5 mg/L, BA 1 mg/L and KT 0.5 mg/L). The culture medium is an important aspect in controlling gynogenesis in vitro, and the composition of culture media components also helps to the advancement of gynogenic procedures. The customised medium is made up of organic nitrogen and carbohydrates that can be changed. They initially tested the optimal media for watermelon ovule enlargement, and the results revealed that M2 (0.02 mg/L TDZ, 0.5 mg/L NAA, 0.5 mg/L 6-B) medium, which was rich in organic matter, was suitable for ovule expansion. In vitro gynogenesis can be triggered as a result of the findings.

The ovary sections of four watermelon cultivars were planted in seven different types of induction medium to see how they affected embryoid. In five types of induction medium (EI1, EI2, EI4, EI5, EI6), the ovary sections could grow into regenerated plants and bud, with EI6 (TDZ 0.02 mg/L, NAA 0.5 mg/L, BA 0.5 mg/L) having the best impact. There were no regenerated plants in the ovaries of EI3 and EI0. In vitro, the hormones TDZ, NAA, BA and KT are commonly employed in gynogenesis. Induction and regeneration media frequently contain TDZ, an active growth regulator. The hormones NAA, BA and KT are important for unpollinated ovule and ovary culture. The concentrations of TDZ, NAA, BA, and KT affect ovule enlargement in the current study, with TDZ and BA having the greatest influence. The findings revealed that the right hormone concentration is required for watermelon ovule growth.

T02 (Zhengkangjufeng, oval shape), T03 (Zhongke NO.2, round shape), and T04 (Xinong NO.8, oval shape) samples were planted in EI5 (TDZ 0.05 mg/L, NAA 0.5 mg/L, BA 1.0 mg/L) and EI6 (TDZ 0.02 mg/L, NAA 0.5 mg/L, BA 0.5 mg/L) medium to see if genotype influences the embryoid T04 had the highest ratio of embryoid induction (1.67%), whereas T02 had no embryoid in EI5 medium. In EI6 medium, T03 had a greater ratio of embryoid induction (1.67%), while T02 and T04 had a lower ratio (0.83%). According to the findings, donor genotype is the most important factor impacting gynogenesis (Gueye and Ndir 2010; Tantasawat et al. 2015). Zhu et al. (2019) discovered that various samples in the same culture media had varying ovule enlargement and embryoid induction ratios. The surrounding environment, such as the embryo's exact month, the culture season, and the culture region, has also been shown to influence the gynogenesis outcome. Summer, for example, has been shown to have more available time for the excision of embryos than spring and winter (Lim and Earle 2008). Furthermore, seedlings that have been treated with culture media have a higher likelihood of doubling. The ploidy level of regenerated plants was determined using a flow cytometer.

Finally, Zhu et al. (2019) found that sterilising with 10% sodium hypochlorite for 10 min is the best method for ovarian expansion. The ovarian expansion was aided by the dark culture. The embryoid differentiation was aided by the MS medium, which contained 0.5 mg/L NAA, 1.0 mg/L 6-BA, and 0.5 mg/L KT. Complete plants may be grown in the M2 medium, which contains 0.02 mg/L TDZ, 0.5 mg/L NAA, and 0.5 mg/L 6-BA. The tetraploid, haploid and diploid plants obtained in this study Seedlings treated with culture medium increased the possibility of doubling. The diploid plants were homozygous double haploid plants.

Lee Hickey and colleagues recently proposed the concept of 'speed breeding,' a non-GMO method that allows researchers to turn over many generations and select plants for desired features among many variations (Voss-Fels et al. 2019). This method uses controlled climatic conditions and long photoperiods to produce four to six generations of long-duration crops per year.

6.10 Speed Breeding a Tool for Accelerated Plant Breeding

Researchers at the University of Queensland established the phrase 'speed breeding' in 2003 to describe a combination of technologies designed to speed up wheat breeding. For numerous crops, speed breeding procedures are now being developed (Ghosh et al. 2018).

Another new technology ideal for watermelon breeding is speed breeding, which can be easily implemented under controlled settings in four to six generations to accelerate genetic gain in breeding and reduce the generation period of a breeding cycle from several years to three to four years.

Watermelon improvement requires eight to ten generations of traditional breeding to achieve a homozygous genotype with desired features. It will be challenging to meet global demand in a short period of time at current rate of improvement. When contrasted to cereal and oilseed crops such as rice, wheat, pearl millet and mustard, speed breeding in vegetable crops, particularly watermelon, has received little attention. Speed breeding has a lot of potential for speeding up the traditional breeding process.

The number of generations that can be cultivated in a year is determined by the length of time it takes for the crop to mature. Each generation starts with the sowing of seeds from the previous generation in order to harvest seeds for the next generation. More generations can be grown in a year if this duration is cut short by getting early flowering and seed set. For example, by manipulating the photoperiod so that early flowering occurs, speed breeding can achieve up to six generations per year for spring wheat (*Triticum aestivum*), durum wheat (*T. durum*), barley (*Hordeum vulgare*), chickpea (*Cicer arietinum*) and pea (*Pisum sativum*) and four generations for canola (*Brassica napus*) instead of two to three under normal glasshouse conditions (Watson et al. 2018). In watermelon, no research has been done on photoperiod alteration to induce earlier flowering.

The principle behind speed breeding is to increase the rate of photosynthesis, which directly stimulates early flowering, by using optimal light intensity, temperature and daytime length control (22 h light, 22 °C day/17 °C night and high light intensity), combined with annual seed harvesting to shorten the generation time (Chiurugwi et al. 2018). Flowering is controlled by the intensity and wavelength of light (Weller et al. 2001). Croser et al. (2016) used various sections of the light spectrum to generate early- and late-flowering genotypes for peas, faba beans and lupins under controlled settings (blue and far red-improved LED lights and metal halide). The proportion of far red-red in these species exhibited a positive correlation with the decreasing red: far red proportion (R:FR). As a result, light with the highest power in the FR region is the most inductive (Ribalta et al. 2017). In general, high R:FR light (e.g. from fluorescent lamps) limits stem enlargement while increasing lateral branching, whereas low R:FR light (e.g. from incandescent lights) greatly enhances stem elongation while inhibiting lateral branching and flowering. FR controls this activity, and phytochrome FR is mediated by blue light (Pfr). Furthermore, phytochrome R (Pr) mediates the action of R light on flowering suppression (Moe and Heins 1990).

Short days or vernalisation, such as RGA, have been created as species-specific procedures to induce early flowering using certain environmental signals (Samineni et al. 2019). Greenhouse tactics were compared to in vitro plus in vivo strategies and fast generation cycling by extending the photoperiod under controlled conditions (Ochatt et al. 2002).

Speed breeding had previously been shown to cut generation time by lengthening photoperiods, while specific crop species, including radish (*Raphanus sativus*), pepper (*Capsicum annum*) and green vegetables like Amaranth (*Amaranthus* spp.), responded well to extended day duration (Stetter et al. 2016). Because of their flowering requirements, short-day crops have been slow to breed. However, Lee Hickey and his research team have recently been working on creating methods for short-day crops.

O'Connor et al. (2013) have already reported positive results in peanut speed breeding (*Arachis hypogaea*). Amaranth (*Amaranthus* spp.) was able to produce more generations per year as the day length extended. It is possible to generate successive generations of improved watermelon crops for field testing via SSD using speed breeding, which is less expensive than producing DHs.

For speed and precision breeding, other existing technologies such as marker-assisted selection, genomic selection and CRISPR gene editing can be combined.

6.11 Conclusion

The mode of plant reproduction, the heritability of the trait(s) to be improved, and the type of cultivar utilised commercially all influence the breeding or selection methods used (e.g. F_1 hybrid cultivar, pureline cultivar, etc.). Selection should be based on mean values obtained from replicated evaluations of families of related plants for traits with low heritability, whereas for highly heritable traits, a choice of superior individual plants evaluated at a single location will be effective. Pedigree selection, modified pedigree selection, mass selection, and recurrent selection are all common selection methods.

A frequent, objective evaluation of the breeding procedure's efficiency should be included in every breeding effort. The number of successful cultivars produced per unit of input should be included in the evaluation criteria, which vary depending on the goal and objectives. Gain from selection per year based on comparisons to an appropriate standard, overall value of advanced breeding lines, and number of Successful cultivars produced per unit of input should all be included (e.g. per year, per dollar expended, etc.). Advanced breeding lines that show promise are rigorously examined and compared to relevant standards for at least three years in environments representative of the commercial target area(s). The best lines are used as parents to develop new populations for further selection; those that are still lacking in a few traits are utilised as parents to develop new populations for further selection. Watermelon plant breeding aims to create new, one-of-a-kind, and superior watermelon inbreds and hybrids. The breeder first chooses and crosses two or more parental lines, then repeats the process of selfing and selection.

From the time the first cross is formed, these processes, which lead to the final stage of marketing and distribution, normally take eight to twelve years. As a result, developing new cultivars is a time-consuming process that necessitates meticulous ahead planning, resource efficiency and a minimum of direction changes. In this regard, speed breeding is a potential method for generating elite watermelon cultivars in a short period of time.

The benefit of conventional plant breeding is that it increases the genetic resources available for crop development by incorporating desired traits. Some plants, on the other hand, are at risk of succumbing to environmental stress and losing genetic diversity (Basey et al. 2016). As a result, traditional agricultural practices are unable to address global food security concerns. Although combining numerous phenotypic traits within a single-plant variety has been shown to increase yield, new breeding procedures are less expensive and will allow for speedier production of genetically enhanced crops (Krimsky 2019).

A trait (e.g., stress tolerance) can be increased by cross-breeding the best hybrid progeny with the desired characteristic (Dolferus et al. 2011). To eliminate undesired phenotypic combinations, desired features can also be introduced into a chosen 'best' recipient line by backcrossing the selected progeny with the recipient line for several generations (Caligari and Brown 2016). The introduction of new genes is essential for the enhancement of desirable features via speed breeding because genetic variability can be reduced by using long-term traditional breeding procedures.

Breeders prefer crops with shorter reproductive cycles because they allow for the production of more generations in a single year, resulting in faster artificial breeding of desirable phenotypes than crops that only reproduce once a year or perennial plants that only reproduce every few years (Abreu et al. 2010). Plant breeding, when integrated with genomic research, improves breeding accuracy and saves time (Doust and Diao 2017).

References

Abbo S, Pinhasi van-Oss R, Gopher A, Saranga Y, Ofner I, Peleg Z (2014) Plant domestication versuscrop evolution: a conceptual framework for cereals and grain legumes. Trends Plant Sci 19(6):351–360. http://doi.org/10.1016/j.tplants.2013.12.002

Abdel-Magid A (1989) Plant domestication in the Middle Nile Basin. BAR International Series 523, Oxford, UK

Abreu GB, Ramalho MAP, Toledo FHRB, De Souza JC (2010) Strategies to improve mass selection in maize. Maydica 55:219–225

Adams MJ, Lefkowitz EJ, King AMQ, Harrach B, Harrison RL, Knowles NJ, Kropinski AM, Krupovic M, Kuhn JH, Mushegian AR, Nibert M (2017) Changes to taxonomy and the international code of virus classification and nomenclature ratified by the international committee on taxonomy of viruses. Arch Virol 162:2505–2538

Akhtar MS, Goldschmidt EE, John I, Rodoni SM, P and Grierson, D. (1999) Altered patterns of senescence and ripening in gf, a stay-green mutant of tomato (*Lycopersiconesculentum* Mill.). J Exp Bot 50:1115–1122

Ali A, Mohammad O, Khattab A (2012) Distribution of viruses infecting cucurbit crops and isolation of potential new virus-like sequences from weeds in Oklahoma. Plant Dis 96:243–248

Andrews FW (1950) The flowering plants of Anglo-Egyptian Sudan. Buncle, Arbroath, UK

Anonymous (2008) Annual report of all India coordinated research project on vegetable crops, 2008–09. ICAR-Indian Institute of Vegetable Research, Varanasi, U.P., India. Available at: https://www.iivr.org.in/annual-reports

Apak R, Guclu K, Demirata B, Ozyurek M, Celik SE, Bektasoglu B, Berker KI, Ozyurt D (2007) Comparative evaluation of various total antioxidant capacity assays applied to phenolic compounds with the CUPRAC assay. Molecules 12:496–547

Artés-Hernández F, Robles PA, Gómez P, Callejas A, Artés F (2010) Low UV-C illumination for keeping overall quality of fresh-cut watermelon. Postharvest Biol Technol 55:114–120

Aveskamp M, De Gruyter J, Woudenberg J, Verkley G, Crous PW (2010) Highlights of the Didymellaceae: a polyphasic approach to characterise *Phoma* and related *Pleosporalean* genera. Stud Mycol 65:1–60

Bangalore DV, McGlynn WG, Scott DD (2008) Effects of fruit maturity on watermelon ultrastructure and intracellular lycopene distribution. J Food Sci 73:S222–S228

Bang H, Davis AR, Kim S, Leskovar DI, King SR (2010) Flesh color inheritance and gene interactions among canary yellow, pale yellow, and red watermelon. J Am Soc Hortic Sci 135(4):362–368

Basey AC, Fant JB, Kramer AT (2016) Producing native plant materials for restoration: 10 rules to collect and maintain genetic diversity. Native Plants J 16:37–53

Ben-Naim Y, Cohen Y (2015) Inheritance of resistance to powdery mildew race 1W in watermelon. Phytopathology 105:1446–1457

Bowen P, Chen L, Stacewicz-Sapuntzakis M, Duncan C, Sharifi R, Ghosh L et al (2002) Tomato sauce supplementation and prostate cancer: lycopene accumulation and modulation of biomarkers of carcinogenesis. Exp Biol Med 227:886–893

Boyhan G, Norton JD, Abrahams BR (1994) Screening for resistance to anthracnose (race 2), gummy stem blight, and root knot nematode in watermelon germplasm. Cucurb Genet Coop Rep 17:106–110

Branham SE, Levi A, Farnham MW, Wechter WP (2016) A GBS-SNP-based linkage map and quantitative trait loci (QTL) associated with resistance to *Fusarium oxysporum* f. sp. *niveum* race 2 identified in *Citrullus lanatus* var. *citroides*. Theor Appl Genet. http://doi.org/10.1007/s00122-016-2813-0

Brar JS, Nandpuri KS (1977) A note on the heritance of total soluble solids in watermelon (*Cirtrullus lantus* (Thunb.) Mansf.). Haryana J Hortic Sci 6:193–195

Braun AE (1942) Resistance of watermelon to the wilt disease. Am J Bot 27:683–684

Brewer MT, Rath M, Li HX (2015) Genetic diversity and population structure of cucurbit gummy stem blight fungi based on microsatellite markers. Phytopathology 105:815–824

Bruton BD, Fish WW, Roberts W, Popham TW (2009) The influence of rootstock selection on fruit quality attributes of watermelon. Open Food Sci J 3:15–34

Burkill HM (1985) The useful plants of west tropical Africa, 2nd edn. Royal Botanic Gardens, Kew

Bush A (1978) Citron melon for cash and condiment. Econ Bot 32:182–184

Caligari PDS, Brown J (2016) Plant breeding, practice. In: Encyclopaedia of applied plant sciences, vol 2. Academic Press, Cambridge, MA, USA, pp 229–235

Chandrika UG, Fernando KSSP, Ranaweera KKDS (2009) Carotenoid content and in vitro bio accessibility of lycopene from guava (*Psidium guajava*) and watermelon (*Citrullus lanatus*) by high-performance liquid chromatography diode array detection. Int J Food Sci Nutr 60:558–566

Charoensiri R, Kongkachuichai R, Suknicom S, Sungpuag P (2009) Beta-carotene, lycopene, and alpha-tocopherol contents of selected. Thai fruits. Food Chem 113:202–207

Chiurugwi T, Kemp S, Powell W, Hickey LT, Powell W (2018) Speed breeding orphan crops. Theor Appl Genet 132(3):607–616

Chomicki G, Renner SS (2015) Watermelon origin solved with molecularphylogenetics including Linnaean material: another example of museomics. New Phytol 205(2):526–532. http://doi.org/10.1111/nph.13163

Choo WS, Sin WY (2012) Ascorbic acid, lycopene and antioxidant activity of red-fleshed and yellow-fleshed watermelons. Adv Appl Sci Res 3(5):2779–2784

Choudhary R, Bowser TJ, Weckler P, Maness NO, McGlynn W (2009) Rapid estimation of lycopene concentration in watermelon and tomato puree by fiber optic visible reflectance spectroscopy. Postharvest Biol Technol 52:103–109

Choudhary BR, Haldhar SM, Maheshwari SK, Bhargava R, Sharma SK (2015) Phytochemicals and antioxidants in watermelon (*Citrullus lanatus*) genotypes under hot arid region. Indian J Agric Sci 85 (3):414–417

Coffey JL, Simmons AL, Shepard BM, Tadmor Y, Levi A (2015) Potential sources of whitefly (Hemiptera: Aleyrodidae) resistance in desert watermelon (*Citrullus colocynthis*) Germplasm. Hort Sci 50:13–17

Collins JK, Davis AR, Adams A, Perkins-Veazie P (2005) Sensory evaluation of low sugar watermelon by consumers. Hort Sci 40:883

Cox A, van der Veen M (2008) Changing foodways: watermelon (*Citrullus lanatus*) consumption in Roman and Islamic Quseir al-Qadim, Egypt. Veg Hist Archaeobot 17(S1):181–189. https://doi.org/10.1007/s00334-008-0164-8

Croser JS, Pazos-Navarro M, Bennett RG, Tschirren S, Edwards K, Erskine W, Creasy R, Ribalta FM (2016)

Time to flowering of temperate pulses in vivo and generation turnover in vivo–in vitro of narrow-leaf lupin accelerated by low red to far-red ratio and high intensity in the far-red region. Plant Cell Tissue Organ Cult 127:591–599

Dahl Jensen B, Maïga Touré F, Ag Hamattal M, Aya Touré F, Dolo Nantoumé A (2011) Watermelons in the Sand of Sahara: cultivation and use of indigenous landraces in the Tombouctou Region of Mali. Ethnobot Res Appl 9:151–162

Dane F, Liu J (2007) Diversity and origin of cultivated and citron type watermelon (*Citrullus lanatus*). Genet Resour Crop Evol 54:1255–1265

Davis AR, Thomas CE, Levi A, Bruton BD, Pair SD (2002) Watermelon resistance to powdery mildew race 1. In: Maynard DN (ed) Cucurbitaceae '02. ASHS Press, Alexandria, VA, pp 192–198

Davis AR, Levi A, Tetteh A, Wehner T, Russo V, Pitrat M (2007) Evaluation of watermelon and related species for resistance to race 1W powdery mildew. J Am Soc Hort Sci 132:790–795

De Vartavan C, Asensi Amoros V (1997) Codex of ancient Egyptian plant remains. TriadeExploration, London

De Winter B (1990) Notes on African plants. Cucurbitaceae. A new species of *Citrullus* (Benincaseae) from the Namib Desert, Namibia. Bothalia 20:209–211

Desbiez C, Lecoq H (1997) Zucchini yellow mosaic virus. Plant Pathol 46:809–829

Dirks RHG, van Dun CMP, Reinink K (2003) Rijk Zwaan Zaadteelt enZaadhandel B.V. Reverse Breeding, WO/2003/017753

Dolferus R, Ji X, Richards RA (2011) Abiotic stress and control of grain number in cereals. Plant Sci 181:331–341

Dou J, Zhao S, Lu X, He N, Zhang L, Ali A, Kuang H, Liu W (2018) Genetic mapping reveals a candidate gene (ClFS1)for fruit shape in watermelon (*Citrullus lanatus* L.). Theor Appl Genet 131:947–958

Doust A, Diao X (2017) Plant genetics and genomics: crops and models. Genet Genom Setaria Ser 19:377

Ebrahimzadeh MA, Pourmorad F, Bekhradnia AR (2008) Iron chelating activity screening, phenol and flavonoid content of some medicinal plants from Iran. Afr J Biotech 32:43–49

Edwards AJ, Vinyard BT, Wiley ER, Brown ED, Collins JK, Perkins-Veazie P et al (2003) Consumption of watermelon juice increases plasma concentrations of lycopene and β-carotene in humans. J Nutr 133:1043–1050

El-Hafez AAA, Gaafer AK, Allam AMM (1981) Inheritance of flesh color, seed coat cracks and total soluble solids in watermelon and their genetic relations. Acta Agron Acad Hung 30:82–86

Feliks Y (2005) Talmud Yerushalmi: Massekhet Ma'asrot. Bar-Ilan, Ramat Gan, Israel

Fenko A, Schifferstein HN, Huang TC, Hekkert P (2009) What makes products fresh: the smell or the colour. Food Qual Prefer 20:372–379

Fish WW, Davis AR (2003) The effects of frozen storage conditions on lycopene stability in watermelon tissue. J Agric Food Chem 51:3582–3585

Forster BP, Heberle-Bors E, Kasha KJ, Touraev A (2007) The resurgence of haploids in higher plants. Trends Plant Sci 12:368–375

Germer R (1988) Katalog der Altägyptischen Pflanzenreste der Berliner Museeum. A gyptologische Abhandlunger, vol 47. Otto Harrassowitz, Wiesbaden, Germany

Ghosh S, Watson A, Gonzalez-Navarro OE, Ramirez-Gonzalez RH, Yanes L, Mendoza-Suárez M, Simmonds J, Wells R, Rayner T, Green P et al (2018) Speed breeding in growth chambers and glasshouses for crop breeding and model plant research. Nat Protoc 13:2944–2963

Gillaspie AG, Wright JM (1993) Evaluation of *Citrullus* sp. germplasm for resistance to watermelon mosaic virus 2. Plant Dis 77:352–354

Grassi S, Piro G, Lee JM, Zheng Y, Fei ZJ, Dalessandro G et al (2013) Comparative genomics reveals candidate carotenoid pathway regulators of ripening watermelon fruit. BMC Genom 14(1):781

Gueye T, Ndir KN (2010) In vitro production of double haploid plants from two rice species (*Oryza sativa* L. and *Oryza glaberrima* Steudt.) for the rapid development of new breeding material. Sci Res Essays 5 (7):709–713

Guner N, Wehner TC (2003) Gene list for watermelon. Cucurbit Genet Cooper Rep 26:76–92

Guner N, Wehner TC (2004) The genes of watermelon. Hort Sci 39:1175–1182

Guner N, Wehner TC (2008) Overview of potyvirus resistance in watermelon. In: M Pitrat (ed) Cucurbitaceae 2008: Proceedings of the IXth EUCARPIA meeting on genetics and breeding of Cucurbitaceae. Avignon, France, pp 445–452

Guner N, Pesic-VanEsbroeck Z, Rivera-Burgos LA, Wehner TC (2018) Inheritance of resistance to Papaya ringspot virus-watermelon strain in watermelon HortScience 53:624–627

Guo S, Zhang J, Sun H, Salse J, Lucas WJ, Zhang H, Zheng Y, Mao L, Ren Y, Wang Z et al (2013) The draft genome of watermelon (*Citrullus lanatus*) and resequencing of 20 diverse accessions. Nat Genet 45 (1):51–58. http://doi.org/10.1038/ng.2470

Guo S, Zhao S, Sun H et al (2019) Resequencing of 414 cultivated and wild watermelon accessions identifies selection for fruit quality traits. Nat Genet 51:1616–1623

Gusmini G, Wehner T (2005) Foundations of yield improvement in watermelon. Crop Sci 45:141–146

Gusmini G, Wehner TC (2006) Qualitative inheritance of rind pattern and flesh color in watermelon. J Hered 97:177–185

Gusmini G, Wehner TC, Jarret RL (2004) Inheritance of egusi seed type in watermelon. J Hered 95(3):268–270. http://doi.org/10.1093/jhered/esh03

Gusmini G, Song R, Wehner TC (2005) New sources of resistance to gummy stem blight in watermelon. Crop Sci 45(2):582–588

Gusmini G, Rivera-Burgos LA, Wehner TC (2017) Inheritance of resistance to gummy stem blight in watermelon. HortScience 52(11):1477–1482
Henderson WR (1992) Corrigenda to 1991 watermelon gene list (CGC 14:129–137). Cucurbit Genet Cooper Rep 15:110
Henderson WR, Scott GH, Wehner TC (1998) Interaction of flesh color genes in watermelon. J Hered 89:50–53
Holkar SK, Choudhary H, Mandal B, Jain RK (2018) Evaluation of watermelon genotypes against bud necrosis disease caused by a distinct watermelon bud necrosis orthotospovirus under field and glasshouse conditions. Sci Hortic 235:106–115
Huh YC, Solmaz I, Sari N (2008) Morphological characterization of Korean and Turkish watermelon germplasm. In: Pitrat M (ed) Cucurbitaceae 2008. Proceedings of the IXth EUCARPIA meeting on genetics and breeding of Cucurbitaceae, Avignon (France), May 21–24, pp 327–333
Hussain AI, Rathmore HA, Sattar MZ, Chatha SA, Sarker SD, Gilani AH (2014) *Citrullus colocynthis*(L.) Schrad. (Bitter apple fruit): a review of its phytochemistry, pharmacology, traditional uses and nutritional potential. J Ethnopharmacol 8:54–66
Ishida BK, Bartley GE (2005) Carotenoids: chemistry, sources, and physiology. In: Caballero B, Allen L, Prentice A (eds) Encyclopedia of human nutrition, 2nd edn., vol 1. Elsevier, Amsterdam, pp 330–338
Jain S, Sopory SK, Veilleux RE (1996) In vitro haploid production in higherplants. In: Current plant science and biotechnology in agriculture. Kluwer Academic Publishers, Dordrecht, Boston, London
Jain RK, Pappu HR, Pappu SSK, Reddy M, Vani A (1998) Watermelon bud necrosis tospovirus is a distinct virus species belonging to serogroup IV. Arch Virol 143:1637–1644
Jain RK, Bag S, Umamaheswaran K, Mandal B (2007) Natural infection by tospoviruses of cucurbitaceous and fabaceous vegetable crops in India. J Phytopathol 155:22–25
Jain RK, Mandal B, Pappu HR, Holkar SK (2014) ICTV taxonomic proposal 2014.005aV.A.v2. Tospovirus_sp. Create 1 new species in the genus Tospovirus, family *Bunyaviridae*. http://www.ictvonline.org/proposals-14/2014.005aV.A.v2.Tospovirus_sp.pdf
Jamatia J, Choudhary H, Saha K, Basavaraj B (2022) Identification of new sources of resistance to watermelon bud necrosis virus (WBNV). Indian Phytopathol. https://doi.org/10.1007/s42360-022-00486-9
Jarret RL, Newman M (2000) Phylogenetic relationships among species of *Citrullus* and the placement of *C. rehmii* De Winter as determined by internal transcribed spacer (ITS) sequence heterogeneity. Genet Resour Crop Evol 47:215–222
Jarret RL, Merrick LC, Holms T, Evans J, Aradhya MK (1997) Simple sequence repeats in watermelon [*Citrullus lanatus* (Thunb.) Matsum. & Nakai]. Genome 40:433–441
Jaskani MJ, Kwon SW, Kim DH (2005) Comparative study on vegetative, reproductive and qualitative traits of seven diploid and tetraploid watermelon lines. Euphytica 145:259–268
Jaskani MJ, Kwon SW, Kim DH (2005) Comparative study on vegetative, reproductive and qualitative traits of seven diploid and tetraploid watermelon lines. Euphytica 145: 259–268
Jaskani MJ, Kwon SW, Kim DH, Abbas H (2006) Seed treatments and orientation affects germination and seedling emergence in tetraploid watermelon. Pak J Bot 38:89
Jeffrey C (1967) *Cucurbitaceae*. In: Milne-Redhead E, Polhill RM (eds) Flora of Tropical East Africa. Published under the Authority of the Minister for Overseas Development. Crown Agents for Oversea Governments and Administration, London
Ji G, Zhang J, Gong G, Shi J, Zhang H, Ren Y, Guo S, Gao J, Shen H, Xu Y (2015) Inheritance of sex forms in watermelon (*Citrullus lanatus*). Sci Hortic 193:367–373
Kang BS, Zhao WE, Hou YB, Tian P (2010) Expression of carotenogenic genes during the development and ripening of watermelon fruit. Sci Hortic 124:368–375
Kaur C, Kapoor HC (2001) Antioxidants in fruits and vegetables—the millennium's health. Int J Food Sci Technol 36:703–725
Keimer L (1924) Die Gartenpflanzen im Alten Agypten. Hoffmann & Campe, Hamburg, Germany
Keinath AP (2011) From native plants in central Europe to cultivated crops worldwide: the emergence of *Didymella bryoniae* as a cucurbit pathogen. Hort Sci 46(4):532–535
Keinath AP, Dubose VB (2004) Evaluation of fungicides for prevention and management of powdery mildew on watermelon. Crop Prot 23:35–42
Kenigsbuch D, Cohen Y (1987) Inheritance of gynoecious sex type in muskmelon. Cucurbit Genet Coop Rep 10:47–48
Kenigsbuch D, Cohen Y (1990) The inheritance of gynoecy in muskmelon. Genome 33(3):317–320
Kihara H (1951) Triploid watermelons. Proc Am Soc Hort Sci 58:217–230
Kim KH, Ahn SG, Hwang JH, Choi YM, Moon HS, Park YH (2013) Inheritance of resistance to powdery mildew in the watermelon and development of a molecular marker for selecting resistant plants. Hort Environ Biotechnol 54(2):134–140
Kousik CS, Shepard BM, Hassell R, Levi A, Simmons AM (2007) Potential sources of resistance to broad mites (*Polyphagotarsonemus latus*) in watermelon germplasm. Hort Sci 42:1539–1546
Kousik CS, Adkins S, Turechek WW, Roberts PD (2009) Sources of resistance in U.S. plant introductions (PI) to watermelon vine decline caused by *Squash Vein Yellowing Virus*. Hort Sci 44:256–262
Kousik CS, Ling K, Adkins ST, Webster CG, Turechek W (2014) *Phytophthora* fruit rot-resistant watermelon germplasm lines: USVL489-PFR,

USVL782-PFR, USVL203-PFR, and USVL020-PFR. Hort Sci 49:101–104

Krimsky S (2019) Traditional plant breeding. In: GMOs decoded. MIT Press, Cambridge, MA, USA

Krishna Kumar NK, Venkatesh N, Kallesh-waraswamy CM, Ranganath HR (2006) Seasonal incidence of thrips and bud necrosis virus on watermelon. Pest Manag Hortic Ecosyst 12(2):85–92

Kubicki B (1969a) Investigations on sex determination in cucumber (*Cucumis sativus* L.). III. Variability of sex expression in the monoecious and gynoecious lines. Genet Pol 10:5–22

Kubicki B (1969b) Investigations on sex determination in cucumber (*Cucumis sativus* L.). V. Genes controlling intensity of femaleness. Genet Pol 10:69–86

Kubicki B (1969c) Investigations on sex determination in cucumber (*Cucumis sativus* L.). VI. Androecism. Genet Pol 10:87–100

Kubicki B (1969d) Investigations on sex determination in cucumber (*Cucumis sativus* L.). VII. Andromonoecism and hermaphroditism. Genet Pol 10

Kumar R, Wehner TC (2011) Discovery of second gene for soliddark green versus light green rind pattern in watermelon. J Hered 102:489–493

Kumar R, Mandal B, Geetanjali AS, Jain RK, Jaiwal PK (2010) Genome organisation and sequence comparison suggest intra-species incongruence in m RNA of watermelon bud necrosis virus. Arch Virol 155:1361–1365

Kunkalikar SR, Sudarsana P, Arun BM, Rajagopalan P, Chen TC, Yeh SD, Naidu RA, Zehr UB, Ravi KS (2011) Importance and genetic diversity of vegetable infecting tospoviruses in India. Phytopathology 101:367–376

Kurtar ES, Balkaya A, Ozer MO (2017) Production of callus mediated gynogenic haploids in WinterSquash (Cucurbita maxima Duch.) and pumpkin (Cucurbita moschata Duch.). Czech J Genet Plant Breed 54 (1):2018. https://doi.org/10.17221/30/2017-CJGPB

Lambel S, Lanini B, Vivoda E, Fauve JP, Wechter W, Harris-Shultz KR, Massey L, Levi A (2014) A major QTL associated with *Fusarium oxysporum* race 1 resistance identified in genetic populations derived from closely related watermelon lines using selective genotyping and genotyping-by-sequencing for SNP discovery. Theor Appl Genet 127:2105–2115. https://doi.org/10.1007/s00122-014-2363-2

Lee WM, Son SH, Park DK, Lee HJ, Huh YC, Cho IH (2010) Selection of watermelon germplasm with disease resistance to powdery mildew (*Sphaerotheca fusca*). Kor J Hort Sci Technol 28(Suppl. II):73 (Abstr.)

Leskovar DI, Bang H, Crosby KM, Maness N, Franco JA, Perkins-Veazie P (2004) Lycopene, carbohydrates, ascorbic acid and yield components of diploid and triploid watermelon cultivars are affected by deficit irrigation. J Hortic Sci Biotechnol 79:75–81

Levi A, Thomas CE, Keinath AP, Wehner TC (2001a) Genetic diversity among watermelon (*Citrullus lanatus* and *Citrullus colocynthis*) accessions. Genet Resour Crop Evol 48:559–566

Levi A, Thomas CE, Wehner TC, Zhang X (2001b) Low genetic diversity indicates the need to broaden the genetic base of cultivated watermelon. Hort Sci 36:1096–1101

Levi A, Thomas CE, Joobeur T, Zhang X, Davis A (2002) A genetic linkage map for watermelon derived from a testcross population: (*Citrullus lanatus* var. *citroides* x *C. lanatus* var. *lanatus*) x *C.colocynthis*. Theor Appl Genet 105:555–563

Levi A, Thomas CE, Newman M, Zhang X, Xu Y, Wehner TC (2003) Massive preferential segregation and non-random assortment of linkage-groups produce quasi-linkage in an F_2 mapping population of watermelon. Hort Sci 38:782

Levi A, Hernandez L, Thimmapuram J, Donthu R, Wright C, Ali C, Wechter WP, Reddy U, Mikel M (2011a) Sequencing the genome of the heirloom watermelon cultivar Charleston Gray. In: XX plant and animal genome conference, San Diego, p P047

Levi A, Wechter WP, Massey LM, Carter L, Hopkins D (2011b) Genetic linkage map of *Citrullus lanatus* var. *citroides* chromosomal segments introgressed into the watermelon cultivar Crimson Sweet (*Citrullus lanatus* var. *lanatus*) genome. Am J Plant Sci 2:93–110

Levi A, Thies JA, Wechter WP, Harrison HF, Simmons AL, Reddy UK, Nimmakayala P, Fei Z (2013) High frequency oligonucleotides: targeting active gene (HFO-TAG) markers revealed wide genetic diversity among *Citrullus* spp. accessions useful for enhancing disease or pest resistance in watermelon cultivars. Genet Resour Crop Evol 60:427–440

Levi A, Coffey J, Massey L, Guner N, Oren E, Tadmor Y, Ling K (2016) Resistance to papaya ringspotvirus-watermelon strain (PRSV-W) in the desert watermelon *Citrullus colocynthis*. Hort Sci 51:4–7

Levi A, Jarret R, Kousik S, Wechter WP, Nimmakayala P, Reddy UK (2017) Genetic resources of watermelon. In: Grumet RKN, Garcia-Mas J (eds) Genetics and genomics of Cucurbitaceae. Springer, Cham, pp 87–110

Lim W, Earle ED (2008) Effect of in vitro and in vivo colchicine treatments on pollen production and fruit set of melon plants obtained by pollination with irradiated pollen. Plant Cell Tissue Organ Cult 95(1):115–124. https://doi.org/10.1007/s11240-008-9422-9

Lisa V, Dellvalle G (1981) Characterization of two potyvirus from zucchini squash. Phytopathology 100:279–286

Liu CH, Zhang HY, Dai ZY, Liu X, Liu Y, Deng XX et al (2012) Volatile chemical and carotenoid profiles in watermelons [*Citrullus vulgaris* (Thunb.) Schrad (Cucurbitaceae)] with different flesh colors. Food Sci Biotechnol 21(2):531–541

Liu J, Guo S, He H, Zhang H, Gong G, Ren Y, Xu Y (2013) Dynamic characteristics of sugar accumulation and related enzyme activities in sweet and non-sweet watermelon fruits. Acta Physiol Plant 35:3213–3222

Liu S, Gao P, Wang XZ, Davis AR, Baloch AM, Luan FS (2015) Mapping of quantitative trait loci for lycopene content and fruit traits in *Citrullus lanatus*. Euphytica 202(3):411–426

Lou LL, Wehner TC (2009) Quantitative inheritance of fruit weight and the total soluble solids content in watermelon. Inheritance of fruit characteristics in watermelon [*Citrullus lanatus* (Thunb.) Matsum. & Nakai], pp 95–119

Lou LL, Wehner TC (2016) Qualitative inheritance of external fruit traits in watermelon. Hort Sci 51:487–496

MacDonald GM (2003) Biogeography: space, time, and life. Wiley, New York, p 518

Maluszynski M, Kasha KJ, Szarejko I (2003) Published protocols for other crop plant species. In: Maluszynski M, Kasha KJ, Forster BP, Szarejko I (eds) Doubled haploid production in crop plants: a manual. Kluwer Academic Publishers, Dordrecht, The Netherlands, pp 309–336

Mandal B, Jain RK, Chaudhary V, Varma A (2003) First report of natural infection of *Luffa acutangula* by watermelon bud necrosis virus in India. Plant Dis 87:598

Mandal B, Jain RKK, Reddy MK, Krishna NK, Ravi KS, Popup HR (2012) Emerging problems of tospoviruses (*Bunyaviridae*) and their management in the Indian subcontinent. Plant Dis 96:468–479

Manniche L (1989) An ancient Egyptian herbal. British Museum, London

Manzano S, Martínez C, García JM, Megías Z, Jamilena M (2014) Involvement of ethylene in sex expression and female flower development in watermelon (*Citrullus lanatus*). Plant Phyiol Biol 85:96–104

Mariod AA, Ahmed YM, Matthäus B, Khaleel G, Siddig A, Gabra AM, Abdelwahab SI (2009) A comparative study of the properties of six Sudanese cucurbit seeds and seed oils. J Am Oil Chem Soc 86 (12):1181–1188. http://doi.org/10.1007/s11746-009-1459-3

Martyn RD (2012) Fusarium wilt of watermelon: a historical review. In: Cucurbitaceae 2012, Proceedings of the 10th EUCARPIA meeting on genetics and breeding of Cucurbitaceae

Martyn RD, Netzer D (1991) Resistance to races 0, 1, and 2 of Fusarium wilt of watermelon in *Citrullus* sp. PI-296341-FR. Hort Sci 26:429–432

Mayr E (1954) Change of genetic environment and evolution. In: Huxley J (ed) Evolution as a process. George Allen & Unwin, London

Maragal S, Rao ES and Reddy DCL (2019) Genetic analysis of fruit quality traits in prebred lines of watermelon derived from a wild accession of *Citrullus amarus*. Euphytica 215:199

Maynard DN, Hopkins DL (1999) Watermelon fruit disorders. HortTechnology 9:155–161

McGregor C (2012) *Citrullus lanatus* germplasm of Southern Africa. Isr J Plant Sci 60:403–414

Meeuse ADJ (1962) The *Cucurbitaceae* of southern Africa. Bothalia 8:1–111

Min Z, Li H, Zou T, Tong L, Cheng J, Sun X (2016) University HA studies of in vitro culture and plant regeneration of unfertilized ovary of pumpkin. Chin Bull Bot 51(01):74–80

Mortensen A, Skibsted LH, Sampson J, Rice-Evans C, Everett SA (1997) Comparative mechanisms and rates of free radical scavenging by carotenoid antioxidants. FEBS Lett 418: 91–97

Moe R, Heins R (1990) Control of plant morphogenesis and flowering by light quality and temperature. Acta Hortic 81–90

Mort A, Zheng Y, Qiu F, Nimtz M, Bell-Eunice G (2008) Structure of xylogalacturonan fragments from watermelon cell-wall pectin. Endopolygalacturonase can accommodate a xylosyl residue on the galacturonic acid just following the hydrolysis site. Carbohydr Res 343(7):1212–1221

Muschler R (1912) A manual flora of Egypt. R. Friedlaender, Berlin

Mutanen M, Pajari A-M (eds) (2004) Vegetables, whole grains, and their derivatives in cancer prevention (Diet and cancer, vol 2). Springer, Dordrecht

Nagal S, Kaur C, Choudhary H, Singh J, Singh BB, Singh KN (2012) Lycopene content, antioxidant capacity and colour attributes of selected watermelon [*Citrullus lanatus* (Thunb.)] cultivars grown in India. Int J Food Sci Nurtition. http://doi.org/10.3109/09637486.2012.694848

Nagendran KS, Mohankumar R, Aravintharaj CG, Balaji SK, Manoranjitham AK, Singh AB, Rai BS, Karthikeyan G (2017) The occurrence and distribution of major viruses infecting cucurbits in Tamil Nadu state, India. Crop Prot 99:10–16

Nagesh GC, Rao ES, Pitchaimuthu M, Varalakshmi B, Lakshmana Reddy DC, Samuel DK, Rekha A, Krishna Reddy M (2018) Genetic analysis of resistance to watermelon bud necrosis orthotospovirus in watermelon [*Citrullus lanatus* (Thunb.) Matsum. & Nakai]. Plant Breed 137:814–822

Netzer C, Weintall C (1980) Inheritance of resistance in watermelon to race 1 of *Fusarium oxysporum* f. sp. *niveum*. Plant Dis 64:853–854

Nimmakayala P, Tomason YR, Jeong J, Ponniah SK, Karunathilake A, Levi A, Perumal R, Reddy UK (2010) Genetic reticulation and interrelationships among *Citrullus* species as revealed by joint analysis of shared AFLPs and species-specific SSR alleles. Plant Genet Res 8:16–25

Nimmakayala P, Levi A, Abburi L, Abburi VL, Tomason YR, Saminathan T, Vajja VG, Malkaram S, Reddy R, Wehner TC, Reddy UK (2014) Single nucleotide polymorphisms generated by genotyping by sequencing used to characterize genome wide diversity, linkage disequilibrium and selection sweep for worldwide cultivated watermelon. BMC Genomics 15:767

Nimmakayala P, Levi A, Abburi L, Abburi VL, Tomason YR, Saminathan T, Vajja VG, Malkaram S, Reddy R, Wehner TC, Mitchell SE, Reddy UK (2014) Single nucleotide polymorphisms generated by genotyping by sequencing to characterize genome-wide diversity, linkage disequilibrium, and selective sweeps in cultivated watermelon. BMC Genom 15:767

Norton JD (1979) Inheritance of resistance to gummy stem blight in watermelon. Hort Sci 14:630–632

Norton J, Boyhan G, Smith D, Abrahams B (1995) "AU-Sweet Scarlet" watermelon. Hort Sci 30(2):393–394

Ochatt SJ, Sangwan RS, Marget P, Assoumou Ndong Y, Rancillac M, Perney P (2002) New approaches towards the shortening of generation cycles for faster breeding of protein legumes. Plant Breed 121:436–440

O'Connor D, Wright G, George D, Hunter M (2013) Development and application of speed breeding technologies in a commercial peanut breeding program. Peanut Sci 40:107–114

Omoni AO, Aluko RE (2005) The anti-carcinogenic and anti-atherogenic effects of lycopene: a review. Trends Food Sci Technol 16:344–350

Orton WA (1911) The development of disease resistant varieties of plants. In: IV international conference on genetique, Paris, pp 247–265

Pantelidis GE, Vasilakakis M, Manganaris GA, Diamantidis G (2007) Antioxidant capacity, phenol, anthocyanin and ascorbic acid contents in raspberries, blackberries, red currants, gooseberries and cornelian cherries. Food Chem 102:777–783

Pardo JE, Gomez R, Tardaguila J, Amo M, Varon R (1997) Quantity evaluation of watermelon varieties (*Citrullus valgaris* S.). J Food Qual 20:547–557

Paris HS (2015) Origin and emergence of the sweet dessert watermelon *Citrullus lanatus*. Ann Bot 116:133–148

Paris HS (2017) Origin of the dessert watermelon *Citrullus lanatus*. Acta Hortic 1151:87–93

Perkins-Veazie P (2007) Carotenoids in watermelon and mango. Acta Hortic 746:259–264

Perkins-Veazie P, Collins JK (2004) Flesh quality and lycopene stability of fresh-cut watermelon. Postharvest Biol Technol 31:159–166

Perkins-Veazie P, Collins JK, Pair SD, Roberts W (2001) Lycopene content differs among red-fleshed watermelon cultivars. J Sci Food Agric 81:983–987

Perkins-Veazie P, Collins JK, Davis AR, Roberts W (2006) Carotenoid content of 50 watermelon cultivars. Agric Food Chem 54:2593–2597

Pitrat (ed) Cucurbitaceae 2008: Proceedings of the IXth EUCARPIA meeting on genetics and breeding of Cucurbitaceae. Avignon, France, pp 445–452

Poole CF (1944) Genetics of cultivated cucurbits. J Hered 35:122–128

Poole CF, Grimball PC (1939) Inheritance of new sex forms in *Cucumis melo* L. J Hered 30(1):21–25

Poole CF, Grimball PC (1944) Interaction of sex, shape, and weight genes in watermelon. J Agric Res 71:533–552

Poole CF, Grimball PC, Porter DR (1941) Inheritance of seed characters in watermelon. J Agric Res 63:433–456

Porter DR (1937) Inheritance of certain fruit and seed characters in watermelons. Hilgardia 10:489–509

Porter DR, Melhus IE (1932) The pathogenicity of Fusarium niveum. (E. F. Sm) and the development of wilt-resistant strains of *Citrullus vulgaris* (Schrad.). Res Bull Iowa Agric Expt Stat (149):183 p

Provvidenti R, Gonsalves D, Humaydan HS (1984) Occurrence of Zucchini yellow mosaicvirus in cucurbits from Connecticut, NewYork, Florida and California. Plant Dis 68:443–446

Rai M, Pandey S, Kumar S (2008) Cucurbit research in India: a retrospect. In: Pitrat M (ed) Cucurbitaceae proceedings of the IXth EUCARPIA meeting on genetics and breeding of Cucurbitaceae. INRA, Avignon (France), May 21–24, 2008

Rajasekharam T (2010) Biological and molecular characterization and management of watermelon bud necrosis virus. Ph.D. thesis, UAS, Dharwad, Karnataka

Rakha MT, Metwally EI, Moustafa SA, Etman AA, Dewir YH (2012) Evaluation of regenerated strains from six Cucurbita interspecific hybrids obtained through anther and ovule in vitro cultures. Aust J Crop Sci 6(1):23–30

Rebijith KB, Asokan R, Krisna Kumar NK, Krishna V, Ramamurthy VV (2012) Development of species-specific markers and molecular differences in mtDNA of *Thrips palmi* Karny and *Scirtothrips dorsalis* Hood (Thripidae: *Thysanoptera*), vectors of tospoviruses *Bunyaviridae*) in India. Entomol News 122:201–213

Rebijith KB, Asokan R, Hande HR, Kumar NK (2016) The first report of miRNAs from a thysanopteran insect, Thrips palmi Karny using high-throughput sequencing. PLoS One 11:e0163635

Reddy UK, Aryal N, Islam-Faridi N, Tomason Y, Levi A, Nimmakayala P (2013) Cytomolecular characterization of rDNA distribution in various *Citrullus* species using fluorescent *in situ* hybridization. Genet Resour Crop Evol. http://doi.org/10.1007/s10722-013-9976-1

Reddy UK, Abburi L, Abburi VL, Saminathan T, Cantrell R, Vajja VG, Reddy R, Tomason YR, Levi A, Wehner TC, Nimmakayala P (2014a) A genome-wide scan of selective sweeps and association mapping of fruit traits using microsatellite markers in watermelon. J Hered. http://doi.org/10.1093/jhered/esu077

Reddy UK, Nimmakayala P, Levi A, Abburi VL, Saminathan T, Tomason YR, Vajja G, Reddy R, Abburi L, Wehner TC, Ronin Y, Korol A (2014b) High-resolution genetic map for understanding the effect of genome-wide recombination rate on nucleotide diversity in watermelon. G3:Genes|Genomes|Genetics. http://doi.org/10.1534/g3.114.012815

Ren Y, McGregor C, Zhang Y, Gong G, Zhang H, Guo S, Sun H, Cai W, Zhang J, Xu Y (2014) An integrated genetic map based on four mapping populations and quantitative trait loci associated with economically important traits in watermelon (*Citrullus lanatus*). BMC Plant Biol 14:33

Ren R, Xu J, Zhang M, Liu G, Yao X, Zhu L, Hou Q (2019) Identification and molecular mapping of a gummy stem blight resistance gene in wild watermelon (*Citrullus amarus*) germplasm PI 189225. Plant Dis 104(1):16–24

Renner SS, Sousa A, Chomicki G (2017) Chromosome numbers, Sudanese wild forms, and classification of the watermelon genus *Citrullus*, with 50 names allocated to seven biological species. Taxon 66 (6):1393–1405

Ribalta FM, Pazos-Navarro M, Nelson K, Edwards K, Ross JJ, Bennett RG, Munday C, Erskine W, Ochatt SJ, Croser JS (2017) Precocious floral initiation and identification of exact timing of embryo physiological maturity facilitate germination of immature seeds to truncate the lifecycle of pea. Plant Growth Regul 81:345–353

Riley DG, Pappu HR (2004) Tactics for management of thrips (Thysanoptera: Thripidae) and tomato spotted wilt virus in tomato. J Econ Entomol 97:1648–1658

Robinson RW, Decker-Walters DS (1997) Cucurbits. CAB International Publishing, Oxon

Romay G, Lecoq H, Desbiez C (2014) Cucurbit crops and their viral diseases in Latin

Rosa JT (1924) Fruit habit and pollination in cantaloupe. Proc Am Soc Hort Sci 21:51–57

Rosa JT (1928) The inheritance of flower types in Cucumis and Citrullus. Hilgardia 3:233–250

Saftner R, Luo Y, McEvoy J, Abbott JA, Vinyard B (2007) Quality characteristics of fresh-cut watermelon slices from non-treated and 1-methylcyclopropene-and/or ethylene-treated whole fruit. Postharvest Biol Technol 44:71–79

Saha K, Choudhary H, Jamatia J, Munshi AD (2022) Genetic diversity analysis among different horticultural groups of indigenous and exotic Citrullus landraces using microsatellite markers. Electron J Plant Breed 13(1):168–174

Sain RS, Joshi P (2003) Pollen fertility of interspecific F_1 hybrids in genus *Citrullus* (*Cucurbitaceae*). Curr Sci 85:431–434

Salman-Minkov A, Levi A, Wolf S, Trebitsh T (2008) ACC synthase genes are polymorphic in watermelon (Citrullus spp.) and differentially expressed in flowers and in response to auxin and gibberellin. Plant Cell Physiol 49(5):740–750

Samineni S, Sen M, Sajja SB, Gaur PM (2019) Rapid generation advance (RGA) in chickpea to produce up to seven generations per year and enable speed breeding. Crop J 8(1):164–169

Sharma RR, Choudhury B (1988) Studies on some quantitative characters in watermelon (*Citrullus Lanatus* Thunb. Masf.) II. Inheritance of total soluble solids and rind thickness. Indian J Hort 45:283–287

Shi J, Maguer ML (2000) Lycopene in tomatoes: chemical and physical properties affected by food processing. Crit Rev Food Sci Nutr 40:1–42

Singh SJ, Krishna Reddy M (1996) Watermelon bud necrosis: a new tospovirus disease. Acta Hortic (Wagening) 431:68–77

Song R, Gusmini G, Wehner TC (2002) Screening the watermelon germplasm collection for resistance to gummy stem blight. In: XXVI international horticultural congress. Advances in vegetable breeding, vol 637, pp 63–68

Sowell G (1975) Additional source of resistance to gummy stem blight in watermelon. Plant Dis Rep 59:413–415

Sowell G, Pointer GR (1962) Gummy stem blight resistance of introduced watermelons. Plant Dis Rep 46:883–885

Stetter MG, Zeitler L, Steinhaus A, Kroener K, Biljecki M, Schmid KJ (2016) Crossing methods and cultivation conditions for rapid production of segregating populations in three grain amaranth species. Front Plant Sci 7:816

Stewart JE, Turner AN, Brewer MT (2015) Evolutionary history and variation in host range of three *Stagonosporopsis* species causing gummy stem blight of cucurbits. Fungal Biol 119(5):370–382

Strange EB, Guner N, Pesic-VanEsbroeck Z, Wehner TC (2002) Screening the watermelon germplasm collection for resistance to Papaya ringspot virus type-W. Crop Sci 42:1324–1330

Tadmor Y, King S, Levi A, Davis A, Meir A, Wasserman B, Hirschberg J, Lewinsohn E (2005) Comparative fruit colouration in watermelon and tomato. Food Res Int 38:837–841

Tantasawat PA, Sorntip A, Pornbungkerd P (2015) Effects of exogenous application of plant growth regulators on growth, yield, and in vitro gynogenesis in cucumber. Hort Sci 50(3):374–382

Tetteh AY, Wehner TC, Davis AR (2010) Identifying resistance to powdery mildew race 2W in the USDA-ARS watermelon germplasm collection. Crop Sci 50:933–939

Thies JA, Levi A (2003) Resistance of watermelon germplasm to the peanut root-knot nematode. Hort Sci 38:1417–1421

Thies J, Ariss JJ, Hassell R, Buckner S, Levi A (2015) Accessions of *Citrullus lanatus* var *citroides* are valuable rootstocks for grafted watermelon in fields infested with root-knot nematodes. Hort Sci 50:4–8

Tiwari AK, Rao GP (2014) Viruses infecting *Cucurbita pepo*: current status and management. In: Kharwar RN, Upadhyay RS, Dubey NK, Raghuwanshi R (eds) Microbial diversity and biotechnology in food security. Springer India, New Delhi, India, pp 357–371

Tlili I, Hdider C, Lenucci MS, Riadh I, Jebari H, Dalessandro G (2011) Bioactive compounds and antioxidant activities of different watermelon [*Citrullus lanatus* (Thunb.)] cultivars as affected by fruit sampling area. J Food Compos Anal 24(3):307–314

Van Zeist W (1983) Fruits in foundation deposits of two temples. J Archaeol Sci 10(4):351–354. http://doi.org/10.1016/0305-4403(83)90072-9

Voss-Fels KP, Stahl A, Hickey LT (2019) Modern crop breeding for future food security. BMC Biol 17:18

Wang C, Qiao A, Fang X, Sun L, Gao P, Davis AR, Liu S, Luan F (2019) Fine mapping of lycopene content and flesh color related gene and development of molecular marker-assisted selection for flesh color in watermelon (*Citrullus lanatus*). Front Plant Sci 10:1–16

Watson A, Ghosh S, Williams MJ, Cuddy WS, Simmonds J, Rey MD, Asyraf Md Hatta M, Hinchliffe A, Steed A, Reynolds D, Adamski NM, Breakspear A, Korolev A, Rayner T, Dixon LE, Riaz A, Martin W, Ryan M, Edwards D, Batley J, Raman H, Carter J, Rogers C, Domoney C, Moore G, Harwood W, Nicholson P, Dieters MJ, DeLacy IH, Zhou J, Uauy C, Boden SA, Park RF, Wulff BBH, Hickey LT (2018) Speed breeding is a powerful tool to accelerate crop research and breeding. Nat Plants 4(1): 23–29

Wechter WP, Kousik CS, McMillan ML, Levi A (2012) Identification of resistance to *Fusarium oxysporum* f. sp. niveum race 2 in *Citrullus lanatus* var. *citroides* plant introductions. Hort Sci 47:334–338

Weetman LM (1937) Inheritance and correlation of shape, size, and color in the watermelon, *Citrullus vulgaris* Schrad. Iowa Agric Exp Stn Res Bull 228:222–256

Wehner T (2008) Watermelon. In: Prohens J, Nuez F (eds) Vegetables I: Asteraceae, Brassicaceae, Chenopodicaceae, and Cucurbitaceae. Springer, New York, pp 381–418

Wehner TC, Barrett C (1996) Vegetable cultivar descriptions for North America, lists 1–26 combined. Am Soc Hortic Sci. http://www2.ashs.org/cultivars/

Wei L, Yongan C, Enhui Z (2012) Study on in vitro regeneration of double haploid plants offspring of pumpkin (Cucurbita moschata Duch.) with cotyledon explants. J Northwest Agric Forest Univ 40(3):141–146

Weller JL, Beauchamp N, Kerckho LHJ, Platten JD, Reid JB (2001) Interaction of phytochromes A and B in the control of de-etiolation and flowering in pea. Plant J 26:283–294

Whitaker TW, Davis GN (1962) Cucurbits. Interscience, New York

Xu YWY, Ge X, Song F, Zheng Z (2000) The relation between the induced constriction resistance and changes in activities of related enzymes in watermelon seedlings after infection by *Fusarium oxysporum* f. sp. *niveum*. J Fruit Sci 17:123–127

Yoshimura K, Masuda A, Kuwano M, Yokota A, Akashi K (2008) Programmed proteome response for drought avoidance/tolerance in the root of a C3 xerophyte (wild watermelon) under water deficits. Plant Cell Physiol 49:226–241

Zhang D, Hamauzu Y (2004) Phenolic compounds and their antioxidant properties in different tissues of carrots (*Daucus carota* L.). J Food Agric Environ 2:95–100

Zhang H, Gong G, Guo S, Ren Y, Xu Y, Ling KS (2011) Screening the USDA watermelon germplasm collection for drought tolerance at the seedling stage. Hort Sci 46:1245–1248

Zhao R (2015) A history of food culture in China. SCPG Publishing Co, New York

Zhao W, Lv P, Gu H (2013) Studies on carotenoids in watermelon flesh. Agric Sci 4(7A):13–20

Zhou XG, Everts KL, Bruton BD (2010) Race 3, a new and highly virulent race of Fusarium oxysporum f. sp. niveum causing Fusarium wilt in watermelon. Plant Dis 94:92–98

Zhu Y, Sun D, Deng Y, Li W, An G, Si W, Gao P, Liu J (2019) Regeneration of double haploid plants from unpollinated ovary cultures of watermelon. Res Square 1–26. http://doi.org/10.21203/rs.2.14098/v1

Zohary D, Hopf M, Weiss E (2012) Domestication of plants in the old world, 4th edn. Clarendon Press, Oxford, UK

Zhang Z, Zhang Y, Sun L, Qiu G, Sun Y, Zhu Z, Luan F, Wang X (2018) Construction of a genetic map for *Citrullus lanatus* based on CAPS markers and mapping of three qualitative traits. Sci Hortic 233:532–538

Zou J, Feng D (2015) Lycopene reduces cholesterol absorption through the downregulation of Niemann-Pick C1-like 1 in Caco-2 cells. Mol Nut 59(11): 2225–2230

Zhao WE, Kang BS, Hu GQ (2008) Advances in research on carotenoids in watermelon flesh. J Fruit Sci 25 (6):908–915

Recent Advances in Genomics, Genetic Resources of Watermelon

7

Akanksha Jaiswar, Nivedita Rai, Devender Arora, Manisha Malhotra, Sarika Jaiswal, and Mir Asif Iquebal

7.1 Introduction

Watermelon (*Citrullus lanatus*) belongs to the Cucurbitaceae family, a xerophytic plant that grows throughout the world. It is among the top five most consumable fruits for which $\sim$ 7% of the world's land is devoted for their production (http://faostat.fao.org/). The center of origin of *Citrullus* is Africa, where wild, feral, and landrace populations of this genus thrive. Later, this crop was introduced to several continents and has become a major crop of all time. They are cultivated in temperate and tropical regions of the world, serving as a source of water and food for animals and humans. Overall, twelve hundred cultivars of watermelon are produced worldwide while the four most promising cultivars are picnic, icebox, yellow flesh, and seedless. In 2017, this annual creeping herb has become the most important global fruit, both in terms of production quantity (118.4 million tons) and production value ($ 33.9 million of US GDP in 2016) (FAO 2017). Although it mainly comprises water (approx. 90%), 6% sugar, and some other important nutritional compounds like, lycopene, vitamin C, health-promoting amino acids, such as citrulline, arginine, and glutathione that play a crucial role in improving individual health (Hayashi et al. 2005; Perkins-Veazie et al. 2006; Collins et al. 2007). The modern fruit variety of watermelon is highly diverse in shape, size, color, texture, ring pattern, flavor, and nutrient composition. The major bottleneck in watermelon improvement is years of cultivation, selection of targeting yield, and desirable fruit qualities, due to which the genetic bases of watermelon are narrowed (Levi et al. 2001a; Mohr et al. 1986).

The nuclear and plastid data reveal that besides *Citrullus lanatus*, the genus *Citrullus* contains six other extant species (Chomicki and Renner 2015). The *Citrullus naudinianus*, is the only dioecious, morphologically unique species found in Sub-Saharan Africa. While other species like *Citrullusamarus*, *Citrullusecirrhosus*, and *Citrullusrehmii* are adapted in southern Africa. *Citrullus mucosospermus* is a highly seedy variety, mainly cultivated for the consumption of their seeds and distributed in western Africa and China (Renner et al. 2017). *Citrulluscolocynthis*

A. Jaiswar · S. Jaiswal · M. A. Iquebal (✉)
Division of Agricultural Bioinformatics, ICAR-Indian Agricultural Statistics Research Institute, New Delhi 110012, India
e-mail: ma.iquebal@icar.gov.in

S. Jaiswal
e-mail: sarika@icar.gov.in

D. Arora
National Institute of Animal Science, Jeonju-si, South Korea

N. Rai
School of Computational and Integrative Sciences, Jawaharlal Nehru University, New Delhi, India

M. Malhotra
CSIR-Indian Institute of Petroleum, Dehradun, India

S. Kr. Dutta et al. (eds.), *The Watermelon Genome*, Compendium of Plant Genomes,
https://doi.org/10.1007/978-3-031-34716-0_8

is a widely distributed watermelon variety in northern and southwestern Africa as well as central Asia, it is mainly grown for its medicinal properties and seed oil (Chomicki and Renner 2015; Paris 2015; Guo et al. 2019).

Watermelons accessible today are not the same as earlier, over time, several genetic and genomic resources have been generated. A high-resolution genetic map is essential to enlighten the evolution and divergence of *Citrullus* species. A high-resolution genetic map was constructed, comprising 11 linkage groups having 698 Simple Sequence Repeat (SSR), 219 Insertion-Deletion (InDel), and 36 Structural Variants (SV) markers (Ren et al. 2012; Yong and Guo 2016). In 2013, the East Asian watermelon cultivar '97103' of 20 representative watermelon accessions genome was sequenced using Illumina sequencing technology (Guo et al. 2013). Further, an improved genome of watermelon cultivar '97103' was de novo assembled using PacBio long read. A total of 20.3 Gb PacBio sequences are generated by PacBio combined with BioNano optical and Hi-C chromatin mapping. The evolution and divergence of *Citrullus* species are much more elucidated by genome resequencing of 414 watermelon accessions that represent all the seven extant species (Guo et al. 2019). Watermelon is prone to various microbial infections like viral, fungal, bacterial, nematodes, etc. Such agricultural problems are difficult to solve through conventional breeding but they can be solved through biotechnology. Transgenic crops are the most familiar face of biotechnology worldwide. These transgenic crops are engineered in such a way that they are resistant against bacteria, viruses, herbicides, nematodes or insects, etc. (Flores et al. 2002; Gaba et al. 2004). In the current breeding program, *C. colocynthis*, *C. amarus*, and *C. mucosospermus* species are used to identify new sources of disease and pest resistance for improving the sweet watermelon variety (Levi et al. 2017). In this chapter, we describe the recent advancement in genomics, genetic resources, and transgenic watermelon to accelerate biological discovery and crop improvement.

7.2 Origin and Evolution in Watermelon

Watermelons, *Citrullus* species (Cucurbitaceae), are among the most widely grown vegetable crops in the warmer parts of the world. The archeological artifacts, iconography, and literature survey evidence that northeastern Africa is the central origin of watermelon. The watermelon has been cultivated in southern Africa whose flesh is watery, hard-in-texture, bland, and bitter. Based on nuclear and plastid data analysis, the sweet watermelon variety originated from west Africa is non-bitter, tender, well-colored flesh suggesting that they originated from a series of selections from their ancestral population (Erickson et al. 2005; Chomicki and Renner 2015). Over the 4000 years, the watermelon has been domesticated for water and food, whereas, the sweet dessert watermelons emerged in Mediterranean land approx 2000 years ago (Paris 2015). The colocynthis watermelon is sparingly cultivated, whereas, the widely cultivated watermelon is citron (*C. amarus* Schrad.), egusi (*C. mucosospermus*), Fursa and the dessert watermelon is *C. lanatus* (Renner et al. 2014). The first and most important trait selected in watermelon domestication is its non-bitterness quality, the evolutionary event for subsequent selection is tender and sweet fruit flesh (Cohen et al. 2014). Similarly, the crossing ability between various watermelon varieties will also help to improve the reproductive barriers of genes. Genomic sequencing has fostered the suggestion that the citron, egusi, and dessert watermelons differ significantly in genome organization, and the possible force for the rapid evolution of reproductive barriers is domestication (Guo et al. 2013). These evolutionary events can be understood by the latest sequencing and genomic technologies where the wild and primitive watermelon from northeastern Africa with modern sweet dessert varieties of other *Citrullus* taxa are compared (Fig. 7.1). The germplasm study of these watermelon varieties exhibits a clear dimorphism between wild/domesticated gene pools as a result of continual selection for improving horticultural traits (Abbo et al. 2014).

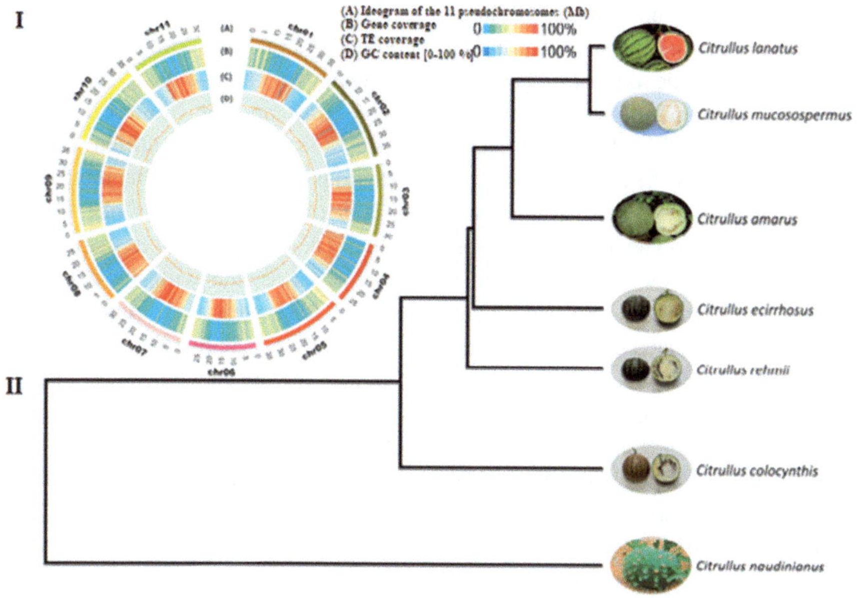

Fig. 7.1 Circos plot showing the genomic landscape of watermelon (I), along with phylogenetic trees of all seven extant species watermelon (II) ref. *Citrullus lanatus* Guo, Shaogui et al. Trends in Genetics, Volume 36, Issue 6, 456–457

7.3 Watermelon Morphology

The watermelon plants are a typical cucurbit with long internodes, the alternate distribution of leaves on the stems with multiple-branching, procumbent tendril with extensive root systems. Watermelons are monoecious, most of the flowers are staminate, and a hermaphroditic flower appears at every seventh or eighth leaf axil. The leaf laminae are pinnatifid in shape (Paris et al. 2013). Ovaries and primordial fruits are lanate, smooth, and glossy as they grow. The most common pollinators are bees. Fruit is usually harvested 25–40 d from anthesis to fruit maturity, but the external indications of fruit ripening are subtle (Gouda 2007; Wehner 2008). In watermelon cultivation, the fruit maturation requires fertile land and a long sunny day (Paris 2015). The fruit of dessert watermelons can weigh from 1 to 100 kg, but commercially it is available in the range of 3–13 kg. The shape ranges from spherical to oval to short oblong while the exterior exocarp ranges from light to dark green color, underneath a thick mesocarp is white and hard. The fruit flesh or endocarp is the main edible portion, however the exocarp is also used in some parts of the world. The fruit flesh of citron watermelon is hard, colorless, and tasteless, whereas the dessert watermelon fruit are tender, juicy, and accumulates sucrose and carotenoid pigment (Xu et al. 2012; Soteriou et al. 2014). Depending on the genotype, the flesh pigmentation of ripe watermelon fruits ranges from red, pink, white, orange, or even yellow color. Along with the fruit flesh many compressed seed are embedded in the middle layer which can be of a variety of colors like, black, brown, white, or yellow (Njoya et al. 2019; Paris et al. 2013; Perkins-Veazie et al. 2012).

7.4 Genomic Assembly to Improve Watermelon Crop

As the rise of genomic research, the watermelon genome of Chinese inbred line '97103' was selected for genome assembly, from which 46.18 Gb of high-quality genomic sequence were generated by using Illumina short reads (Guo et al. 2013). Based on 17-mer depth distribution analysis and an earlier flow cytometry analysis of the sequenced reads, the entire watermelon genome was estimated at approximately 425 Mb. 83.2% of the watermelon genome was represented by the de novo assembly of the Illumina reads, generating 353.5 Mb genome assembly (Guo et al. 2013). The assembled watermelon genome quality was evaluated by approximately one million Expressed Sequence tags (ESTs), 4 completely sequenced Bacterial Artificial Chromosomes (BACs), and paired-end sequences of BACs clones. Later on, to improve its quality, the watermelon cultivar '97103' genome was assembled by de novo PacBio long reads (Guo et al. 2019). All the seven extant species are represented in this resequencing of 414 accessions collected from various geographic regions. This improved genome size is 365.1 Mb with 31 scaffolds. A total of 22,596 high-confidence genes were predicted in the assembly and the repeat content of the assembly is 55.5%. The identified SNPs were 19,725,853 out of which 1,100,803 were located in coding regions. Altogether, the watermelon genome assembly is substantially more complete and a contiguous, almost 100-fold improvement than the previous one. This high-quality genome assembly accelerates to identify the regions under artificial selection during domestication, tightly linked with fruit quality traits, and disease resistance (Guo et al. 2019; Subburaj et al. 2019).

From the mid-nineteenth century, numerous watermelon cultivars have been developed in the US including 'Charleston Gray' (Levi et al. 2001a). The 'Charleston Gray' genomes is an American dessert watermelon cultivar developed by Charles Fredrick Andrus. Over the decades, it was the most commercially produced large oblong fruit with a thick and tough rind, which is required for long-distance shipping in the US. This cultivar is well-known for its resistance against both soil-borne disease Fusarium wilt and the foliar disease anthracnose (Wu et al. 2019). The Charleston Gray genome was sequenced and de novo assembled with a total size of 396 Mb, and 22,546 protein-coding genes were predicted among which 18,982 genes (84.2%) were assigned with biological functions. The assembly consists of 21,498 contigs and an N50 scaffold length of 7.47 Mb. Comparative genome analyses between 'Charleston Gray' and the Chinese East Asian watermelon line '97103' identified various genomic variations overlap with Quantitative Trait Locus (QTLs) of important traits like fruit weight and shape. The elongated fruit shape of 'Charleston Gray' is associated with the deletion of 159 bp in the ClFS1 gene (Wu et al. 2019). By using Genotyping-By-Sequencing (GBS), 1365 watermelon Plant Introduction (PI) lines were maintained at the US National Plant Germplasm System (NPGS; https://www.ars-grin.gov/npgs/index.html). These PI lines mainly belong to three *Citrullus* species, *C. lanatus*, *C. mucosospermus,* and *C. amarus*. From this GBS data, approximately 25,000 high-quality Single Nucleotide Polymorphism (SNPs) were derived and population genomic analysis of these SNPs shows a close relationship between *C. lanatus* and *C. mucosospermus* and a distinct genetic makeup of *C. amarus*. By utilizing resequencing data, identification of SNPs associated with different traits can be performed, which will be useful for the rapid development of diagnostic markers for Marker Assisted Selection (MAS) and fine mapping. These genomics approaches could also help in studying population genetics and identifying markers associated with producing lycopene-enriched watermelon (Wang et al. 2019; Branham et al. 2017a, b; Lee et al. 2021).

7.5 Genetic Resources of Watermelon

Frequently increasing demands for highly nutritive, good yielding, and disease resistant food crops, including watermelon by the constantly escalating human population have intensified the research on collection and conservation of genetically diverse watermelon germplasm. Moreover, only limited genetic resources of watermelon are available at present (Assefa et al. 2020; Bai et al. 2016). Although a RNA-Seq gene expression atlas in melon, Melonet-DB exists (Yano et al. 2018) but that of watermelon is warranted. This scarcity of genomic resources along with low genetic diversity among *Citrullus lanatus* cultivars have further concerned the plant breeders and researchers for incorporation of new genetic resources of *Citrullus lanatus* in gene banks and also necessitates the requirement of efficient and effective conservation of watermelon genetic resources, along with better understanding of genetic diversity between and within the population (Xiao et al. 2004; Gusmini and Wehner 2005; Guo et al. 2011). Also, it has been evident from previous studies that the narrow genetic base among watermelon cultivars is due to excessive long-term selective domestication of watermelon for desirable fruit quality, resulting in genetic bottleneck and loss of alleles contributing to resistance against pests, disease, and stress (Levi et al. 2001b, c; Guo et al. 2013). Hence, the assembly and the conservation of genetically and morphologically diverse watermelon germplasm are essential activities to ensure the current and future success of watermelon breeding programs. Various centers involved in prioritizing the collection and conservation of watermelon germplasms include the Southern African Development Community (SADC) Plant Genetic Resources Centre (SPGRC), National Plant Resources Centre Regional Network, National Botanical Research Institute (NBRI), National Plant Genetic Resources Centre (Namibia), Zambia Agriculture Research Institute (ZARI), and N.I. Vavilov Research Institute of Plant Industry (Russian Federation) (Mujaju et al. 2010; Munyenyembe 2009; McGregor 2012). Turkey is the second largest watermelon producer after China and is involved in maintaining an extensive watermelon genetic resource (Sari et al. 2007). Currently, National Agrobiodiversity Cnter (NAC) at Rural Development Administration in South Korea collected around 1100 watermelon accessions from all over the world (Lee et al. 2019), around 67 morphologically different watermelon accessions obtained from Spain were sent to National seed storage laboratory, USDA, Fort Collins, Colorado, USA (Nuez et al. 1987) and around 400 USA-developed watermelon cultivars and 1800 accessions of Citrullus germplasm are maintained in the USDA/ARS Plant Genetic Resources Unit (PGRU), Griffin, GA (Levi et al. 2017). About 400 watermelon accessions were collected across most of Turkey and maintained by Cukurova University in 1993 (Levi et al. 2017) (Table 7.1).

Wide knowledge regarding watermelon genetic resources and availability of genetic diversity help overcome the challenges related to crop improvement in terms of yield, fruit quality, early maturation, pests and disease resistance, adaptation, and tolerance against stress (Gusmini and Wehner 2005). As evident from previous research that Genetic resources within Citrullus contain genes conferring resistance to a broad range of fungal diseases such as Fusarium wilt, anthracnose, gummy stem blight (Sowell et al. 1980; Boyhan et al. 1994; Sowell 1975; Sowell and Pointer 1962; Gusmini et al. 2005; Netzer and Martyn 1989; Dane et al. 1998; Wechter et al. 2012); oomycete diseases including Phytophthora fruit rot, powdery mildew, downy mildew (Kousik et al. 2012a, 2014; Davis et al. 2007; Tetteh et al. 2010) viruses such as the potyviruses, and Squash vein yellowing virus (SqVYV) (Kousik et al. 2009, 2012b; Guner 2005; Strange et al. 2002; Levi et al. 2016); and insect pests such as root-knot nematodes, white flies, and mites. (Thies and Levi 2003, 2007; Kousik et al. 2007; Coffey et al. 2015). Moreover, various Citrullus species, exhibiting resistance against pests and diseases, are also a potential source of genes, providing tolerance against cold, drought, and high temperature (Akashi et al. 2001; Huh et al. 2001; Rivero et al. 2001; Zhang et al. 2011) which includes *C. amarus* (Yoshimura et al. 2008), *C. colocynthis, C. mucosopermus* (McGregor 2012), *C. rehmii, C. naudinianus, C. ecirrhosus* (Levi et al. 2017). Collection and conservation of Citrullus germplasm in Gene banks help identifying the sources of disease, pest, and stress resistance to gain knowledge about trait inheritance and in studying the Citrullus taxonomy and evolution (Reddy et al. 2014, 2015; Levi et al. 2013, 2017; Branham et al. 2017a, b), hence potentially contributing to the present and future success of

Table 7.1 Databases and sites holding genetic resources of watermelon

Databases/centers	Resource link	References
Watermelon Genome Database	http://www.iwgi.org/	Guo et al. (2013)
Indian Council of Agricultural Research (ICAR)–National Bureau of Plant Genetic Resources (NBPGR)	http://www.nbpgr.ernet.in/	Grumet et al. (2021)
USDA/ARS/NPGS, Germplasm Resources Information Network (GRIN)	http://www.ars-grin.gov/npgs	Levi et al. (2017)
Cucurbit genomics database (CuGenDB)	http://www.icugi.org/	Guo et al. (2013)
FAO statistics database	http://faostat.fao.org/	Guo et al. (2013)
National Centre for Biotechnology Information (NCBI) database	www.ncbi.nlm.nih.gov	Verma and Arya (2008)
National Agriculture and Food Research Organization (NARO)	https://www.naro.affrc.go.jp/archive/nias/eng/genresources/index.html	Grumet et al. (2021)
The Vavilov Institute of Plant Industry	http://www.vir.nw.ru	Grumet et al. (2021)
National Agrobiodiversity Center	http://genebank.rda.go.kr/	Grumet et al. (2021)
The Uzbek Research Institute of Plant Industry, the Uzbek Research Institute of Vegetables, Melons, and Potato, the Karakalpak Research Institute of Agriculture, Tashkent, and the World Vegetable Center	https://avrdc.org http://seed.worldveg.org/	Grumet et al. (2021)
The GeneSys system	https://www.genesys-pgr.org/	Grumet et al. (2021)

watermelon breeding programs (Levi et al. 2017; Zhang et al. 2019). Recently, watermelon rootstocks with enhanced resistance against nematodes and other diseases have been developed utilizing the germplasm collected in Plant Genetic Resources Unit (PGRU) (Levi et al. 2013; Thies et al. 2015).

7.6 Watermelon Genome in the Age of Next-Generation Sequencing

Narrow knowledge about the genetic resources of Citrullus and the urgent need for identifying and incorporating new sources of disease and pest resistance genes into the cultivated crop for improving the present levels of production, certain high throughput, and cost-effective Next-Generation Sequencing (NGS) technologies including Roche/454 and Illumina/Solexa sequencing technologies are required (Levi et al. 2017). Next-generation sequencing (NGS) technologies may potentially be used as an important tool for understanding the genetic diversity, population structure, and also for identifying the gene loci responsible for heterosis and those involved in resistance to biotic and abiotic stresses. These technologies may also play an important role in selective incorporation and use of diverse germplasm into watermelon breeding programs (Reddy et al. 2015; Ren et al. 2014; Lambel et al. 2014; Branham et al. 2017a, b). Utilizing these technologies, genome sequencing of various closely related cucurbit species has been carried out, for instance genome sequencing of watermelon by International Watermelon Genomics Initiative (Arumuganathan and Earle 1991) and that of melon by Spanish Genomics Initiative (MELONOMICS). However, whole genome sequencing still includes huge investments and efforts, therefore, discovery of gene, its function, and expression analysis has been carried out by a relatively low cost and efficient transcriptome sequencing (Guo et al. 2011).

Recently, Roche/454 sequencing and traditional Sanger sequencing technologies have been utilized for the generation of large Expressed Sequence Tag (EST) datasets in cucurbit species such as 1.2 million, 350,000 and 500,000 ESTs have been generated from melon (Portnoy et al. 2011), cucumber (Guo et al. 2010) and cucurbita pepo (Blanca et al. 2011) respectively. However, it is evident that the availability of watermelon ESTs in Gene banks is only about 12,000 which is why genetic resources of watermelon needed to be expanded to carry out gene discoveries, functional and expression analysis, and molecular breeding (Guo et al. 2011).

7.7 Databases and Bioinformatics for *Citrullus* Species

As the rise of genomic research, the watermelon genomes were sequenced and annotated well, as a result the genetic and transcriptomic data have also increased exponentially in the last two decades. To provide access, storage, and also manage the vast amount of genetics and genomics data it is necessary to develop a web-based database. The genomics and their functional annotation of watermelon were previously proposed in the Cucurbit crop as shown in Fig. 7.2. The most current database is cucurbit genomics database (CuGenDB, http://www.icugi.org) where four major economically important cucurbit crops data have been deposited. CuGenDB is a repository of genetics, genomics, transcriptomes, annotations, genetic maps, and markers as well as data mining tools that provide browsing, search, and downloading facilities for the cucurbit family. In CuGenDB, the four major economically important cucurbit are cucumber (*Cucumis sativus* L.), watermelon [*C. lanatus* (Thunb.) Matsum. and Nakai], melon (*Cucumis melo* L.), and squash/pumpkin (*Cucurbita* spp.), along with some other cucurbits like pumpkin, bottle gourd, and wax gourd genome sequence are also available. Initially, the watermelon genome sequences were deposited in Watermelon Genome Database (http://www.iwgi.org) as well as in CuGenDB, later on all genetic resources are only maintained by CuGenDB.

CuGenDB contains two types of watermelon cultivars, '97103' genomes (Guo et al. 2013, 2019) and 'Charleston Gray' genomes (Wu et al. 2019). The 'Charleston gray' is a high-quality genome assembly of three *Citrullus* species collected throughout the world. The 97103 cultivar had two versions, version 1 is the East Asian watermelon cultivar obtained from resequencing of 20 watermelon accessions representing three different *C. lanatus* subspecies (Guo et al. 2013). In this version, the 11 watermelon chromosomes had 23,440 genes that were predicted in the assembly. The version 2 of watermelon was de novo assembly of all seven extant species by resequencing of 414 accessions from various geographic regions (Guo et al. 2019). In this version, 22,596 genes were predicted in the assembly. Overall, the CuGenDB database is engaged in providing valuable information regarding genome synteny blocks, homologous gene pairs, gene expression profiles, biochemical pathways, and genetic datasets of cucurbit species (Zheng et al. 2019). Therefore, bioinformatics databases are valuable tools for deriving large-scale phenotypic data and a number of molecular markers for studying phenotypic and genotypic information.

7.8 Exploration of Genetic Resources for Transgenic Watermelon

Transgenic/genetic modifications are adopted more rapidly than any other agricultural techniques, it offers artificial insertion of one or more genes from an unrelated plant or from different species and generates agronomically important mutations to crops (Compton et al. 2004a, b). In case of watermelon fruit cultivars which vary in

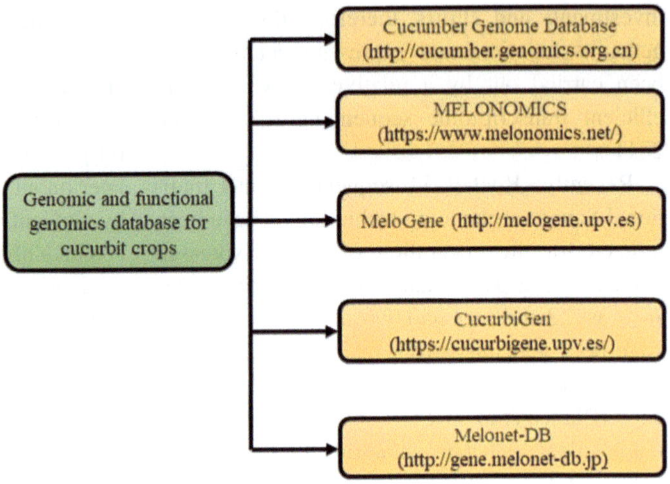

Fig. 7.2 Various genomic and functional database for cucurbit crops

size, shape, color and rind pattern, flesh and seed color and maturity time etc., Various improvements like, fruit quality (especially shelf-life), the herbicide resistance, male sterility, high-sugar content are more prevalent tasks which are achieved by the genetic transformation. For example, the long shelf-life has been achieved by genetic engineering [4–5]. Apart from improving the breeding techniques to gain high-yield varieties, the watermelon plants are more susceptible to a wide range of pathogens such as bacterial, fungal, nematode, insect, and virus etc. (Compton et al. 2004a, b). By using conventional breeding techniques, watermelon cultivars are resistant to Fusarium wilt, gummy stem blight, and anthracnose (Compton et al. 2004a, b).

7.9 Conclusion

Whole genome sequence serves as the blueprint for crop improvement by facilitating the scientific community toward understanding the functional genomics mostly related to yield, stress-tolerance, engineering the crops with important agronomic traits, etc. With the recent advancement in genome assembly of such high quality, it provides the basis of investigating molecular markers, studying species, cultivar and clonal diversity, researching evolution, and domestication events.

References

Abbo S, Pinhasi van-Oss R, Gopher A, Saranga Y, Ofner I, Peleg Z (2014) Plant domestication versus crop evolution: a conceptual framework for cereals and grain legumes. Trends Plant Sci 19(6):351–360

Akashi K, Miyake C, Yokota A (2001) Citrulline, a novel compatible solute in drought-tolerant wild watermelon leaves, is an efficient hydroxyl radical scavenger. FEBS Lett 508(3):438–442

Arumuganathan K, Earle E (1991) Nuclear DNA content of some important plant species. Plant Mol Biol Rep 9:208–218

Assefa AD, Hur OS, Ro NY, Lee JE, Hwang AJ, Kim BS, Rhee JH, Yi JY, Kim JH, Lee HS, Sung JS (2020) Fruit morphology, citrulline, and arginine levels in diverse watermelon (*Citrullus lanatus*) germplasm collections. Plants 9(9):1054

Bai Y, Zhang Z, Fei Z (2016) Databases and bioinformatics for cucurbit species. In: Grumet R, Katzir N, Garcia-Mas J (eds) Genetics and genomics of Cucurbitaceae. Plant genetics and genomics: crops and models, vol 20. Springer, Cham. http://doi.org/10.1007/7397_2016_27

Blanca J, Cañizares J, Roig C, Ziarsolo P, Nuez F, Picó B (2011) Transcriptome characterization and high throughput SSRs and SNPs discovery in Cucurbita pepo (Cucurbitaceae). BMC Genom 12(1):1–15

Boyhan G, Norton JD, Abrahams BR (1994) Screening for resistance to anthracnose (race 2), gummy stem blight and root knot nematode in watermelon germplasm. Cucurbit Genet Coop 17:106–110

Branham S, Vexler L, Meir A, Tzuri G, Frieman Z, Levi A, Wechter WP, Tadmor Y, Gur A (2017a) Genetic mapping of a major codominant QTL associated with β-carotene accumulation in watermelon. Mol Breed 37(12):1–13

Branham SE, Levi A, Farnham MW, Wechter WP (2017b) A GBS-SNP-based linkage map and quantitative trait loci (QTL) associated with resistance to *Fusarium oxysporum* f. sp. niveum race 2 identified in *Citrullus lanatus* var. citroides. Theor Appl Genet 130 (2):319–330

Chomicki G, Renner SS (2015) Watermelon origin solved with molecular phylogenetics including Linnaean material: another example of museomics. New Phytol 205(2):526–532

Coffey JL, Simmons AM, Shepard BM, Tadmor Y, Levi A (2015) Potential sources of whitefly (Hemiptera: Aleyrodidae) resistance in desert watermelon (Citrullus colocynthis) germplasm. Hortic Sci 50 (1):13–17

Cohen S, Itkin M, Yeselson Y et al (2014) The *PH* gene determines fruit acidity and contributes to the evolution of sweet melons. Nat Commun 5:4026. https://doi.org/10.1038/ncomms5026

Collins JK et al (2007) Watermelon consumption increases plasma arginine concentrations in adults. Nutrition 23:261–266

Compton ME, Gray DJ, Gaba VP (2004a) Use of tissue culture and biotechnology for the genetic improvement of watermelon. Plant Cell Tissue Organ Cult 77 (3):231–243

Compton ME, Gray DJ, Gaba VP (2004b) Genetic transformation of watermelon. In: Curtis IS (ed) Transgenic crops of the world. Springer, Dordrecht. http://doi.org/10.1007/978-1-4020-2333-0_31

Crall JM, Elmstrom GW, McCuistion FT Jr (1994) SSDL: a high-quality icebox watermelon breeding line resistant to fusarium wilt and anthracnose. Hortic Sci 29:707–708

Dane F, Hawkins LK, Norton JD, Kwon YS, Om YH (1998) New resistance to race 2 of Fusarium oxysporum f. sp. niveum in watermelon. Report Cucurbit Genetics Cooperative 21:37–39

Davis AR, Levi A, Tetteh A, Wehner T, Russo V, Pitrat M (2007) Evaluation of watermelon and related species for resistance to race 1W powdery mildew. J Am Soc Hortic Sci 132(6):790–795

Erickson DL, Smith BD, Clarke AC, Sandweiss DH, Tuross N (2005) An Asian origin for a 10,000-year-old domesticated plant in the Americas. Proc Natl Acad Sci U S A 102(51):18315–18320

FAO (2017) FAOSTAT. Food and Agriculture Organization of the United Nations, Rome, Italy. http://www.fao.org/faostat/en/#data. Accessed 18 Jan 2019

Flores F, El Yahyaoui F, de Billerbeck G, Romojaro F, Latche A, Bouzayen M, Pech JC, Ambid C (2002) Role of ethylene in the biosynthetic pathway of aliphatic ester aroma volatiles in Charentais Cantaloupe melons. J Exp Bot 53:201–206

Fukushima A, Wada M, Kanaya S, Arita M (2008) SVD-based anatomy of gene expressions for correlation analysis in Arabidopsis thaliana. DNA Res 15(6):367–374

Gaba V, Zelcer A, Gal-on A (2004) Cucurbit biotechnology-the importance of virus resistance. In Vitro Cell Dev Biol-Plant 40:346–358. https://doi.org/10.1079/IVP2004554

Garcia-Mas J, Benjak A, Sanseverino W, Bourgeois M, Mir G, González VM, Hénaff E, Câmara F, Cozzuto L, Lowy E, Alioto T (2012) The genome of melon (*Cucumis melo* L.). Proc Natl Acad Sci 109 (29):11872–11877

Gouda M (2007) Diversity of local genetic resources of watermelon, *Citrullus lanatus* (Thunb.) Matsum. and Nakai, in Sudan. MS thesis, University of Uppsala

Grumet R, Katzir N, Garcia-Mas (eds) (2017) Genetics and genomics of Cucurbitaceae. Springer International Publishing, Cham

Grumet R, McCreight JD, McGregor C, Weng Y, Mazourek M, Reitsma K, Labate J, Davis A, Fei Z (2021) Genetic resources and vulnerabilities of major cucurbit crops. Genes 12: 1222. https://doi.org/10.3390/genes12081222

Guner N (2005) Papaya ringspot virus watermelon strain and Zucchini yellow mosaic virus resistance in watermelon. North Carolina State University

Guo S, Zheng Y, Joung JG, Liu S, Zhang Z, Crasta OR, Sobral BW, Xu Y, Huang S, Fei Z (2010) Transcriptome sequencing and comparative analysis of cucumber flowers with different sex types. BMC Genom 11 (1):1–13

Guo S, Liu J, Zheng Y, Huang M, Zhang H, Gong G, He H, Ren Y, Zhong S, Fei Z, Xu Y (2011) Characterization of transcriptome dynamics during watermelon fruit development: sequencing, assembly, annotation and gene expression profiles. BMC Genomics 12(1):1–13

Guo S, Zhang J, Sun H, Salse J, Lucas WJ, Zhang H, Zheng Y, Mao L, Ren Y, Wang Z, Min J, Guo X, Murat F, Ham BK, Zhang Z, Gao S, Huang M, Xu Y, Zhong S, Bombarely A, Mueller LA, Zhao H, He H, Zhang Y, Zhang Z, Huang S, Tan T, Pang E, Lin K, Hu Q, Kuang H, Ni P, Wang B, Liu J, Kou Q, Hou W, Zou X, Jiang J, Gong G, Klee K, Schoof H, Huang Y, Hu X, Dong S, Liang D, Wang J, Wu K, Xia Y, Zhao X, Zheng Z, Xing M, Liang X, Huang B, Lv T, Wang J, Yin Y, Yi H, Li R, Wu M, Levi A, Zhang X, Giovannoni JJ, Wang J, Li Y, Fei Z, Xu Y (2013) The draft genome of watermelon (*Citrullus lanatus*) and resequencing of 20 diverse accessions. Nat Genet 45 (1):51–58. http://doi.org/10.1038/ng.2470. Epub 2012 Nov 25. PMID: 23179023

Guo S, Zhao S, Sun H et al (2019) Resequencing of 414 cultivated and wild watermelon accessions identifies selection for fruit quality traits. Nat Genet 51:1616–1623

Gusmini G, Wehner TC (2005) Foundations of yield improvement in watermelon. Crop Sci 45(1):141–146

Gusmini G, Song RH, Wehner TC (2005) New sources of resistance to gummy stem blight in watermelon. Crop Sci 45:582–588

Hayashi T et al (2005) l-citrulline and l-arginine supplementation retards the progression of high-cholesterol-diet–induced atherosclerosis in rabbits. Proc Natl Acad Sci U S A 102:13681–13686

Huang S, Li R, Zhang Z, Li L, Gu X, Fan W, Lucas WJ, Wang X, Xie B, Ni P, Ren Y (2009) The genome of the cucumber (*Cucumis sativus*) L. Nat Genet 41 (12):1275–1281

Huh YC, Om YH, Lee JM (2001) Utilization of Citrullus germplasm with resistance to fusarium wilt (*Fusarium oxysporum* f. sp. niveum) for watermelon rootstocks. In: II international symposium on cucurbits, vol 588, pp 127–132

Kousik CS, Shepard BM, Hassell R, Levi A, Simmons AM (2007) Potential sources of resistance to broad mites (Polyphagotarsonemus latus) in watermelon germplasm. Hortic Sci 42(7):1539–1544

Kousik CS, Adkins S, Turechek WW, Roberts PD (2009) Sources of resistance in US plant introductions to watermelon vine decline caused by Squash vein yellowing virus. Hortic Sci 44(2):256–262

Kousik CS, Ikerd JL, Wechter P, Harrison H, Levi A (2012a) Resistance to Phytophthora fruit rots of watermelon caused by *Phytophthora capsici* in US plant introductions. Hortic Sci 47(12):1682–1689

Kousik CS, Adkins S, Turechek WW, Webster CG, Roberts PD (2012b) 392291-VDR, a watermelon germplasm line with resistance to squash vein yellowing virus-caused watermelon vine decline. Hortic Sci 47(12):1805–1807

Kousik CS, Ling KS, Adkins S, Webster CG, Turechek WW (2014) Phytophthora fruit rot-resistant watermelon germplasm lines USVL489-PFR, USVL782-PFR, USVL203-PFR and USVL020-PFR. Hortic Sci 49(1):101–104

Lambel S, Lanini B, Vivoda E, Fauve J, Wechter WP, Harris-Shultz KR, Massey L, Levi A (2014) A major QTL associated with *Fusarium oxysporum* race 1 resistance identified in genetic populations derived from closely related watermelon lines using selective genotyping and genotyping-by-sequencing for SNP discovery. Theor Appl Genet 127(10):2105–2115

Lee KJ, Lee JR, Sebastin R, Shin MJ, Kim SH, Cho GT, Hyun DY (2019) Genetic diversity assessed by Genotyping by Sequencing (GBS) in watermelon germplasm. Genes 10(10):822

Lee ES, Kim DS, Kim SG, Huh YC, Back CG, Lee YR, Siddique MI, Han K, Lee HE, Lee J (2021) QTL mapping for gummy stem blight resistance in watermelon (*Citrullus* spp.) Plants 10(3):500

Levi A, Thomas CE, Wehner TC, Zhang X (2001a) Low genetic diversity indicated the need to broaden the genetic base of cultivated watermelon. Hortic Sci 36:1096–1101

Levi A, Thomas CE, Keinath AP, Wehner TC (2001b) Genetic diversity among watermelon (*Citrullus lanatus* and *Citrullus colocynthis*) accessions. Genet Resour Crop Evol 48(6):559–566

Levi A, Thomas CE, Wehner TC, Zhang X (2001c) Low genetic diversity indicates the need to broaden the genetic base of cultivated watermelon. Hortic Sci 36 (6):1096–1101

Levi A, Thies JA, Wechter WP, Harrison HF, Simmons AM, Reddy UK, Nimmakayala P, Fei Z (2013) High frequency oligonucleotides: targeting active gene (HFO-TAG) markers revealed wide genetic diversity among *Citrullus* spp. accessions useful for enhancing disease or pest resistance in watermelon cultivars. Genet Resour Crop Evol 60(2):427–440

Levi A, Coffey J, Massey L, Guner N, Oren E, Tadmor Y, Ling KS (2016) Resistance to papaya ringspot virus-watermelon strain (PRSV-W) in the desert watermelon *Citrullus colocynthis*. Hortic Sci 51(1):4–7

Levi A, Jarret R, Kousik S, Wechter WP, Nimmakayala P, Reddy UK (2017) Genetic resources of watermelon. In: Genetics and genomics of Cucurbitaceae. Springer, Cham, pp 87–110

Lin WD, Liao YY, Yang TJ, Pan CY, Buckhout TJ, Schmidt W (2011) Co expression-based clustering of Arabidopsis root genes predicts functional modules in early phosphate deficiency signaling. Plant Physiol 155(3):1383–1402

Mantegazza O, Gregis V, Chiara M, Selva C, Leo G, Horner DS, Kater MM (2014) Gene co expression patterns during early development of the native Arabidopsis reproductive meristem: novel candidate developmental regulators and patterns of functional redundancy. Plant J 79(5):861–877

McGregor C (2012) *Citrullus lanatus* germplasm of southern Africa. Isr J Plant Sci 60(4):403–413

Mohr HC (1986) Watermelon breeding. In: Bassett MJ (ed) Breeding vegetable crops. AVI Publishing Co., Inc, Westport, Connecticut, USA

Montero-Pau J, Blanca J, Bombarely A, Ziarsolo P, Esteras C, Martí-Gómez C, Ferriol M, Gómez P, Jamilena M, Mueller L, Picó B (2018) De novo assembly of the zucchini genome reveals whole-genome duplication associated with the origin of the Cucurbita genus. Plant Biotechnol J 16(6):1161–1171

Mujaju C (2009) Diversity of landraces and wild forms of watermelon (*Citrullus lanatus*) in southern Africa

Mujaju C, Sehic J, Werlemark G, Garkava-Gustavsson L, Fatih M, Nybom H (2010) Genetic diversity in watermelon (*Citrullus lanatus*) landraces from Zimbabwe revealed by RAPD and SSR markers. Hereditas 147(4):142–153

Munyenyembe P (2009) Conservation of plant genetic resources of southern Africa: prospects and challenges. In: 36th session of the FAO conference, Rome, Italy

Nagel DH, Doherty CJ, Pruneda-Paz JL, Schmitz RJ, Ecker JR, Kay SA (2015) Genome-wide identification of CCA1 targets uncovers an expanded clock network in Arabidopsis. Proc Natl Acad Sci 112(34):E4802–E4810

Netzer D, Martyn RD (1989) PI 296341, a source of resistance in watermelon to race 2 of *Fusarium oxysporum* f. sp. niveum. Plant Dis 73(6)

Njoya Hk et al (2019) Estimation of some phytoconstituents in the aqueous extract of the endocarp, seeds and exocarp of watermelon (*Citrullus lanatus*) fruit. J Pharmacogn Phytochem 8:4750–4757

Noh JJ, Hur OS, Ro NY, Lee JE, Hwang AJ, Kim BS, Rhee JH, Yi JY, Kim JH, Lee HS, Sung JS (2020)

Lycopene content and fruit morphology of red, pink, orange and yellow fleshed watermelon (*Citrullus lanatus*) germplasm collections. Korean J Plant Res 33(6):624–637

Nuez F, Diez MJ, Palmares G, Ferrandol C, Cuartero J, Costa J (1987) Germplasm resources of *Citrullus lanatus* (Thunb./Matsum and Nakai). Cucurbit Genet Coop Rep 10:64

Obayashi T, Kinoshita K, Nakai K, Shibaoka M, Hayashi S, Saeki M, Shibata D, Saito K, Ohta H (2007) ATTED-II: a database of co-expressed genes and cis elements for identifying co-regulated gene groups in Arabidopsis. Nucleic Acids Res 35(-suppl_1):D863–D869

Obayashi T, Hayashi S, Saeki M, Ohta H, Kinoshita K (2009) ATTED-II provides co expressed gene networks for Arabidopsis. Nucleic Acids Res 37(-suppl_1):D987–D991

Paris HS (2015) Origin and emergence of the sweet dessert watermelon *Citrullus lanatus*. Ann Bot 116 (2):133–148

Paris HS, Daunay MC, Janick J (2013) Medieval iconography of watermelons in Mediterranean Europe. Ann Bot 112(5):867–879. https://doi.org/10.1093/aob/mct151

Park G, Kim J, Jin B, Yang HB, Park SW, Kang SC, Chung SM, Park Y (2018) Genome-wide sequence variation in watermelon inbred lines and its implication for marker-assisted breeding

Perez-de-Castro AM, Vilanova S, Cañizares J, Pascual L, Blanca JM, Diez MJ, Prohens J, Picó B (2012) Application of genomic tools in plant breeding. Curr Genom 13(3):179–195

Perkins-Veazie P, Collins JK, Davis AR, Roberts W (2006) Carotenoid content of 50 watermelon cultivars. J Agric Food Chem 54:2593–2597

Perkins-Veazie P, Davis A, Collins JK (2012) Watermelon: from dessert to functional food. Isr J Plant Sci 60:395–402

Portnoy V, Diber A, Pollock S, Karchi H, Lev S, Tzuri G, Harel-Beja R, Forer R, Portnoy VH, Lewinsohn E, Tadmor Y (2011) Use of non-normalized, non-amplified cDNA for 454-based RNA sequencing of fleshy melon fruit. Plant Genome 4(1)

Qi J, Liu X, Shen D, Miao H, Xie B, Li X, Zeng P, Wang S, Shang Y, Gu X, Du Y (2013) A genomic variation map provides insights into the genetic basis of cucumber domestication and diversity. Nat Genet 45(12):1510

Reddy UK, Nimmakayala P, Levi A, Abburi VL, Saminathan T, Tomason YR, Vajja G, Reddy R, Abburi L, Wehner TC, Ronin Y (2014) High-resolution genetic map for understanding the effect of genome-wide recombination rate on nucleotide diversity in watermelon. G3 Genes Genome Genet 4(11):2219–2230

Reddy UK, Abburi L, Abburi VL, Saminathan T, Cantrell R, Vajja VG, Reddy R, Tomason YR, Levi A, Wehner TC, Nimmakayala P (2015) A genome-wide scan of selective sweeps and association mapping of fruit traits using microsatellite markers in watermelon. J Hered 106(2):166–176

Ren Y, Zhao H, Kou Q, Jiang J, Guo S, Zhang H et al (2012) A high-resolution genetic map anchoring scaffolds of the sequenced watermelon genome. PLoS One 7(1):e29453

Ren Y, McGregor C, Zhang Y, Gong G, Zhang H, Guo S, Sun H, Cai W, Zhang J, Xu Y (2014) An integrated genetic map based on four mapping populations and quantitative trait loci associated with economically important traits in watermelon (*Citrullus lanatus*). BMC Plant Biol 14(1):1–11

Renner SS, Chomicki G, Greuter W (2014) Proposal to conserve the name *Momordica lanata* (*Citrullus lanatus*) (watermelon, *Cucurbitaceae*), with a conserved type against *Citrullus battich*. Taxon 63:941–942

Renner SS, Sousa A, Chomicki G (2017) Chromosome numbers, Sudanese wild forms and classification of the watermelon genus Citrullus with 50 names allocated to seven biological species. Taxon 66:1393–1405

Rivero RM, Ruiz JM, Garcıa PC, Lopez-Lefebre LR, Sánchez E, Romero L (2001) Resistance to cold and heat stress: accumulation of phenolic compounds in tomato and watermelon plants. Plant Sci 160(2):315–321

Sari N, Solmaz I, Yetisir H, Unlu H (2005) Watermelon genetic resources in Turkey and their characteristics. In: III international symposium on Cucurbits, vol 731, pp 433–438

Sari N, Solmaz I, Yetisir H, Unlu H (2007) Watermelon genetic resources in Turkey and their characteristics. Acta Hortic 731: 433–438. https://doi.org/10.17660/ActaHortic.2007.731.59

Silva AT, Ribone PA, Chan RL, Ligterink W, Hilhorst HW (2016) A predictive co expression network identifies novel genes controlling the seed-to-seedling phase transition in Arabıdopsis thaliana. Plant Physiol 170(4):2218–2231

Soteriou GA, Kyriacou MC, Siomos AS, Gerasopoulos D (2014) Evolution of watermelon fruit physicochemical and phytochemical composition during ripening as affected by grafting. Food Chem 165:282–289

Sowell G Jr (1975) Additional source of resistance to gummy stem blight in watermelon. In: Plant disease reporter

Sowell G, Pointer GR (1962) Gummy stem blight resistance of introduced watermelons

Sowell G Jr, Rhodes BB, Norton JD (1980) New sources of resistance to watermelon anthracnose. J Am Soc Hortic Sci 105(2):197–199

Strange EB, Guner N, Pesic Van Esbroeck Z, Wehner TC (2002) Screening the watermelon germplasm collection for resistance to Papaya Ringspot Virus Type-W. Crop Sci 32:1324–1330

Subburaj S, Lee K, Jeon Y, Tu L, Son G, Choi S, Lim YP, McGregor C, Lee GJ (2019) Whole genome resequencing of watermelons to identify single nucleotide polymorphisms related to flesh color and lycopene content. PLoS One 14(10):e0223441

Tetteh AY, Wehner TC, Davis AR (2010) Identifying resistance to powdery mildew race 2W in the USDA-ARS watermelon germplasm collection. Crop Sci 50 (3):933–939

Thies JA, Levi A (2003) Resistance of watermelon germplasm to the peanut root-knot nematode. Hortic Sci 38(7):1417–1421

Thies JA, Levi A (2007) Characterization of watermelon (*Citrullus lanatus* var. citroides) germplasm for resistance to root-knot nematodes. Hortic Sci 42(7):1530–1533

Thies JA, Ariss JJ, Hassell RL, Buckner S, Levi A (2015) Accessions of *Citrullus lanatus* var. citroides are valuable rootstocks for grafted watermelon in fields infested with root-knot nematodes. Hortic Sci 50(1):4–8

Verma M, Arya L (2008) Development of EST-SSRs in watermelon (*Citrullus lanatus* var. *lanatus*) and their transferability to *Cucumis* spp. J Hortic Sci Biotechnol 83(6): 732–736. https://doi.org/10.1080/14620316.2008.11512452

Wang C, Qiao A, Fang X, Sun L, Gao P, Davis AR, Liu S, Luan F (2019) Fine mapping of lycopene content and flesh color related gene and development of molecular marker–assisted selection for flesh color in watermelon (*Citrullus lanatus*). Front Plant Sci 10:1240

Wechter WP, Kousik C, McMillan M, Levi A (2012) Identification of resistance to *Fusarium oxysporum* f. sp. niveum race 2 in Citrullus *lanatus* var. citroides plant introductions. Hortic Sci 47(3):334–338

Wehner TC (2008) Watermelon. In: Prohens J, Nuez F (eds) Handbook of plant breeding, vegetables I. Springer, New York, pp 381–418

Wu S, Wang X, Reddy U, Sun H, Bao K, Gao L, Mao L, Patel T, Ortiz C, Abburi VL, Nimmakayala P, Branham S, Wechter P, Massey L, Ling KS, Kousik C, Hammar SA, Tadmor Y, Portnoy V, Gur A, Katzir N, Guner N, Davis A, Hernandez AG, Wright CL, McGregor C, Jarret R, Zhang X, Xu Y, Wehner TC, Grumet R, Levi A, Fei Z (2019) Genome of 'Charleston Gray', the principal American watermelon cultivar, and genetic characterization of 1,365 accessions in the U.S. National Plant Germplasm System watermelon collection. Plant Biotechnol J 17 (12):2246–2258

Xiao LQ, Ge XJ, Gong XUN, And HG, Zheng SX (2004) ISSR variation in the endemic and endangered plant Cycas guizhouensis (Cycadaceae). Ann Bot 94 (1):133–138

Xu Y, Guo S, Liu J et al (2012) Dynamic characteristics of enzymes and transcriptome related to sugar metabolism and accumulation in sweet and non-sweet watermelon fruits. In: Sari N, Solmaz I, Aras V (eds) Cucurbitaceae 2012, Proceedings of the Xth Eucarpia meeting on genetics and breeding of Cucurbitaceae. Cukurova University, Adana, pp 103–111

Yano R, Nonaka S, Ezura H (2018) Melonet-DB, a grand RNA-Seq gene expression atlas in melon (*Cucumis melo* L.). Plant Cell Physiol 59(1):e4

Yong X, Guo S (2016) The watermelon genome. In: Grumet R, Katzir N, Garcia-Mas J (eds) Genetics and genomics of Cucurbitaceae, vol 20. Springer, Cham

Yoshimura K, Masuda A, Kuwano M, Yokota A, Akashi K (2008) Programmed proteome response for drought avoidance/tolerance in the root of a C3 xerophyte (wild watermelon) under water deficits. Plant Cell Physiol 49(2):226–241

Zhang H, Gong G, Guo S, Ren Y, Xu Y, Ling KS (2011) Screening the USDA watermelon germplasm collection for drought tolerance at the seedling stage. Hortic Sci 46(9):1245–1248

Zhang H, Li X, Yu H, Zhang Y, Li M, Wang H, Wang D, Wang H, Fu Q, Liu M, Ji C (2019) A high-quality melon genome assembly provides insights into genetic basis of fruit trait improvement. iScience 22:16–27

Zheng Y, Wu S, Bai Y, Sun H, Jiao C, Guo S, Zhao K, Blanca J, Zhang Z, Huang S, Xu Y (2019) Cucurbit Genomics Database (CuGenDB): a central portal for comparative and functional genomics of cucurbit crops. Nucleic Acids Res 47(D1):D1128–D1136

Health Properties and Breeding for Phytonutrients in Watermelon (*Citrullus lanatus* L.)

8

Gograj Singh Jat and Umesh K. Reddy

8.1 Introduction

Watermelon (*Citrullus lanatus*, 2n = 2x = 22) is one of the most popular vegetable fruit crop worldwide. It belongs to the Cucurbitaceae family and originated in Africa (Robinson 1997). It has been domesticated for more than 4000 years and improved by domestication and modern breeding techniques from its wild relatives which are harboring hard, pale-colored, and bitter-or bland-tasting flesh into modern sweet watermelon cultivars which possess red flesh, large-sized fruits with crispy sweet taste and thin rind (Paris 2015). The annual production of watermelon is estimated to be around 90 million tons and ranks in the top 10 fruit crops produced globally (FAOSTAT 2020). China is the leading producer of watermelon, contributing 60 million tons per annum from 1.4-million-hectare area (FAOSTAT 2020).

In recent years, the economic and nutritional value of watermelon has been recognized globally, which has created an excellent opportunity to breed new cultivars combining phytonutrients and high yield (Yang et al. 2016). Nutraceuticals have received considerable attention because they are safe, efficacious, and have potential nutritional as well as therapeutic values (Manivannan et al. 2020). Nutraceuticals provide several health benefits including prevention and treatment of several deadly diseases including certain cancers and cardiovascular diseases (Connolly et al. 2019; Tarazona-Diaz et al. 2011). Watermelon is also known as a functional food because it provides several types of phytochemicals and nutrients, which play an important role in the human diet and promote health. The global trade in nutraceuticals and functional foods is growing at an exponential rate. These compounds also determine eating quality, fruit ripening, and post-harvest shelf life (Wang et al. 2021; Mashilo et al. 2022). Therefore, modern breeding focuses on the development of watermelon cultivars with improved fruit quality and bioactive health compounds or nutraceuticals such as lycopene, β-carotene, flesh color, sugar content, TSS, amino acids (citrulline and arginine), organic acids (citric and glutamic acids), and volatile compounds (Davis et al. 2011; Liu et al. 2012; Ren et al. 2014; Jawad et al. 2020; Mashilo et al. 2022). Fruit sugar determines sweetness, volatile compounds impart flavor and aroma, whereas organic acids enhance fruit acidity (Liu et al. 2012; Dima et al. 2014).

G. S. Jat
Texas A&M University, College Station, Texas, USA

Division of Vegetable Science, ICAR-Indian Agricultural Research Institute, New Delhi 110012, India

U. K. Reddy (✉)
Department of Biology, Gus R. Douglass Institute, West Virginia State University, Institute, WV, USA
e-mail: ureddy@wvstateu.edu

S. Kr. Dutta et al. (eds.), *The Watermelon Genome*, Compendium of Plant Genomes,
https://doi.org/10.1007/978-3-031-34716-0_9

There is a need to enhance the bioactive health compounds of fresh products, which in turn will add value and create new opportunities for producers and processors by developing health-oriented markets. Therefore, plant bioactive compounds important for human nutrition must be identified well and intended to breed new watermelon cultivars with improved nutraceuticals or bioactive health compounds through available rich genetic resources using conventional as well as molecular breeding (Reddy et al. 2014a, b). This chapter aims to present the human health properties, genetic resources, and genetic inheritance of quality traits, as well as breeding behavior, methods, and objectives for phytonutrients in watermelon.

8.2 Health Properties of Watermelon Phytochemicals

8.2.1 Lycopene

Watermelon is an excellent source of red-pigmented strong antioxidant compound known as lycopene which also serves as an intermediate for the biosynthesis of many other carotenoids with the highest degree of unsaturation (Kehili et al. 2017; Maoto et al. 2019). Watermelon is also known as the "lycopene leader" as it contains the highest bioavailable lycopene which is about 60% more than that found in tomatoes (Oberoi and Sogi 2017). Watermelon pomace contains 20–24 mg/100 g lycopene which makes it an excellent functional food (Oberoi and Sogi 2017). Lycopene is an effective free radical scavenger and O_2 quencher among all carotenoids including β-carotene and tocopherol (Oberoi and Sogi 2017; Maoto et al. 2019). In terms of single O_2 quenching ability activity, lycopene is twice as potent as β-carotene and ten times more potent than tocopherol (Naz et al. 2014; Kulczynski et al. 2017; Maoto et al. 2019). In developed countries, it is estimated that the intake of lycopene is 5–7 mg/day and only 10–30% is absorbed by the human body (Petyaev 2016). Lycopene contributes about 70–80 of the total carotenoids in red-fleshed cultivars. The other carotenoids in these cultivars include phytofluene, β-carotene, lutein, phytoene, nerosporene, and £-carotene. The quantity of lycopene in watermelon also depends on cultivar type and growing conditions (Soteriou et al. 2014; Kyriacou et al. 2018). In recent times, the demand for natural lycopene has increased due to its several health benefits such as reduced cancer cell growth and induced cell death in malignant leukemia, endometrial, mammary, lung, breast, colon, stomach, and prostate cancer cells (Oberoi and Sogi 2017). It is also effective at curbing destructive free radicals including sulfide, nitrogen dioxide, and singlet O_2 and inhibiting DNA and cellular membrane damage (Kulczynski et al. 2017; Kehili et al. 2017; Kyriacou et al. 2018). Lycopene protects against lipid peroxidation and foam cell production, which are implicated in the initiation of atherosclerosis (Elumalai et al. 2013). A diet consisting of fruits and vegetables rich in lycopene can protect against stroke and cardiovascular diseases. Studies have shown that the content of lycopene and carotenoids increases rapidly and accumulates 10–12 days after pollination in diploid watermelons and continues to accumulate as the fruit matures. Watermelon juice containing lycopene and citrulline may improve athletes recovery and performance (Maoto et al. 2019).

8.2.2 β-carotene

B-carotene is a precursor of vitamin A in the human body, which is an insoluble vitamin in water due to its highly conjugated long chain (Kong et al. 2017). β-carotene cannot be synthesized by the human body; therefore, it is essential to take it from plant-based foods including watermelon which is a good source of β-carotene (Shao et al. 2017). The quantity of β-carotene depends on cultivar types and growing conditions, but on average in the fresh flesh of watermelon, it is 4.82 mg/g (Kim et al. 2014; Maoto et al. 2019). β-carotene has several health benefits such as possessing antioxidant and prooxidant properties (Chen et al. 2017); reconstructing blood vessel walls (Kulczynski et al. 2017); maintaining the immune system; supporting cell growth and differentiation; playing a role in the formation and maintenance of heart, kidney, and other organs (Shao et al. 2017); possessing anticancer properties (Barkura et al. 2017); inhibiting tumor progression in some cancers (Kulczynskiet al. 2017); reducing the risk of type-2 diabetes; and lowering the metabolic syndrome in middle-aged adults (Chen et al. 2017; Maoto et al. 2019).

8.2.3 Vitamin C

Vitamin C is a water-soluble vitamin that is important for the biosynthesis of collagen and certain hormones (Rodríguez-Roque et al. 2015). Vitamin C cannot be synthesized by the human body; therefore, it is essential to acquire it from plant-based foods including watermelon (Oberoi and Sogi 2015; Maoto et al. 2019) which is a good source of vitamin C (Jumde et al. 2015). It is an important chain-breaking antioxidant that inhibits lipid peroxidation (Doll and Ricou 2013).The quantity of vitamin C depends on cultivar types and growing conditions, but on average, fresh watermelon juice contains 3.72 mg/100 g vitamin C (Oberoi and Sogi 2015). Twenty percent of the daily vitamin C requirement can be fulfilled by drinking one cup of watermelon juice and at least 10 mg daily will prevent scurvy (Pacier and Martirosyan 2015). Vitamin C can improve the quality of life for cancer patients by reducing high oxidative stress (Takahashi et al. 2012; Maoto et al. 2019), inhibiting cancer cell growth by cutting off the blood supply to growing cancers (Lemos et al. 2017), and possessing anticancer properties by suppressing cancer cells in patients (Ijah et al. 2015). It has long been reported to be beneficial in the prevention and treatment of a variety of ailments, including scurvy, simple cold, as well as being stress resistant (Vaccaa et al. 2016; Doll and Ricou 2013; Maoto et al. 2019).

8.2.4 Citrulline

Citrulline is a nonprotein amino acid abundantly present in watermelon ranging from 0.7 to 3.6 mg/g fresh FW (Sonteriou et al. 2014; Joshi et al. 2021), which is higher in canary yellow watermelon. The natural source of citrulline is more bioavailable than a synthetic source (Barba et al. 2015; Maoto et al. 2019). A diet rich in citrulline is associated with several health benefits due to its strong antioxidant activities and works as an efficient hydroxyl radical scavenger (Soteriou et al. 2014; Maoto et al. 2019). Citrulline is found to be effective in sections such as pharmacology, immunology, and neurology (Odewunmi et al. 2015). It has been known for its effectiveness in improving sexual stamina and erectile functions (Soteriou et al. 2014; Maoto et al. 2019). It has several other health benefits such as muscle recovery during exercise, reduced blood pressure, and increased vasodilation in many tissues of the body, which help mitigate cardiovascular disease. Ijah et al. (2015) reported its importance for young adults in trauma, burn injury, and renal failure (Joshi et al. 2019).

8.2.5 Polyphenolic Compounds

Polyphenolic compounds are the most dominant antioxidants in a diet derived from fruits and vegetables (Barba et al. 2015; Maoto et al. 2019). In fresh watermelon juice, the total polyphenols range from 16.94 to 20.23 mg GAE/100 ml (Feng et al. 2013). Daily consumption of polyphenol-rich beverages induces positive effects on human health such as the ability to stop the formation of ROS in the human body (Choudhary et al. 2014; Maoto et al. 2019). In vitro and in vivo studies have proven that polyphenols possess anticancer and anti-inflammatory properties (Choudhary et al. 2015) and prevent "psoriasis disease" (Garcia-Perez et al. 2017), diabetes, neurodegenerative disorders, osteoporosis, inflammation, arthritis, and headaches (Choudhary et al. 2014; Maoto et al. 2019).

8.3 Germplasm Resources

The USDA-ARS Plant Genetic Resources Unit of National Plant Germplasm System, Griffin, Georgia, maintains 400 watermelon cultivars and 1800 accessions of all 7 *Citrullus* spp. (Levi et al. 2017). Detailed information on germplasms is accessed at National Plant Germplasm System (NPGS) and Germplasm Resources Information Network (GRIN). The National Plant Genetic Resources Centre of Namibia, Zambia Agriculture Research Institute (ZARI), and the N. I. Vavilov Research Institute of Plant Industry, Russian Federation, maintain many *Citrullus* germplasms collected from southern African countries (McGrgor 2012). Other countries such as China, Turkey, South Africa, Zimbabwe, and India also maintain watermelon germplasm collections at their genetic resource institutions. The TILLING (Targeting Induced Local Lesions in Genomes) population of Charleston Grey is a valuable source for genetic improvement of watermelon cultivars (Levi et al. 2017).

Interspecific crosses among different watermelon species have been achieved successfully (Robinson and Decker-Walters 1997) but with very few seed settings and pollen viability (Sain and Joshi 2003). *C. lanatus* and *Citrullus mucosospermus* can develop fertile hybrids (Levi et al. 2011b). *Citrullus amarus* can hybridize with *C. lanatus* and *C. mucosospermus* (Guo et al. 2013). These watermelon accessions and species have been used for new cultivar development for useful traits including fruit quality (Reddy et al. 2014a, b) and can broaden the genetic base of cultivated *C. lanatus* (Nimmakayala et al. 2022). The seeds of egusi watermelon (*C. mucosospermus)* are used for edible oil extraction (Nimmakayala et al. 2011; Dahl Jensen et al. 2011). Citron (*C. amarus)* has hard flesh, which is used as a cooked vegetable and pickle (Bush 1978), whereas colocynth (*Citrullus colocynthis*) has several medicinal properties and its seeds possess 17–18% edible oil with 80–85% unsaturated fatty acids, colocynthin, and citrulline (Dane et al. 2007; Nimmakayala et al. 2011; Hussain et al. 2014). The several accessions of desert watermelon (*C. lanatus*) and other species might be excellent sources of phytochemicals such as L-citrulline, lycopene, ascorbic acids, potassium, flavonoids, and carotenoids (α- and β-carotene) (Davis et al. 2007).

8.3.1 Screening of Germplasm

The primary objective of watermelon breeding is to screen a large number of germplasm lines for superior parental lines rich in nutraceuticals and other health compounds. Laboratory analysis of these bioactive compounds using high-throughput metabolomic approaches is preferred over visual observations. The information from earlier work done and new sources identified as rich in phytonutrients in watermelons is presented in Table 8.1, which may be useful for the introgression of these quality traits into a widely adapted cultivar of watermelon.

Analysis of quality traits in watermelon is crucial to developing new cultivars with enhanced bioactive compounds and shelf life. Watermelon is an excellent source of potential natural antioxidants such as lycopene and β-carotene. Davis et al. (2004) identified accession PI 288232 (81.0 μg/g FW) as the richest source of lycopene. Wehner (2017) also screened a large number of watermelon germplasm for lycopene content and recorded Dixielee (59.26 μg/g FW), Sugar Baby (53.37 μg/g FW), and All Sweet (41.77 μg/g FW) as good sources of lycopene, whereas Levi et al. (2017) identified new sources of lycopene (red and pink fleshed) such as Chinese97, Chinese103, Charleston Grey, Crimson Sweet, and All Sweet. Wehner (2017) identified the germplasm sources for β-carotene in yellow and orange watermelon accessions. The accession, PI 629111 (13.0 μg/g FW), was rich in β-carotene, whereas NC-517 (4.35 μg/g FW) and Yellow Doll (4.18 μg/g FW) were excellent sources of citrulline and both citrulline + arginine, whereas Yellow Doll and Yellow Crimson were rich in total soluble solids (11.60 and 11.37%, respectively).

Table 8.1 Germplasm source for higher bioactive compounds/phytonutrients in watermelons

S. No	Traits	Range/sources	Maximum in germplasm	References
1	Lycopene	10.0–81.0 μg/g FW	PI 288232	Davis et al. (2004)
		59.26 > 53.37 > 41.77 g/kg FW	Dixielee > Sugar Baby > All Sweet	Wehner (2017)
2	Carotenoid	3.0–13.0 μg/g FW	PI 629111	Davis et al. (2004)
3	Citrulline and Citrulline + arginine	4.35 > 4.18 μg/g FW	NC-517 > Yellow Doll	Wehner (2017)
4	Soluble solids	11.60% > 11.37%	Yellow Doll > Yellow Crimson	
5	Flesh color and small seeds	Rich in lycopene	Crimson Sweet, All Sweet	Levi et al. (2017)
6	Red flesh	Rich in lycopene	Chinese 97 and 103	
7	Pink flesh	Rich in lycopene	Charleston Grey	

8.4 Traditional Breeding

Mendelian genetics has played a crucial role in understanding different aspects of plant biology, which facilitates plant breeders in the development of many cultivars and hybrids for improved traits including fruit quality. Mendelian genetics using classical techniques has also facilitated the discovery of many useful genes and their inheritance in watermelon for fruit quality (Fall et al. 2019; Pei et al. 2021). These findings will help crop breeders to imply appropriate breeding procedures and methods for genetic improvement of watermelon, especially for quality improvement. The major objective of traditional breeding of watermelons is the development of high-yielding varieties rich in lycopene, β-carotene, high TSS, acidity, vitamins, and minerals. In recent years, many cultivars have been developed in watermelons using different breeding techniques that are rich in phytonutrients (Dia et al. 2016). Positive heterosis has been reported for several quality traits in watermelons (Singh and Dadwadia 2009).

8.4.1 Inheritance Studies for Quality Traits

Improvement of any crop species for quality traits requires information about genetic inheritance of these traits. Understanding the genetic makeup for various quality traits enables plant breeders to develop or introgress desirable traits into widely adapted cultivars. The knowledge of the genetics of important quality traits such as lycopene, β-carotene, TSS, flavor and pale-yellow flesh, red flesh in watermelon is indispensable for genetic improvement of watermelon. Poole and Grimbell (1945) reported that 2–3 genes along with modifiers influence the inheritance of flesh color in watermelon. Bang et al. (2010) reported that several genes are responsible for producing watermelon with flesh colors of coral red, scarlet red, orange, salmon yellow, canary yellow, or white. Genes such as *B*, *C*, *i-C*, *Wf*, *y*, and *y-o* are responsible for flesh color in watermelon cultivars. Canary yellow (*C*) is dominant over other colored flesh (c). Coral red flesh (*Y*) is dominant over salmon yellow (*y*). Orange flesh (*y–o*) is a member of multiple allelic system at that locus; hence, coral red flesh (*Y*) is dominant over both orange flesh (*y–o*) and salmon yellow (*y*), and orange flesh (*y–o*) is dominant over salmon yellow (*y*). *YY* (red flesh) is dominant over *yy* (yellow), *WY* (white flesh) over *C* (canary yellow), and golden yellow flesh over red flesh. *Wf* (white flesh) is dominant over *wf* (red flesh) in *C. colosynthesis* × *C. lanatus* (Singh et al. 2020). Bitterness in *C. colosynthesis* is dominant over nonbitterness. Incomplete dominant genes are responsible for total soluble solids in watermelon (Singh et al. 2020). The production of mixed colorations would have been caused by recombination of these genes. It may be useful to have a separate rating of the color of different parts of the flesh to determine whether there are genes controlling the color of each part: the endocarp between the carpel walls and the mesocarp (white rind); the flesh within the carpels, originating from the stylar column; and the carpel walls. A novel chromoplast phosphate transporter (*CLPHT4; 3*) was identified for flesh color development in watermelon (Zhang et al. 2016). Several quantitative trait loci (QTLs) have been identified and are well characterized for fruit quality traits such as lycopene (Fall et al. 2019; Li et al. 2020), citrulline, and arginine (Fall et al. 2019); flesh color (Liu et al. 2015; Pei et al. 2021); β-carotene (Branham et al. 2017); and total soluble solids including glucose, fructose, and sucrose (Ren et al. 2014; Cheng et al. 2016; Fall et al. 2019; Table 8.2).

8.5 Breeding Behavior and Floral Biology

Watermelon is a highly cross-pollinated vegetable crop of the Cucurbitaceae family. It is basically a monoecious (presence of both male and female flowers on the same plant but in different positions), but different sex forms such as gynoecious, gynomonoecious, andromonoecious, hermaphrodite, and trimonoecious (Zhang et al. 2017) have been reported. It is entomophilous in nature and pollination is done by honeybee species such as *Apisflorea*, *Apisdorsata*, and *Apismelifera* in most of the watermelon-growing regions. The watermelon crop is different from other cross-pollinated vegetables such as root, onion, and cole crops. Hand pollination without emasculation in monoecious cultivars is preferred for crossing program, because the flowers of watermelons are attractive, orange yellow, and large. Bagging female flowers in female parents is a must in monoecious cultivars for hybrid development (Munshi et al. 2017). Controlled pollination can be done easily in a greenhouse as there is no need to cover individual flowers to protect them from pollinating agents. Hand pollination is more effective between 6 to 9 am. In monoecious cultivars, many seeds can be obtained from a single cross fruit. Although watermelon is outcrossing in nature, the extent of the inbreeding depression is very negligible.

8.6 Breeding Methods and Objectives for Phytonutrients

Single plant selection from F_2 segregating population is generally practiced for the selection of superior genotypes due to low inbreeding depression in watermelon. Therefore, different modified breeding methods of self-pollinated as well as cross-pollinated crops are practiced based on breeding objectives. The traditional breeding methods practiced in watermelon include plant introduction, mass selection, pure line selection, pedigree method followed by hybridization, recurrent selection, single-seed descent method, backcrossing, and heterosis breeding (Delannay et al. 2010; Munshi et al. 2017). In watermelon, several F_1 hybrids have been developed by private and public institutions for quality traits such as attractive flesh color, texture, and high sugar content using heterosis breeding. The different breeding methods adopted for the genetic improvement of watermelon for quality traits are as follows:

Pedigree Breeding: This is the most common method used in watermelon breeding. The objective of this method is to develop new lines

Table 8.2 Inheritance or quantitative trait loci (QTLs) of quality traits in watermelons

S. No	Traits	Genes/OTLs	Gene action/segregation	References
1	Red flesh	Monogenic	*Wf-Y*—white flesh *Wfwfy*—red flesh *Wfwfyy*—red flesh *C*—canary yellow *B*—yellow flesh *Y*—coral red flesh *y—orange flesh* *y—salmon yellow*	Wehner (2007); Bang et al. (2010)
2	Lycopene	*LCYB4.1*	Chromosome 4	Liu et al. (2015)
		FC4.1	Chromosome 4	Zhang et al. (2017)
		Qlyc3-1, Qlyc11-1	Chromosomes 3 and 11	Fall et al. (2019)
		Yscr	Chromosome 6	Li et al. (2020)
3	Flesh color	Single recessive gene	LSW-177 (red flesh) × Saskatchewan (pale yellow flesh)	Wang et al. (2019)
		FC4.1, FC3.1, FC6.1, FC11.1	Chromosomes 4, 3, 6, and 11	Liu et al. (2015)
		FCR1, FCR2, CC	Chromosome 10	Pei et al. (2021)
4	Pale yellow flesh	Quantitative traits	Cream of Saskatchewan (pale yellow) × PI 186,490 (white flesh)	Wang et al. (2021)
5	B-carotene	*qFC.1*	Chromosome 1	Branham et al. (2017)
6	Citrulline	*Qcit1-1, Qcit8-1, Qcit11-1*	Chromosomes 1, 8, and 11	Fall et al. (2019)
7	Arginine	*Qarg2-1, Qarg5-1*	Chromosomes 2 and 5	
8	Sucrose	*Qsur5-2, Qsur8-1,*	Chromosomes 5 and 8	
		Qsur2-1, Qsur2-2, Qsur5	Chromosomes 2, 2, and 5	Ren et al. (2014)
9	Glucose	*Qglu6*	Chromosome 6	
		Qglu5-1, Qglu5-2, Qglu5-3, Qglu8 1, Qglu8-2	Chromosomes 5 and 8	Fall et al. (2019)
10	Fructose	*Qfru2-1, Qfru2-2, Qfru3, Qfru6, Qfru8*	Chromosomes 2, 3, 6, and 8	Ren et al. (2014)
		FCE10.10	Chromosome 10	Cheng et al. (2016)
		Qfru5-1, Qfru5-2	Chromosome 5	Fall et al. (2019)
11	TSS	*Brx7.1, Brx8.1 Brx9.2 Brx9.1*	Chromosomes 7, 8, and 9	Sandlin et al. (2012)
		Qbrix2-1, Qbrix2-2, Qbrix6, Qbrix8	Chromosome 2, 6, and 8	Ren et al. (2014)
		BCC8.1, BCE2.1, BCE8.1		Liu et al. (2015)
		BCC2.1 2	Chromosome 2	Cheng et al. (2016)
		Qbrx5-1, Qbrx8-2	Chromosomes 5 and 8	Fall et al. (2019)

for high yield, early maturity, and high fruit quality. In this method, breeders select two widely adapted parental lines, which is complement each other for some traits. These parental lines are crossed to develop the F_1 generation, which is then self-pollinated to generate a segregating F_2 population. The F_2 population is sib- or self-pollinated to develop F_3 generation while selecting for highly heritable characters. The breeder concentrates on selecting superior plants in each of the best F_3 families. Plants within family rows should be selected for the next generation that have excellent fruit quality traits. After six generations of selfing (F_5), plants become more uniform as inbred lines. The number of plants handled might decrease from 100 F_2 plants to a few F_5 lines.

Single-seed descent method: This method of breeding is also known as the modified pedigree method where inbred lines are developed by self-pollination under protected cultivation and winter nurseries, and selection is practiced after F_4–F_6 generations. The single-seed descent method requires less record keeping and fruit quality traits such as flesh color can be improved. The traditional pedigree method is more useful for watermelon where several quality traits can be selected in early generations, and plants or families with undesirable traits that are governed by simple inheritance like poor fruit flesh color can be eliminated in early generations.

Backcross breeding: This method is used to transfer highly heritable traits into a superior inbred line (recurrent parent). Six generations of selection and backcrossing to the recurrent parent are used to recover the similar genotypes of the recurrent parent without other undesirable traits of the nonrecurrent parent (donor) except for the new traits that are to be incorporated. Two different backcross methods are used depending on whether genetic inheritance is recessive or dominant. For recessive genes, recurrent parent is crossed with the donor parent, and the resulting F_1 is backcrossed to the recurrent parent and self-pollinated to develop the F_2 population which will segregate for the desired trait. Plants having desired traits are backcrossed to the recurrent parent to develop the BC_1. The BC_1 generation is grown for the traits of interest and then desired individuals are self-pollinated once again to develop a segregating population for selection and backcrossing to the recurrent parent. This procedure is repeated until BC_6 where the best progenies are self-pollinated, and selection is done for the trait of interest to develop inbred lines. For the traits inherited by dominant genes, F_1 is backcrossed to the recurrent parent to develop BC_1 generation. BC_1 generation is grown for trait, and individuals having trait of interest are backcrossed to recurrent parent. This entire process is repeated for the BC_6 generation until the individual does not attain homozygosity using progeny testing.

Heterosis breeding: This method requires a good number of inbred lines to make all possible combinations to exploit heterosis for quality traits. We should select the inbred lines based on their traits and they should be complementary to each other. In watermelon, many open pollinated lines have been developed and recommended for cultivation and to be used in breeding programes with desirable quality traits. Most of the F_1 hybrids developed are derived from monoecious $\times$ monoecious sex forms having attractive flesh color, high lycopene, TSS, sugar, and other quality traits since dominant gene controls these traits.

8.7 Limitations of Traditional Breeding and Rationale for Molecular Breeding

In traditional breeding, the trait of interest for selection in a segregating population has relied directly or indirectly only on morphological markers. Traditional breeding follows a direct measure of phenotypes (e.g., flesh color) or an association of one phenotype with another (dried tendrils with high sugar content). Breeding cultivars for bioactive health compounds using classical genetics and traditional breeding is often challenging in watermelon crops. Therefore, molecular markers have great potential to overcome the limitations of traditional breeding, as markers are nondestructive, eliminate environmental variation, and can be evaluated for several traits simultaneously. However, the use of molecular markers for breeding bioactive compounds requires segregating populations for a particular trait which needs to be properly identified for marker-assisted selection. There are two important issues regarding the use of molecular markers: time and cost. The development of molecular markers and their validation in multiple populations take significant time and involve a high cost of molecular markers. Genomics approaches including genome-wide association studies (GWAS), genotyping by sequencing (GBS), and re-sequencing along with metabolomics will be useful for better understanding the complex biosynthetic pathways of phytochemicals for the development of phytonutrient-dense cultivars in watermelon.

8.8 Conclusion

The bioactive compounds of watermelon have been reported for their several pharmacological activities and therapeutic properties such as antioxidant, anti-inflammatory, antifungal, antimicrobial, antibacterial, antiulcer, gastroprotective, and laxative, and they play a vital role in both prevention and amelioration of chronic diseases by suppressing the free radicals and decreasing the oxidative stress, leading to a decrease in the risk of dreadful diseases such as cancers, diabetes, CVDs, hypertension, and asthma. Traditional breeding has made enormous progress in understanding and improving watermelon crops. Classical genetics has successfully been used to enhance our understanding of taxonomy and phylogenetic relationships in sweet watermelon. Watermelon breeders have identified several genes associated with quality traits. Based on traditional breeding for the past ten decades, several F_1 hybrids/cultivars have been developed in watermelon. However, genetic improvement for quality traits is limited using traditional breeding; hence, the use of molecular marker technology and genomics will play a significant role in watermelon. Therefore, the integration of traditional breeding with molecular breeding will possibly accelerate improvement in watermelon's quality traits such as lycopene, β-carotene, total soluble solids, sugar content, and essential amino acids (citrulline and arginine).

References

Bang H, Davis AR, Kim S, Leskovar DI, King SR (2010) Flesh color inheritance and gene interactions among canary yellow, pale yellow, and red watermelon. J Am Soc Horti Sci 135(4):362–368

Barba FJ, Terefe NS, Buckow R, Knorr D, Orlien V (2015) New opportunities and perspectives of high-pressure treatment to improve health and safety attributes of foods. A review. Food Res Int 77:725–742

Barkura S, Bankapura A, Chidangila S, Mathura D (2017) Effect of infrared light on live blood cells: role of β carotene. J Photochem Photobiol 171:104–116

Branham SE, Vexler L, Meir A, Tzuri G, Frieman Z, Levi A et al (2017) Genetic mapping of a major codominant QTL associated with β-carotene accumulation in watermelon. Mol Breed 37:146

Bush A (1978) Citron melon for cash and condiment. Eco Bot, 182–184

Chen J, Li F, Li Z, McClements DJ, Xiao H (2017) Encapsulation of carotenoids in emulsion-based delivery systems: enhancement of β-carotene water-dispersibility and chemical stability. Food Hydro 69:49–55

Cheng Y, Luan F, Wang X, Gao P, Zhu Z, Liu S et al (2016) Construction of a genetic linkage map of watermelon (*Citrullus lanatus*) using CAPS and SSR markers and QTL analysis for fruit quality traits. Sci Hortic 202:25–31

Choudhary BR, Jayaprakasha GK, Porat R, Patil BS (2014) Low temperature conditioning reduces chilling injury while maintaining quality and certain bioactive compounds of star ruby grapefruit. Food Chem 153:243–249

Choudhary BR, Haldhar SM, Maheshwari SK, Bhargava R, Sharma SK (2015) Phytochemicals and antioxidants in Watermelon (*Citrullus Lanatus*) genotypes under hot arid region. Indian J Agric Sci 85:414–417

Connolly M, Lum T, Marx A, Hooshmand S, Kern M, Liu C, Hong MY (2019) Effect of fresh watermelon consumption on risk factors for cardiovascular disease in overweight and obese adults (P06-102-19). Curr Dev Nutr 3

Dane F, Liu J (2007) Diversity and origin of cultivated and citron type watermelon (*Citrullus lanatus*). Genet Resour Crop Evol 54:1255–65

Davis AR, King SR, Jarret RL, Levi A, Tadmor YA (2004) Lycopene and total carotenoid content as descriptors for *Citrullus lanatus*: limitations and preliminary trials. Rep Cucurbit GenetCoop 27:34

Davis AR, Levi A, Tetteh A, Wehner T, Russo V, Pitrat M (2007) Evaluation of watermelon and related species for resistance to race 1W Powdery Mildew. J Am Soc Hort Sci 132:790–5

Davis AR, Webber CL, Fish WW, Wehner TC, King S, Perkins-Veazie P (2011) L-Citrulline levels in watermelon cultigens tested in two environments. J Am Soc Hortic Sci 46:1572–1575

Dia M, Wehner TC, Perkins-Veazie P, Hassell R, Price DS, Boyhan GE, Olson SM, King SR, Davis AR, Tolla GE, Bernier J, Juarez B (2016) Stability of fruit quality traits in diverse watermelon cultivars tested in multiple environments. Horti Res 3:16066

Dima G, Tripodi G, Condurso C, Verzera A (2014) Volatile constituents of mini-watermelon fruits. Essent Oil Res 26:323–327

Doll S, Ricou B (2013) Severe vitamin c deficiency in a critically Ill adult: a case report. Eur J Clin Nutr 67:881–882

Elumalai M, Karthika B, Usha V (2013) Lycopene—role in cancer prevention. Int J Pharm Bio Sci 4:371–378

Fall LA, Perkins-Veazie P, Ma G (2019) QTLs associated with flesh quality traits in an elite × elite watermelon population. Euphytica 215:30

FAOSTAT (2020) Available online at http://www.fao.org/faostat/en/#data/QC. Accessed 10 Jan 2022

Feng M, Ghafoor K, Seo B, Yang K, Park J (2013) Effects of ultraviolet-C treatment in Teflon®-coil on microbial populations and physico-chemical characteristics of watermelon juice. Innov Food Sci Emerg Technol 19:133–139

García-Pérez ME, Kasangana PB, Stevanovic T (2017) Chapter 6-bioactive polyphenols for diabetes and inflammation in psoriasis disease. Stud Nat Prod Chem 52:231–268

Guo S, Zhang J, Sun H, Salse J, Lucas WJ, Zhang H, Zheng Y, Mao L, Ren Y, Wang Z, Min J, Guo X, Murat F, Ham BK, Zhang Z, Gao S, Huang M, Xu Y, Zhong S, Bombarely A, Mueller LA, Zhao H, He H, Zhang H, Zhang Z, Huang S, Tan T, Pang E, Lin K, Hu Q, Kuang H, Ni P, Wang B, Liu J, Kou Q, Hou W, Zou X, Jiang J, Gong G, Klee K, Schoof H, Huang Y, Hu X, Dong S, Liang D, Wang J, Wu K, Xia Y, Zhao X, Zheng Z, Xing M, Liang X, Huang B, Lv T, Wang J, Yin Y, Yi H, Li R, Wu M, Levi A, Zhang X, Giovannoni JJ, Wang J, Li Y, Fei Z, Xu Y (2013) The draft genome of watermelon (*Citrullus lanatus*) and resequencing of 20 diverse accessions. Nat Genet 45:51–8

Hussain AI, Rathmore HA, Sattar MZ, Chatha SA, Sarker SD, Gilani AH (2014) Citrullus colocynthis (L.) Schrad. (Bitter apple fruit): a review of its phyochemistry, pharmacology, traditional uses and nutritional potential. J Ethnopharmacol 8:54–66

Ijah UJJ, Ayodele HS, Aransiola SA (2015) Microbiological and some sensory attributes of watermelon juice and watermelon-orange juice mix. J Food Resour Sci 4:49–61

Jawad UM, Gao L, Gebremeskel H, Safdar LB, Yuan P, Zhao S et al (2020) Expression pattern of sugars and organic acids regulatory genes during watermelon fruit development. Sci Hortic 265:109102

Jensen BD, Touré FM, Hamattal MA, Touré FA, Nantoumé AD (2011) Watermelons in the sand of Sahara: cultivation and use of indigenous landraces in the Tombouctouregion of Mali. Ethnobot Res Appl 9:151–62

Joshi V, Shinde S, Nimmakayala P, Abburi VL, Alaparthi SB, Lopez-Ortiz C, Levi A, Panicker G, Reddy UK (2019) Haplotype networking of GWAS hits for citrulline variation associated with the domestication of watermelon. Int J Mol Sci 20:5392

Joshi V, Nimmakayala P, Song Q, Abburi V, Natarajan P, Levi A, Crosby K, Reddy UK (2021) Genome-wide association study and population structure analysis of seed-bound amino acids and total protein in watermelon. Peer J 9:e12343

Jumde AD, Shukla RN, Gousoddin (2015) Development and chemical analysis of watermelon blends with beetroot juice during storage. Int J Sci Eng Technol 4:2395–4752

Kehili M, Kammlottb M, Chouraa S, Zammelc A, Zetzlb C, Smirnovab I, Allouched N, Sayadia S (2017) Supercritical CO_2 extraction and antioxidant activity of lycopene and β-carotene-enriched Oleoresin from tomato (*Lycopersicum Esculentum* L.) peels by-product of a Tunisian industry. Food Bioprod Process 102:340–349

Kong L, Bhosale R, Ziegler GR (2018) Encapsulation and stabilization of β-carotene by amylose inclusion complexes. Food Res Int 105:446–452

Kulczynski B, Gramza-Michałowska A, Kobus-Cisowska J, Kmiecik D (2017) The role of carotenoids in the prevention and treatment of cardiovascular disease-current state of knowledge. J Funct Foods 38:45–65

Kyriacou MC, Leskovar DI, Colla G, Rouphael Y (2018) Watermelon and melon fruit quality: the genotypic and agro-environmental factors implicated. Sci Hortic 30:8–12

Lemos AT, Ribeiro AC, Fidalgo LG, Jorge ID, Saraiva A (2017) Extension of raw watermelon juice shelf-life up to 58 days by hyperbaric storage. Food Chem 231:61–69

Levi A, Simmons AM, Massey L, Coffey J, Wechter WP, Jarret RL, Tadmor Y, Nimmakayala P, Reddy UK (2017) Genetic diversity in the desert watermelon Citrullus colocynthis and its relationship with Citrullus species as determined by high-frequency oligonucleotides-targeting active gene markers. J Am Soc Hortic Sci 142(1):47–56

Levi A, Wechter WP, Massey LM, Carter L, Hopkins D (2011b) Genetic linkage map of Citrullus lanatus var. citroides chromosomal segments introgressed into the watermelon cultivar Crimson Sweet (Citrullus lanatus var. lanatus) genome. Am J Plant Sci 2:93–110

Li N, Shang J, Wang J, Zhou D, Li N, Ma S (2020) Discovery of the genomic region and candidate genes of the scarlet red flesh color (*Yscr*) locus in watermelon (*Citrullus Lanatus* L.). Front Plant Sci 11:116

Liu C, Zhang H, Dai Z, Liu Y, Deng X, Chen F, et al (2012) Volatile chemical and carotenoid profile in watermelon [*Citrullus vulgaris* (Thunb.) Schrad (Cucurbitaceae)] with different flesh colors. Food Sci Biotechnol 21:531–541

Liu S, Gao P, Wang X, Davis A, Baloch AM, Luan F (2015) Mapping of quantitative trait loci for lycopene content and fruit traits in *Citrullus lanatus*. Euphytica 202:411–426

Manivannan A, Lee ES, Han K, Lee HE, Kim DS (2020) Versatile nutraceutical potentials of watermelon—a modest fruit loaded with pharmaceutically valuable phytochemicals. Molecules 25:5258

Maoto MM., Beswa D, Jideani AIO (2019) Watermelon as a potential fruit snack. Int J Food Prop 22(1):355–370

Mashilo J, Shimelis H, Ngwepe RM, Thungo Z (2022) Genetic analysis of fruit quality traits in sweet watermelon (*Citrullus lanatus* var. *lanatus*): a review. Front Plant Sci 13:834696

Mcgregor C (2012) *Citrullus lanatus* germplasm of southern Africa. Israel J Plant Sci 60(4):403–413

Munshi AD, Tomar BS, Jat GS, Singh J (2017) Quality seed production of open pollinated varieties and F 1 hybrids in cucurbitaceous vegetables. Compendium-ICAR Sponsored Ten days short course on "Advances in variety maintenance and quality seed production for entrepreneurship, pp 107–125

Naz A, Butt MS, Sultan MT, Qayyum MMN, Niaz RS (2014) Watermelon lycopene and allied health claims. Excli J 13:650–666

Nimmakayala P, Natarajan P, Lopez-Ortiz C, Dutta SK, Levi A, Reddy UK (2022) Population genomics of sweet watermelon. In: Rajora OP (ed) Population Genomics: Crop plants. Springer Nature

Nimmakayala P, Vajja G, Gist RA, Tomason YR, Levi A, Reddy UK (2011) Effect of DNA methylation on molecular diversity of watermelon heirlooms and stability of methylation-specific polymorphisms across the genealogies. Euphytica 177:79–89

Oberoi DPS, Sogi DS (2015) Prediction of lycopene degradation during dehydration of watermelon pomace (Cv Sugar Baby). J Saudi Soc Agric Sci 16:97–103

Oberoi DPS, Sogi S (2017) Utilization of watermelon pulp for lycopene extraction by response surface methodology. Food Chem232:1–7

Odewunmi NA, Umorena SA, Gasema ZM, Ganiyub SA, Muhammad Q (2015) L-Citrulline: an active corrosion inhibitor component of watermelon rind extract for mild steel in HCL medium. J Taiwan Inst Chem Eng 51:177–185

Pacier C, Martirosyan DM (2015) Vitamin C: optimal dosages, supplementation and use in disease prevention. Funct Foods Health Dis 5:89–107

Paris HS (2015) Origin and emergence of the sweet dessert watermelon *Citrullus lanatus*. An Bot 116:133–148

Pei S, Liu Z, Wang X, Luan F, Dai Z, Yang Z et al (2021) Quantitative trait loci and candidate genes responsible for pale green flesh colour in watermelon (*Citrullus lanatus*). Plant Breed 140:349–359

Petyaev IM (2016) Lycopene deficiency in ageing and cardiovascular disease. Oxid Med Cell Longevity, 1–6

Poole CF, Grimball PC (1945) Interaction of sex, shape, and weight genes in watermelon. J Agric Res 71:533–552

Reddy UK, Abburi L, Abburi VL, Saminathan T, Cantrell R, Vajja VG, Reddy R, Tomason YR, Levi A, Wehner TC, Nimmakayala P (2014) A genome-wide scan of selective sweeps and association mapping of fruit traits using microsatellite markers in watermelon. J Hered 106(2):166–176

Reddy UK, Nimmakayala P, Levi A, Abburi VL, Saminathan T, Tomason YR, Vajja G, Reddy R, Abburi L, Wehner TC, Ronin Y, Korol A (2014b) High-resolution genetic map for understanding the effect of genome-wide recombination rate on nucleotide diversity in watermelon. G3: Genes|Genomes|Genetics 1;4(11):2219–30

Ren Y, Mcgregor C, Zhang Y, Gong G, Zhang H, Guo S et al (2014) An integrated genetic map based on four mapping populations and quantitative trait loci associated with economically important traits in watermelon (*Citrullus lanatus*). BMC Plant Biol 14(1):1–1

Robinson RW, Decker-Walters DS (1997) Cucurbits. CAB International Publishing, Oxon

Rodríguez-Roque MJ, de Ancos B, Sánchez-Moreno C, Cano MP, Elez-Martínez P, Martín-Belloso O (2015) Impact of food matrix and processing on the in vitro bioaccessibility of Vitamin C. Phenolic. J. Funct. Foods. 14:33–43. https://doi.org/10.1016/j.jff.2015.01.020

Sain RS, Joshi P (2003) Pollen fertility of interspecific F1 hybrids in genus *Citrullus* (Cucurbitaceae). Curr Sci 85:431–434

Shao P, Qiu Q, Iao J, Sun P (2017) Chemical stability and in vitro release properties of β-carotene in emulsions stabilized by Ulva fasciata polysaccharide. Int J Appl Sci Technol 102:225–323

Singh S, Selvakumar R, Manisha M, Kalia P (2020) Breeding and genomic investigations for quality and nutraceutical traits in vegetable crops-a review. Indian J Hort 77(1):1–40

Singh SP, Dadwadia G (2009) Analysis of heterosis and combining ability status among diallel set of hybrids for yield and quality traits in watermelon (*Citrullus lanatus*Thunb). Veg Sci 36(3s):323–326

Soteriou GA, Kyriacou MC, Gerasopoulos ASD (2014) Evolution of watermelon fruit physicochemical and phytochemical composition during ripening as affected by grafting. Food Chem 165:282–289

Takahashi H, Mizuno H, Yanagisawa A (2012) High-dose intravenous vitamin C improves quality of life in cancer patients. Pers Med Universe 1:49–53

Tarazona-Díaz MP, Viegas J, Moldao-Martins M, Aguayo E (2011) Bioactive compounds from flesh and by-product of fresh-cut watermelon cultivars. J Sci Food Agric 91:805–812

Vaccaa RA, Valenti D, Caccamese S, Dagliac M, Braidy N, Nabavi SM (2016) Plant polyphenols as natural drugs for the management of down syndrome and related disorders. Neurosci Biobehav Rev 71:865–877

Wang Y, Wang J, Guo S (2021) CRISPR/Cas9-mediated mutagenesis of CLBG1 decreased seed size and promoted seed germination in watermelon. Hortic Res. 8:70

Wehner TC (2017) Heritability and genetic variance components associated with citrulline, arginine, and lycopene content in diverse watermelon cultigens. HortSci 52(7):936–940

Yang X, Ren R, Ray R, Xu J, Li P, Zhang M et al (2016) Genetic diversity and population structure of core watermelon (*Citrullus lanatus*) genotypes using dart-seq based SNPs. Plant Genet Resour 14:226–233

Zhang J, Guo S, Ren Y, Zhang H, Gong G, Zhou M, Wang G, Zong M (2017) High-level expression of a novel chromoplast phosphate transporter ClPHT4;2 is required for flesh color development in watermelon. New Phytol 213(3):1208–1221

Zhang J, Jianting SH, Gaojie JI, Zhang H, Guoyi GO, Shaogui GU, Yi RE, Jianguang FA, Shouwei TI, Yong XU (2017) Modulation of sex expression in four forms of watermelon by gibberellin, ethephone and silver nitrate. Hortic Plant J 1;3(3):91–100

Genomic Resources for Disease Resistance in Watermelon

9

Brahma Induri, Padma Nimmakayala,
and Umesh K. Reddy

9.1 Introduction

Watermelon belonging to Cucurbitaceae family is one of the very healthy, fleshy, and economically important fruits in the world. It is produced worldwide in several countries with a yearly production of 117 million tons in 2016 (FAOSTAT, http://www.fao.org/faostat/en/). It promotes individual health with its rich lycopene, vitamin C, and health-promoting amino acids including citrulline, arginine, and glutathione (Hayashi et al. 2005). It is also rich in vitamins A, C, B1, and B6 (Naz et al. 2014). Watermelon shows extensive phenotypic variation in size, color, texture, rind patterns, and other morphological features. Availability of watermelon genome sequence (Guo et al. 2013) is a valuable genomic resource that enables us to study the genomic regions responsible for our target traits. By studying the interaction between phenotypic variation and among different genotypes, we can dissect the genetic basis of this extensive phenotypic variation. One drawback is that most of the watermelon varieties have similar genetic base (Levi et al. 2001a, b) because of the founder effect (Nimmakayala et al. 2014b) caused due to introduction of genetically similar cultivars. Intensive selection for desirable fruit qualities for several years resulted in low levels of genetic diversity (Nimmakayala et al. 2010; Reddy et al. 2014a) resulting in accumulation of deleterious alleles in the process, leading to susceptibility to pests, pathogens, viruses, fungi, and nematodes. Hence, there is a pressing need to increase the allelic diversity conferring resistance to pathogens, nematodes, and fungi in watermelon cultivars (Levi et al. 2017a).

Fusarium wilt, gummy stem blight, *Phytophthora* fruit rot, powdery mildew, downey mildew, bacterial fruit blotch (*Acidovorax citrulli*), and several viruses including the watermelon strain of Papaya ringspot virus (PRSV-W), Zucchini yellow mosaic virus (ZYMV), and Cucumber green mottle mosaic virus (CGMMV) are the major diseases affecting watermelons currently (Grumet et al. 2021).

Citrullus genus from Africa has a total of seven known species (Chomicki and Renner 2015). Though citron *C. amarus* and egusi melon *C. mucosospermus* were until recently reported to be as subspecies of *C. lanatus* (Chomicki and Renner 2015), they are readily crossable with cultivated watermelon. *Citrullus lanatus, C. mucosospermus, and C. amarus* are considered to be native to North Africa, sub-Saharan West Africa, and South Africa, respectively, with the latter two species conferring resistance to *Phytophthora* fruit rot, powdery mildew, *Fusarium*

B. Induri · P. Nimmakayala · U. K. Reddy (✉)
Department of Biology, Gus R. Douglass Institute,
West Virginia State University, Institute, WV, USA
e-mail: ureddy@wvstateu.edu

B. Induri
e-mail: brahma.induri@wvstateu.edu

S. Kr. Dutta et al. (eds.), *The Watermelon Genome*, Compendium of Plant Genomes,
https://doi.org/10.1007/978-3-031-34716-0_10

wilt, gummy stem blight, anthracnose, and various viruses (Levi et al. 2017a).

Presence of landraces and different germplasm accessions in gene banks across the world as shown in Fig. 9.1 is a good genetic resource to explore the genetic diversity and sequence variations using the next-generation sequencing (NGS) technologies and develop cultivars with improved resistance to biotic and abiotic stresses.

The first published genome is from Chinese cultivar name "97103" and is an important genomic resource, but further needs to enrich with the pan-genome studies (Gao et al. 2019; Zhang et al. 2015), as it is important to note that depending on just one cultivar genome for mapping and genome-wide association studies will potentially miss important variations affecting any target trait of interest including disease resistance. "Charleston Gray," the American premier cultivar, that has resistance to *Fusarium* wilt and foliar anthracnose disease is also fully sequenced (Wu et al. 2019) (Fig. 9.2).

The single-nucleotide polymorphisms (SNPs) between the genome sequences of 97103 and "Charleston Gray" were very low (1SNP per 1300 bp) (Levi et al. 2017b). Reduced genetic diversity is the major challenge for the breeders to develop cultivars with disease resistance (Levi et al. 2017b). Using NGS technologies, it is now possible to understand and utilize useful genetic variation to improve various traits of interest (Reddy et al. 2014a, b). The gene banks in Tables 9.1 and 9.2 have a wide collection of accessions that can be explored for widening genetic base in watermelon cultivars (Reddy et al. 2014a). The USDA-SCRI 'CucCAP' project is a collaborative effort by the US cucurbit community to develop SNPs to facilitate the introgression and pyramiding of disease resistance loci. Genotyping-by-sequencing (GBS) of the watermelon collections provided ~11,000 SNPs across the eleven watermelon chromosomes. The SNPs were used to characterize population differentiation of the collections and perform genome-wide association studies (GWAS) for many diseases (Wu et al. 2019).

These collections of watermelon accessions/germplasm/landraces are primarily collected from Africa and well adapted to Europe and the USA are very powerful germplasm resources to widen genetic diversity by enabling researchers to screen for unexplored resources for disease resistance, thereby recovering the rare alleles lost due to the founder effects causing bottleneck. All the seven extant species of *Citrullus* as shown in Table 9.1 are a great source because of cross-compatibility among them facilitating introgression of new alleles. Global germplasm centers of watermelon are excellent resource for watermelon resistant breeding (Table 9.3).

Databases that are freely accessible to watermelon researchers is of immense use for researchers

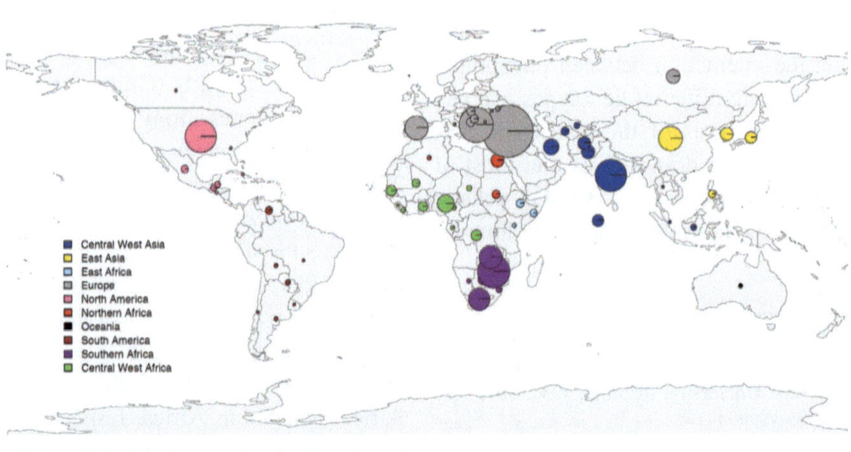

Fig. 9.1 Geographical distribution of the 1365 Citrullus spp. accessions in the US national plant germplasm system. Size of the circle is proportional to the number of accessions from each country (Wu et al. 2019)

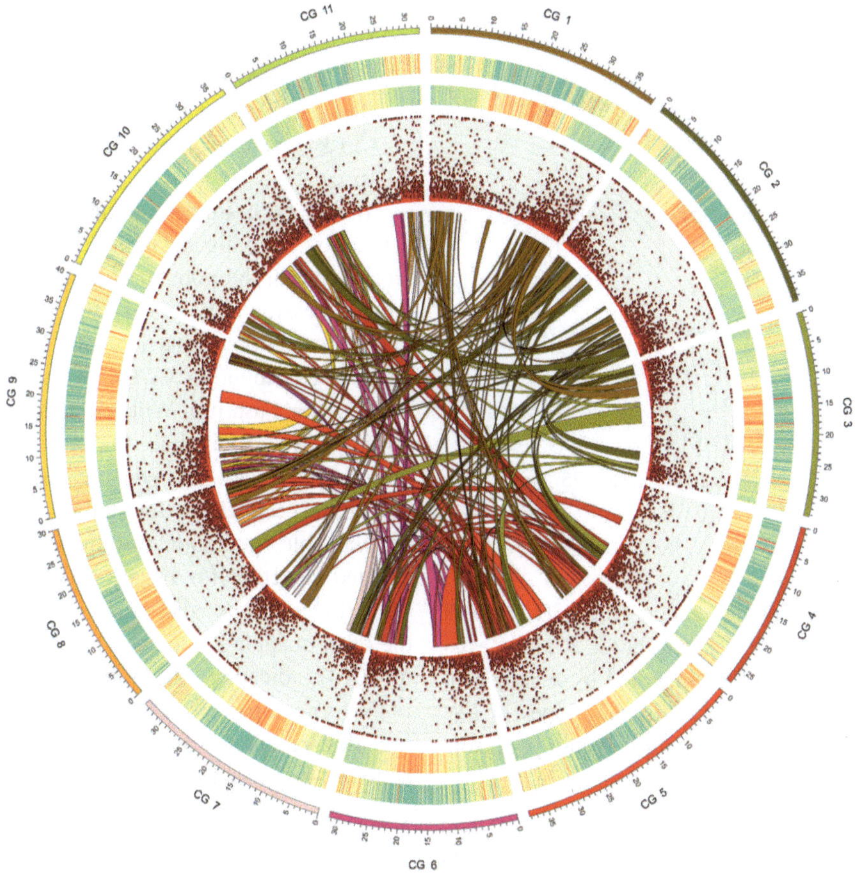

Fig. 9.2 Genomic landscape of watermelon, "Charleston Gray." The outermost circle is the ideogram of 11 chromosomes in Mb scale, followed by circles of gene density and TE density represented by percentage of genomic regions covered by genes and repeat sequences in 200 kb windows, respectively (green to red, low to high), gene expression levels (RPKM; Maximum = 200), and syntenic blocks within the genome depicted by lines. Figure from (Wu et al. 2019)

Table 9.1 Active Citrullus accessions in the NPGC

Species	Number of accessions
C. amarus	151
C. colocynthis	24
C. ecirrhosus	3
C. lanatus	1613
C. mucosospermus	75
C. naudinianus	7
C. rehmii	4
Citrullus spp.	1

https://npgsweb.ars-grin.gov/gringlobal/search.aspx. Accessed 8 March 2021 (Grumet et al. 2021)

to extract phenotypic and genotypic datasets from the listed websites.

9.2 Genes Involved in Disease Resistance in Watermelon

9.2.1 Fusarium Wilt

Since 19th centuary, in the US, Fusarium Wilt (FW) is the most economically important disease of watermelon, a soil-borne pathogen *Fusarium oxysporum f. sp. niveum* (FON) (Martyn 2014), resulting in huge crop losses. The crop losses due

Table 9.2 The major watermelon germplasm collections recorded in the GeneSys system

Country	Number of accessions (*C. lanatus//Citrullus sp.*)
Brazil	2007//2010
US	1922//2211
Sudan	469//471
Spain	428//435
Ukraine	395//452
Germany	249//266
Hungary	240//253
Bulgaria	–//242
Poland	101//101

https://www.genesys-pgr.org. Accessed 25 April 2021 (Grumet et al. 2021)

to Fusarium Wilt can range from 30 to 80% (Rahman et al. 2021, Zhang et al. 2021). Resistance to FW is well documented by several watermelon breeders. There are many races of FW affecting all the cucurbit crops. Notably, Kasote et al. (2020) (Fig. 9.3).

Among the genetic studies, phytosulfokine (PSK), a disulfated pentapeptide signaling pathway is reported to be involved in conferring resistance to FW as validated by CRISPR editing of Clpsk1 gene in watermelon (Zhang et al. 2020).

Among QTL studies, an analysis using a F8 generation recombinant inbred lines (RILs) of a cross between the popular cultivar 97,103 and a wild accession PI 296341-FR resolved QTL FON-1, on chromosome 1, with a LOD score of 13.2, and accounting 48.1% of the phenotypic variance for Fusarium Wilt resistance (Ren et al. 2015). Two additional QTLs Qfon2.1 and Qfon2.2 were also mapped on chromosomes 9 and 10, respectively in this study. These loci will be valuable genomic resources for marker-assisted selection for Fusarium Wilt resistance. QTL FON-1 was also previously reported in the other mapping studies (Lambel et al. 2014; Yi 2013) indicating it is an important QTL. QTL FON-1 encompasses a total of 84 genes that includes receptor kinase (Cla004916), glucan endo-1,3-bglucosidase precursor (Cla004990),

Table 9.3 Databases on genetic resources of watermelon

Databases/centers	Resource link	Reference
Watermelon Genome Database	http://www.iwgi.org/	Guo et al. (2013)
Indian Council of Agricultural Research (ICAR)–National Bureau of Plant Genetic Resources (NBPGR)	http://www.nbpgr.ernet.in/	Grumet (2021)
USDA/ARS/NPGS, Germplasm Resources Information Network (GRIN)	http://www.ars-grin.gov/npgs	Levi et al. (2017)
Cucurbit Genomics Database (CuGenDB)	http://www.icugi.org/	Guo et al. (2013)
FAO statistics database	http://faostat.fao.org/	Guo et al. (2013)
National Centre for Biotechnology Information (NCBI) database	www.ncbi.nlm.nih.gov	
National Agriculture and Food Research Organization (NARO)	https://www.naro.affrc.go.jp/archive/nias/eng/genresources/index.html	Grumet (2021)
The Vavilov Institute of Plant Industry	http://www.vir.nw.ru	Grumet (2021)
National Agrobiodiversity Center	http://genebank.rda.go.kr/	Grumet (2021)
The Uzbek Research Institute of Plant Industry, the Uzbek Research Institute of Vegetables, Melons, and Potato, the Karakalpak Research Institute of Agriculture, Tashkent, and the World Vegetable Center	https://avrdc.org http://seed.worldveg.org/	Grumet (2021)
The GeneSys SYSTEM	https://www.genesys-pgr.org/	Grumet (2021)

Fig. 9.3 Wilting disease symptoms of watermelon plants caused by *Fusarium oxysporum f. sp. niveum* under field condition. **a** A brown necrotic spot occurs first at the base of the stem and then **b** the whole plant wilts and dies (Rahman et al. 2021)

and three duplicate loci of acidic chitinase (Cla004914, Cla004920, and Cla004921), suggesting the need for fine mapping to resolve the causative resistant gene.

The QTL on chromosome 9 (Qfon2.1) has a lipoxygenase (Cla014845) gene, five receptor-like kinases (Cla014853, Cla014955, Cla014972, Cla014725, and Cla014728), and four glutathione S-transferase (Cla014674–Cla014677). Lipoxygenase is reported to be involved in defense responses to biotic and abiotic stresses (Yan et al. 2013). Glutathione S-transferase also known playing active role in detoxification of toxins caused by FW in watermelons (Mazzeo et al. 2014). The QTL on chromosome 10, Qfon2.2, has genes encoding arginine biosynthesis bifunctional protein (Cla017879), two receptor kinase proteins (Cla017918 and Cla017919), and lipid-transfer protein (Cla018045). Upregulation of arginase activity was reported to increase resistance in tomato roots infected with Fusarium wilt (Mazzeo et al. 2014).

9.2.2 Miscellaneous Disease Resistance Mechanisms Against FON

Jasmonic acid-isoleucine (JA-Ile), methyl jasmonate (MeJA), indole-3-acetic acid (IAA), melatonin, and lysine can also induce plant immunity to FON by affecting different signaling pathways in watermelon as reported by Kasote et al. (2020). Grafting of watermelon on a wide-ranging rootstocks has been reported to improve the plant defense mechanisms against FON as shown by the different differentially expressed proteins and through pathogen-induced proteases (Zhang et al. 2021). It is also demonstrated that by using bacterial strain *Bacillus velezensis* F21, we can control and suppress the spore formation of FON in watermelon cultivation to control Fusarium wilt (Jiang et al. 2019). Gene ontology (GO) classification and Kyoto Encyclopedia of Genes and Genomes (KEGG) pathway suggested that several transcription factors (TFs) and disease resistance genes in watermelon show higher expression levels of MAPK, phytohormones, and enzymes including defense enzymes, such as CAT, POD, and SOD (Jiang et al. 2019).

9.2.3 Gummy Stem Blight

A QTL study using the F2 population from a cross between "920533" (*C. lanatus*) and the paternal-resistant line "PI 189225" (*C. amarus*) for Gummy Stem Blight (GSB) resistance in watermelon, reported qLL8.1, qSB8.1, and qSB6.1, on chromosomes 8 and 6, explaining 10.5, 10.0, and 9.7% of the phenotypic variation observed for these traits, and the four candidate genes identified from this study are an important resource for further studies of GSB in watermelon (Lee et al. 2021) (Fig. 9.4).

In another study involving GSB resistance in watermelons examined 178 segregants resulting from a cross between Crimson Sweet (*C. lanatus*) and GSB-resistant PI 482276 (*C. amarus*) to map three QTLs (ClGSB3.1, ClGSB5.1, and ClGSB7.1) with a phenotypic variance ranging between 6.4 and 21.1% (Gimode et al. 2021). The gene underlying the QTL ClGSB5.1 is of NBS-LRR (ClCG05G019540). The QTL locus ClGSB7.1 encompasses ClCG07G013230 that encodes for Avr9/Cf-9, a rapidly elicited disease resistance protein with a non-synonymous point mutation in its DUF761 domain (Gimode et al. 2021).

NBS-encoding R genes comprise Toll/interleukin-1 receptor(TIR), NBS, leucine-rich repeat (LRR), and coiled coil (CC) domains. Six NBSLRR genes (Cla001821, Cla019863, Cla020705, Cla012430, Cla012433, and Cla012439) were highly expressed in the resistant line than that of the susceptible one to D. *bryoniae* infection, suggesting that these genes will be a valuable improve GSB resistance in watermelon (Hassan et al. 2019). An F2 population of a cross between PI189225 and K3 was screened for GSB resistance in watermelon, and *Qgsb8.1* was identified that has 19 genes, of which two genes Cla001017 and Cla 001019 were further characterized as useful for building resistance (Ren et al. 2019). In another study of GSB resistance screening involving intermating of PI 482283, PI 526233, and PI 189225, reported that the GSB inheritance is manifested by multiple genes (Gusmini et al. 2017).

9.2.4 Anthracnose

Anthracnose disease caused by *Colletotrichum orbiculare* is one of the severe diseases in watermelons and the resulting losses due to this disease can range from 5 to 20% (Guo et al. 2022) or even 100% (Prusky 1996). This disease can also result in severe losses during post harvesting operations including transportation and storage (Dean et al. 2012) (Fig. 9.5).

A SNP, CL14-27-9 on Chromosome 8, in the exon of gene Cla001017 encoding a CC–NBS–LRR resistance protein, has been reported as a resistance marker to the race 1 of Anthracnose in watermelon (Jang et al. 2019). Watermelon has a cluster of 18 NBSLLR resistance genes on chromosome 8 and one of these is Cla001017 (Guo et al. 2013). This SNP conferred resistance to Anthracnose in cucumber cultivars also when analyzed using CAPS marker analysis (Matsuo

Fig. 9.4 The symptoms of gummy stem blight, GSB, in watermelon accessions (Hassan et al. 2019)

Fig. 9.5 Anthracnose symptoms in watermelon leaves, stems, and fruits (Guo et al. 2022)

et al. 2022), suggesting its role in the other cucurbit crops.

9.2.5 Powdery Mildew

Powdery Mildew (PM) is one of the most devastating diseases of watermelon and causes low productivity (up to 50% yield losses) of watermelon and in the other cucurbits by affecting as early as cotyledon stage to matured leaves in watermelons causing chlorotic spots on leaves at initial stages to fully formed lesions (Fig. 9.6 and Table 9.4) (Yadav et al. 2021; Levi et al. 2017b). Kousik and Adkins (2020) identified USVL531-MDR that is multiple disease-resistant (MDR) watermelon that exhibits high levels of resistance to a broad range of isolates of cucurbit powdery mildew (Podosphaera xanthii) and Phytophthora capsica that cause Phytophthora fruit rot.

A bulked segregant analysis of different F2 population of watermelon cultivars revealed a single incomplete dominant gene *Pm1.1* responsible for powdery mildew resistance in watermelon and is a valuable locus to target for further studies on disease resistance to powdery mildew studies (Kim et al. 2013).

9.2.6 Downey Mildew

Downey mildew is a sporadic yet dangerous disease affecting watermelon, reducing yield and fruit quality with its defoliation of vines. The oldest leaves are always attacked the first, with symptoms appearing pale green to yellow at the beginning and then to turning dark brown to black in circular to irregular-shaped spots (Fig. 9.7).

The American commercial cultivar, Sakata F1-Charleston Gray 243, was found to be

Fig. 9.6 Powdery mildew symptoms on watermelon leaves and stems (Han et al. 2016)

Table 9.4 United States' watermelon plant introductions (PIs) with disease resistance

Disease	Resistant plant introductions
Fusarium wilt races 1 and 2	PI 482246, PI 482252, PI 271769, PI 299378, PI 482273, PI 482299, PI 482308, PI 296341
Gummy stem blight	PI 271778, PI 244019, PI 271771, PI 482276, PI 482379, PI 500334, PI 500335, PI 296332, PI 505590, PI 526233, P1 189225
Anthracnose	PI 248774, PI 270562, PI 270563, PI 271775, PI 271778, PI 271779, PI 299379, PI 500315, PI 500329, PI 505592, PI 505593
Phytophthora capsici	PI 306782, PI 595203, PI 560020, PI 560002, PI 494531, PI 482328, PI 189225 and PI 532738
Powdery mildew	PI 494531, PI 482323, PI 482326, PI 482328, PI 189225, PI 532738, PI 482312 and PI525088
Watermelon mosaic virus (WMV)	PI 244018, PI 244019, PI 255137 and PI 482252
Papaya ringspot virus	PI 244018, PI 244019, PI 255137, PI 482252, PI 244017, PI 482318, PI 244017, PI 244019, and PI 485583
Watermelon strain (PRSV-W)	PI 482342, PI 482299, PI 482315, PI 482322, PI 482379, PI 485583 and PI 595,203
Squash vein yellowing virus (SqVYV)	PI 392291, PI 482266, PI 386015, PI 386024, PI 459074, PI 500354
Zucchini yellow mosaic virus (ZYMV-FL)	PI 482322, PI 482299, PI 482261, PI 482308, PI 244018, PI 482276, PI 485580, PI 596662, PI 537277, PI 560016, PI 386016, PI 386019, PI 485580, PI 494529, PI 595200, PI 494528, PI 595201, PI 386025, PI 494530, PI 386015, PI 386021, PI 386026, PI 596662, PI 386018, PI 595202 and PI 595203

(Levi et al. 2017b)

resistant to downey mildew in two consecutive seasons of infection compared to the other cucurbits that showed variable levels of downey mildew infection, and this genotype will be useful for further research and watermelon breeding for disease resistance (Zakeri et al. 2022).

9.2.7 Bacterial Fruit Blotch (BFB)

Bacterial fruit blotch, BFB, caused by *Acidovorax citrulli,* is a devastating bacterial disease caused by bacteria in watermelon with no known cure or prevention for this disease (Klomchit et al. 2021). Resistance to BFB will reduce the cost of watermelon production by manifold with integrated pest management practices, but as of now, there are no known resistance cultivars to this deadly seed-borne disease of watermelon (Branham et al. 2019). It infects, seeds, cotyledons, leaves, and fruits and can result in fruit blotching, cracking, and internal necrosis of the fruit parts (Daley and Wehner 2021), and screening of different PIs resulted in resistant PIs for BFB in watermelon including PI 494819, PI 596659, PI 596670, PI 490384, and PI 596656. In a genotype screening study for BFB tolerance study, BGCIA 979, BGCIA 34, and Sugar Baby showed high levels of resistance to BFB (Carvalho et al. 2013).

9.2.8 Viruses

The following are the different viruses affecting watermelon crop in the US and around the world.

- Papaya ringspot virus-Watermelon—(PRSV-W)
- Zucchini yellow mosaic virus—(ZYMV)
- Cucumber green mottle mosaic virus—(CGMMV)

Fig. 9.7 Downey mildew symptoms of dark brown spots and inward curling of leaves in watermelon

- Cucumber mosaic virus—(CMV)
- Cucurbit yellow stunting disorder virus (CYSDV)
- Squash vein yellowing virus—(SqVYV)
- Cucurbit leaf crumple virus—(CuLCrV).

9.2.9 Papaya Ringspot Virus-Watermelon (PRSV-W)

Viruses are a major limiting factor for watermelon production worldwide and around 10 viruses are affecting watermelon with two pathotypes of this potyvirus, affecting watermelon production, growth, and fruit quality (Guner et al. 2018a). The two types of pathotypes are: PRSV-P (papaya strain) that infects papaya, *Carica papaya,* and a potyvirus strain PRSV-W (watermelon strain) that exclusively infects cucurbits (Romay et al. 2014).

The wide genetic distance and genetic diversity between *C. lanatus and C. colocynthis* can be explored by creating backcross populations, and the disease resistance genes can be introgressed into the watermelon cultivars (Levi et al. 2016). Extensive genetic diversity present in all the seven extant species of watermelon is a valuable genetic resource that researchers can exploit for disease resistance alleles and then introgression of these alleles into the existing elite watermelon cultivars.

Papaya ringspot virus—Watermelon (PRSV-W) was formerly known as Watermelon mosaic virus 1 (WMV-1 and WMV-2) and is transmitted through aphids. Mosaic symptoms can appear in the form of blisters and deformations on leaves with malformed fruit and its color and plant showing stunted growth. A single recessive gene, *prv*, controlled the PRSV—W disease resistance to papaya ring spot virus in a screening of different watermelon genotypes (Guner et al. 2018a).

9.2.9.1 Zucchini Yellow Mosaic Virus—(ZYMV)

ZYMV is aphid transmitted and infects all cucurbits and its symptoms include stunting, yellowing, leaf deformations with dark green blisters, mosaic, and fruit deformation (Kaldis et al. 2018). Exogenous application of dsRNAs that were derived from HC-Pro and CP genes of ZYMV improved the disease resistance to ZYMV in watermelon and cucumber, and this resistance is more of a sequence-specific and dose-dependent method (Kaldis et al. 2018). These dsRNAs activated the plant's RNA-silencing mechanism, resulting in degradation of ZYMV genome conferring disease resistance to watermelon plants. Regular disease control methods including chemical controls have been rendered inefficient, and cultural controls such as application of mineral oils and exudates of ZYMV resulted in increasing the production cost

Fig. 9.8 Watermelon leaves inoculated with ZYMV 14 and 20 days after inoculation and the control leaves at the bottom of the picture (Kaldis et al. 2018)

of the crop and without much improvement in disease resistance and genetic resistance are the most effective and efficient method of disease control (Guner et al. 2018b) (Fig. 9.8).

In one screening study for resistance against ZYMV, the very high level of resistance to ZYMV observed in PI 595203 was because of a single recessive gene *zym*-FL (Guner et al. 2018b).These genes will be valuable tools for future studies on disease resistance to ZYMV.

9.2.9.2 Cucumber Green Mottle Mosaic Virus (CGMMV)

CGMMV's spread in 30 countries has been a global threat to cucurbit-based industries worldwide and, with a 6.4 kb—single-stranded and positive-sense RNA genome, CGMMV produces stiff rod-shaped particles and its symptoms include mottling and mosaic on leaves, necrotic lesions on stems, and dirty red discolorations with spongy and rotting pulp in peduncles and flesh (Sun et al. 2019) (Fig. 9.9).

CGMMV infection changed the gene expression levels of genes in watermelon, and functional analysis of these genes showed that these genes were involved in plant–pathogen interactions, secondary metabolism, plant hormone signal transduction, and photosynthesis. Different transcription factor families including WRKY, NAC, MYB, bZIP, and zinc finger levels are affected upon CGMMV infection in watermelon (Sun et al. 2019).

Watermelon blood flesh disease caused by CGMMV can be controlled by exogenous Boron application, and virus-induced gene silencing methods with silencing of SPDS resulted in the inhibition of CGMMV infection and no watermelon blood flesh disease symptoms (Bi et al. 2022), with RAP2-3, MYB6, WRKY12, H2A, and DnaJ11 contributing to host antiviral resistance. This new knowledge of boron application can be used to design better control methods under field conditions and also to breed watermelon and other cucurbit cultivars resistant to CGMMV in future.

9.2.9.3 Cucurbit Yellow Stunting Disorder Virus (CYSDV), Cucurbit Leaf Crumple Virus—(CuLCrV), Squash Vein Yellowing Virus—(SqVYV)

CYSDV was first reported in 2000 in the US, and this virus caused outbreaks in the US in 2006, leading to 100% losses of the watermelon crops. This virus is transmitted by whiteflies along with other viruses including Cucurbit leaf crumple virus—(CuLCrV) and Squash vein yellowing

Fig. 9.9 Control leaf and CGMMV-inoculated leaf and CGMMV accumulation after 24 h as can be seen in the RT-PCR gel image in **c** Figure adopted from (Sun et al. 2019)

Fig. 9.10 Watermelon leaves affected by Cucurbit yellow stunting disorder virus (CYSDV) and Cucurbit leaf crumple virus—(CuLCrV). Figure from (Kousik and Adkins 2020)

virus—(SqVYV) (Kousik and Adkins 2020) (Fig. 9.10).

Symptoms include bright yellow and interveinal chlorosis of older and mid-growth leaves, stunting, necrosis of leaf margins, and leaf crumpling (Gadhave et al. 2018) (Table 9.5).

9.3 Next-Generation Sequencing in Watermelon

Modern sequencing technologies are needed for simultaneous genome sequencing of large number of diverse accessions to help us to generate vast genetic information, and molecular markers developed using this information will help the breeders in marker-assisted selection for disease resistance in watermelon (Wu et al. 2019; Levi et al. 2017b; Reddy et al. 2014a). Understanding genetic diversity and population structure will provide more insights in the genes responsible for biotic and abiotic stresses. SSR markers and SNP markers were developed to construct a linkage map to understand the genetic diversity in watermelon through whole-genome sequencing (Nimmakayala et al. 2014a) and to understand the effect of genome-wide recombination rate on nucleotide diversity in watermelons (Reddy et al. 2014b; Wu et al. 2019). A large number of SSR markers were developed by using next-generation sequencing technologies on watermelon draft genome sequence to study

Table 9.5 Genes involved in disease resistance in watermelon

Disease name	Causative agent	Chromosome	Resistance gene name or genotype name	Location	Author and year
Fusarium Wilt					
	F. oxysporums.sp. niveum (FON)	7, 11	*Clpsk1*	Clpsk1	Zhang et al. (2020)
	FON race 1	1		Locus Fo-1.3	Meru and McGregor (2016)
	FON race 2	2		Locus qFon2-2	Branhan et al. (2017)
	FON race 1	1	*Cla014674, Cla014677, Cla014728, Cla014853, Cla014955, Cla014972, Cla014725, Cla014845, SNP marker Chr1SNP_502124*	Qfon2.1 QTL	Ren et al. (2015)
	FON race 2	9, 10	*Cla017879, Cla017918 and Cla017919, Cla018045*	Qfon2.2 QTL	Ren et al. (2015)
	FON race 1	1, 3, 4, 9, 10	*Cla004884, Cla004990, Cla004914, Cla004920, Cla004921*	QTL Fo-1.1	Lambel et al. (2014)
	FON race 1	1	*Fon—1*	7716_fon, 7419_fon and 4451_fon	Yi et al. (2013)
Gummy stem blight					
	Didymella bryoniae	6, 8		qLL8.1, qSB8.1 on 8, qSB6.1 on 6	Lee et al. (2021)
	Didymella bryoniae	3, 5, 7	*NBS-LRR gene (ClCG05G019540) and 22 other candidate genes including ClCG07G013230, encoding an Avr9/Cf-9 rapidly elicited disease resistance protein contains a non-synonymous point mutation in the DUF761 domain*	Three QTLs (ClGSB3.1, ClGSB5.1 and ClGSB7.1)	Gimode et al.
	Didymella bryoniae		Genotype 189225 PI	Cla 001017 and Cla 001019	Ren et al. (2020)
	Didymella bryoniae		*Nucleotide binding site (NBS)-encoding resistance (R) genes Cla019863, Cla020705, Cla012430, Cla012433 and Cla012439)*		Hassan et al. (2019)

(continued)

Table 9.5 (continued)

Disease name	Causative agent	Chromosome	Resistance gene name or genotype name	Location	Author and year
Anthracnose					
	Colletotrichum orbiculare race 1	8	*Cla001017 encoding variable site in NBS–LRR protein*	Cla001017	Jang et al. (2019)
Powdery mildew					
	Podosphaera xanthii race 2F		M16, M11 and M49 resistant lines		Yadav et al. (2021)
	Race 1		BIU 19 AND PI 189225, PI 482312 and M16, M11 and M49 resistant lines		Ben-Naim and Cohen (2015)
	P. xanthii 1W-AN		*CAPS and HRM-type molecular markers, 254PMR-Nco and 254PMR-HRM3*		Bal-Kum Han et al. (2016)
	Race 1W		The Indian open pollination cultivar "ArkaMank"	Locus *Pm1.1*	Kim et al. (2013)
Downey mildew					
	Pseudoperonospora cubensis		Resistant genotypes screening	Resistant genotypes	
	Pseudoperonospora cubensis		Sakata F1-Charleston Gray 243	Cucurbits resistance testing	Zakeri et al. (2022)
Viruses					
	Papaya ring spot virus—PRSV-W		Griffin 14201, PI 537277, PI 652554 and PI 525080	Resistant genotypes identified	Levi et al. (2016)
	Papaya ring spot virus—PRSV-W	–	PI 244017, PI 244019, and PI 485583	*prv*	Guner et al. (2018a, b)
	Zucchini yellow mosaic virus—ZYMV		PI 595203	*zym-FL*	Guner et al. (2018b)
	Zucchini yellow mosaic virus—ZYMV		Resistant genotypes screening	*dsRNA molecules derived from the helper component-proteinase (HC-Pro) and coat protein (CP) genes of the ZYMV_DE_2014*	Kaldis et al. (2018)

(continued)

Table 9.5 (continued)

Disease name	Causative agent	Chromosome	Resistance gene name or genotype name	Location	Author and year
	Cucumber green mottle mosaic virus—CGMMV		Differentially expressed genes (DEGs)	Transcription factor families, including WRKY, MYB, HLH, bZIP and NAC	Sun et al. (2019)
	Cucumber green mottle mosaic virus—CGMMV		Micronutrient exogenous Boron application	RAP2-3, MYB6, WRKY12, H2A, and DnaJ11	Bi et al. (2022)
	Cucurbit yellow stunting disorder virus (CYSDV)				Kousik and Adkins (2020)
	Squash vein yellowing virus (SqVYV)				
	Cucurbit leaf crumple virus (CuLCrV)				
	Cucumber mosaic virus		Micronutrient exogenous Boron application		
Bacterial fruit blotch	*Acidovorax citrulli*		Genotypes BGCIA 979, BGCIA 34 and sugar baby		Carvalho et al. (2013)
	Acidovorax citrulli		*ClMPK1, ClMPK4-2 and ClMPK7 positively but ClMPK6 and ClMKK2-2 negatively regulate the*		
	Acidovorax citrulli		PI 494819, PI 596659, PI 596670, PI 490384, PI 596656		
	Acidovorax citrulli	1,2,3 and 8	Resistant Genotype—USVL246-FR2	Six QTLs	Branham et al. (2019)

genetic diversity among different sweet and citron watermelons (Zhu et al. 2016).

9.4 CuGenDB and Bioinformatics

With the advent of genome sequencing and genomics era, it is imperative to have web-based databases to store and readily access the extensive information of genes, accessions, and genetic diversity of different crops. Genome sequence and resequencing data of two watermelon cultivars 97103 (Guo et al. 2013, 2019) and Charleston Gray (Wu et al. 2019) were stored in CuGenDB and the data are readily available for extraction from these databases of watermelons and other economically important cucurbits. Information regarding biochemical pathways, gene expression profiles, and large genetic datasets of watermelon and other cucurbits (Zheng et al. 2018). The latest version of CuGenDB database was released and running since 2022, thus confirming that bioinformatics databases are important genomic tools for obtaining phenotypic data and molecular markers, thus facilitating research studies on phenotype and genotype interaction for traits of interest.

CucCAP, a multi-institutional project aimed at introgression of disease resistance into cucurbit cultivars (https://cuccap.org/about/overview/), is another valuable genomic resource that can be utilized for developing watermelon cultivars with improved disease resistance.

9.5 Conclusion

The availability of resequencing technologies of multiple accessions should enable us to conduct further research on evolutionary genomics and comparative genomics, which provides more insights on disease resistance in watermelon and the other cucurbits. The presence of vast genomic resources and germplasm cultivars now should help us to broadly understand the genomic variation across different cultivars of different species around the world and help us understand the disease-resistant mechanisms in watermelons. With the high-throughput whole-genome sequencing available, disease-resistant wild species of *Citrullus* and the vast germplasm/landraces collection in the world and genetic diversity have the potential to provide valuable alleles through allele mining to increase the disease resistance and expand the limited genetic base of the current elite watermelon cultivars. Conducting a genome-wide association study for disease resistance in watermelon on the genes and genomic areas identified so far, and using gene editing through CRISPR/Cas9-mediated system is one of the immediate and practical genomic approaches to dissect disease resistance in watermelon. Even the recently concluded International Conference on Cucurbitaceae at Naples, Florida, USA (https://conference.ifas.ufl.edu/cucurbits/), stressed more on the development of disease and pest-resistant watermelon varieties using the novel genomic tools and resources currently available.

References

Bi X, Guo H, Li X, Zheng L, An M, Xia Z, Wu Y (2022) A novel strategy for improving watermelon resistance to cucumber green mottle mosaic virus by exogenous boron application. Mol Plant Pathol 23:1361–1380

Branham SE, Levi A, Katawczik ML, Wechter WA-O (2019) QTL mapping of resistance to bacterial fruit blotch in Citrullus amarus

Carvalho FCQ, Santos LA, Dias RCS, Mariano RLR, Souza EB (2013) Selection of watermelon genotypes for resistance to bacterial fruit blotch. Euphytica 190:169–180

Chomicki G, Renner SS (2015) Watermelon origin solved with molecular phylogenetics including Linnaean material: another example of museomics. New Phytol 205:526–532

Daley J, Wehner TC (2021) Screening for bacterial fruit blotch resistance in watermelon fruit. Crop Sci 61:1228–1240

Dean R, Van Kan JAL, Pretorius ZA, Hammond-Kosack KE, Di Pietro A, Spanu PD, Rudd JJ, Dickman M, Kahmann R, Ellis J, Foster GD (2012) The top 10 fungal pathogens in molecular plant pathology. Mol Plant Pathol 13:414–430

Dong W, Wu D, Li G, Wu D, Wang Z (2018) Next-generation sequencing from bulked segregant analysis identifies a dwarfism gene in watermelon. Sci Rep 8:2908

FAO (2012) United nations international treaty on plant genetic resources for food and agriculture. Available online: http://www.fao.org/plant-treaty/overview/texts-treaty/en/. Accessed on 25 April 2021

Gadhave KR, Dutta B, Coolong T, Sparks AN, Adkins S, Srinivasan R (2018) First report of cucurbit yellow stunting disorder virus in cucurbits in Georgia, United States. Plant Health Prog 19:9–10

Gao L, Gonda I, Sun H, Ma Q, Bao K, Tieman DM, Burzynski-Chang EA, Fish TL, Stromberg KA, Sacks GL, Thannhauser TW, Foolad MR, Diez MJ, Blanca J, Canizares J, Xu Y, van der Knaap E, Huang S, Klee HJ, Giovannoni JJ, Fei Z (2019) The tomato pan-genome uncovers new genes and a rare allele regulating fruit flavor. Nat Genet 51:1044–1051

Gimode W, Bao K, Fei Z, McGregor C (2021) QTL associated with gummy stem blight resistance in watermelon. Theor Appl Genet 134:573–584

Grumet R, McCreight JD, McGregor C, Weng Y, Mazourek M, Reitsma K, Labate J, Davis A, Fei Z (2021) Genetic resources and vulnerabilities of major cucurbit crops. Genes 12:1222

Guner N, Pesic-VanEsbroeck Z, Rivera-Burgos LA, Wehner TC (2018a) Inheritance of resistance to papaya ringspot virus-watermelon strain in watermelon. HortScience Horts 53:624–627

Guner N, Rivera-Burgos LA, Wehner TC (2018b) Inheritance of resistance to zucchini yellow mosaic virus in watermelon. HortScience Horts 53:1115–1118

Guo S, Zhang J, Sun H, Salse J, Lucas WJ, Zhang H, Zheng Y, Mao L, Ren Y, Wang Z, Min J, Guo X, Murat F, Ham B-K, Zhang Z, Gao S, Huang M, Xu Y, Zhong S, Bombarely A, Mueller LA, Zhao H, He H, Zhang Y, Zhang Z, Huang S, Tan T, Pang E, Lin K, Hu Q, Kuang H, Ni P, Wang B, Liu J, Kou Q, Hou W, Zou X, Jiang J, Gong G, Klee K, Schoof H, Huang Y, Hu X, Dong S, Liang D, Wang J, Wu K, Xia Y, Zhao X, Zheng Z, Xing M, Liang X, Huang B, Lv T, Wang J, Yin Y, Yi H, Li R, Wu M, Levi A, Zhang X, Giovannoni JJ, Wang J, Li Y, Fei Z, Xu Y (2013) The draft genome of watermelon (Citrullus lanatus) and resequencing of 20 diverse accessions. Nat Genet 45:51–58

Guo S, Zhao S, Sun H, Wang X, Wu S, Lin T, Ren Y, Gao L, Deng Y, Zhang J, Lu X, Zhang H, Shang J, Gong G, Wen C, He N, Tian S, Li M, Liu J, Wang Y, Zhu Y, Jarret R, Levi A, Zhang X, Huang S, Fei Z, Liu W, Xu Y (2019) Resequencing of 414 cultivated and wild watermelon accessions identifies selection for fruit quality traits. Nat Genet 51:1616–1623

Guo Z, Luo C-X, Wu H-J, Peng B, Kang B-S, Liu L-M, Zhang M, Gu Q-S 2022 Colletotrichum species associated with anthracnose disease of watermelon (Citrullus lanatus) in China. J Fungi https://doi.org/10.3390/jof8080790

Gusmini G, Rivera-Burgos LA, Wehner TC (2017) Inheritance of resistance to gummy stem blight in watermelon. HortScience Horts 52:1477–1482

Han BK, Rhee S-J, Jang YJ, Sim TY, Kim YJ, Park TS, Lee G (2016) Identification of a causal pathogen of watermelon powdery mildew in Korea and development of a genetic linkage marker for resistance in watermelon (Citrullus lanatus). Horticultural Sci Technol 34:912–923

Hassan MZ, Rahim MA, Jung H-J, Park J-I, Kim H-T, Nou I-S (2019) Genome-wide characterization of NBS-encoding genes in watermelon and their potential association with gummy stem blight resistance. Int J Mol Sci 20:902

Hayashi T, Packiasamy J, Matsui-Hirai H, Miyazaki A, Fukatsu A, Funami J, Iguchi A, Ignarro L (2005) L-Citrulline and L-arginine supplementation retards the progression of high-cholesterol-diet-induced atherosclerosis in rabbits. Proc Natl Acad Sci USA 102:13681–13686

Jang YJ, Seo M, Hersh CP, Rhee S-J, Kim Y, Lee GP (2019) An evolutionarily conserved non-synonymous SNP in a leucine-rich repeat domain determines anthracnose resistance in watermelon. Theor Appl Genet 132:473–488

Jiang C-H, Yao X-F, Mi D-D, Li Z-J, Yang B-Y, Zheng Y, Qi Y-J, Guo J-H (2019) Comparative transcriptome analysis reveals the biocontrol mechanism of Bacillus velezensis F21 against fusarium wilt on watermelon. Front Microbiol 10

Kaldis A, Berbati M, Melita O, Reppa C, Holeva M, Otten P, Voloudakis A (2018) Exogenously applied dsRNA molecules deriving from the Zucchini yellow mosaic virus (ZYMV) genome move systemically and protect cucurbits against ZYMV. Mol Plant Pathol 19:883–895

Kasote DM, Jayaprakasha GK, Ong K, Crosby KM, Patil BS (2020) Hormonal and metabolites responses in Fusarium wilt-susceptible and -resistant watermelon plants during plant-pathogen interactions. BMC Plant Biol 20:481

Kim K-H, Ahn S-G, Hwang J-H, Choi Y-M, Moon H-S, Park Y-H (2013) Inheritance of resistance to powdery mildew in the watermelon and development of a molecular marker for selecting resistant plants. Hortic Environ Biotechnol 54:134–140

Klomchit A, Calderin JD, Jaidee W, Watla-iad K, Brooks S (2021) Napthoquinones from Neocosmospora sp.—antibiotic activity against Acidovorax citrulli, the Causative Agent of Bacterial Fruit Blotch in Watermelon and Melon. J Fungi https://doi.org/10.3390/jof7050370

Kousik CS, Adkins S (2020) Detection of Cucurbit yellow stunting disorder virus infecting watermelon in South Carolina. Plant Health Prog 21:133–134

Lambel S, Lanini B, Vivoda E, Fauve J, Patrick Wechter W, Harris-Shultz KR, Massey L, Levi A (2014) A major QTL associated with Fusarium oxysporum race 1 resistance identified in genetic populations derived from closely related watermelon lines using selective genotyping and genotyping-by-sequencing for SNP discovery. Theor Appl Genet 127:2105–2115

Lee ES, Kim D-S, Kim SG, Huh Y-C, Back C-G, Lee Y-R, Siddique MI, Han K, Lee H-E, Lee J (2021)

QTL Mapping for Gummy Stem Blight Resistance in watermelon (Citrullus spp.). Plants 10:500

Levi A, Thomas CE, Keinath AP, Wehner TC (2001a) Genetic diversity among watermelon (Citrullus lanatus and Citrullus colocynthis) accessions. Genet Resour Crop Evol 48:559–566

Levi A, Thomas CE, Wehner TC, Zhang X (2001b) Low genetic diversity indicates the need to broaden the genetic base of cultivated watermelon. HortScience HortSci 36:1096–1101

Levi A, Coffey J, Massey L, Guner N, Oren E, Tadmor Y, Ling K-s (2016) Resistance to Papaya ringspot virus-watermelon strain (PRSV-W) in the desert watermelon Citrullus colocynthis. HortScience 51:4–7

Levi A, Jarret R, Kousik S, Patrick Wechter W, Nimmakayala P, Reddy UK (2017a) Genetic resources of watermelon. In: Grumet R, Katzir N, Garcia-Mas J (eds) Genetics and genomics of cucurbitaceae. Springer International Publishing, Cham, pp 87–110

Levi A, Jarret R, Kousik S, Wechter W, Nimmakayala P, Reddy UK (2017b) Genetic resources of watermelon

Martyn RD (2014) Fusarium wilt of watermelon: 120 years of research. Hortic Rev 42:349–442

Matsuo H, Mine K, Sano T, Yoneda H, Yoshioka Y (2022) Resistance of a previously screened old Japanese watermelon cultivar to diverse anthracnose strains and its inheritance. Hortic J 91:375–381

Mazzeo MF, Cacace G, Ferriello F, Puopolo G, Zoina A, Ercolano MR, Siciliano RA (2014) Proteomic investigation of response to forl infection in tomato roots. Plant Physiol Biochem 74:42–49

Naz A, Butt MS, Sultan MT, Qayyum MMN, Niaz RS (2014) Watermelon lycopene and allied health claims. EXCLI J 13:650–660

Nimmakayala P, Tomason YR, Jeong J, Ponniah SK, Karunathilake A, Levi A, Perumal R, Reddy UK (2010) Genetic reticulation and interrelationships among citrullus species as revealed by joint analysis of shared AFLPs and species-specific SSR alleles. Plant Genet Resour 8:16–25

Nimmakayala P, Abburi VL, Bhandary A, Abburi L, Vajja VG, Reddy R, Malkaram S, Venkatramana P, Wijeratne A, Tomason YR, Levi A, Wehner TC, Reddy UK (2014a) Use of VeraCode 384-plex assays for watermelon diversity analysis and integrated genetic map of watermelon with single nucleotide polymorphisms and simple sequence repeats. Mol Breeding 34:537–548

Nimmakayala P, Levi A, Abburi L, Abburi V, Tomason Y, Saminathan T, Vajja V, Ma, Sr., Reddy R, Wehner T, Mitchell S, Reddy UK (2014b) Single nucleotide polymorphisms generated by genotyping by sequencing to characterize genome-wide diversity, linkage disequilibrium, and selective sweeps in cultivated watermelon. BMC Genomics. https://doi.org/10.1186/1471-2164-15-767

Prusky D (1996) PATHOGEN QUIESCENCE IN POSTHARVEST DISEASES. Annu Rev Phytopathol 34:413–434

Rahman MZ, Ahmad K, Bashir Kutawa A, Siddiqui Y, Saad N, Geok Hun T, Hata EM, Hossain MI (2021) Biology, diversity, detection and management of Fusarium oxysporum f. sp. niveum causing vascular wilt disease of watermelon (Citrullus lanatus): a review. Agron 11:1310

Reddy UK, Abburi L, Abburi VL, Saminathan T, Cantrell R, Vajja VG, Reddy R, Tomason YR, Levi A, Wehner TC, Nimmakayala P (2014a) A genome-wide scan of selective sweeps and association mapping of fruit traits using microsatellite markers in watermelon. J Hered 106:166–176

Reddy UK, Nimmakayala P, Levi A, Abburi V, Saminathan T, Tomason Y, Vajja G, Reddy R, Abburi L, Wehner T, Ronin Y, Karol A (2014b) High-resolution genetic map for understanding the effect of genome-wide recombination rate on nucleotide diversity in watermelon. G3 (Bethesda, Md.) 4

Ren R, Xu J, Zhang M, Liu G, Yao X, Zhu L, Hou Q (2019) Identification and molecular mapping of a gummy stem blight resistance gene in wild watermelon (Citrullus amarus) germplasm PI 189225. Plant Dis 104:16–24

Ren Y, Di J, Gong G, Zhang H, Guo S, Zhang J, Xu Y (2015) Genetic analysis and chromosome mapping of resistance to Fusarium oxysporum f. sp. niveum (FON) race 1 and race 2 in watermelon (Citrullus lanatus L.). Mol Breed 35:183.

Romay G, Lecoq H, Desbiez C (2014) Cucurbit crops and their viral diseases in Latin America and the Caribbean islands: a review. J Plant Pathol 96:227–242

Sanderson L-A, Ficklin SP, Cheng C-H, Jung S, Feltus FA, Bett KE, Main D (2013) Tripal v1.1: a standards-based toolkit for construction of online genetic and genomic databases. Database

Subburaj S, Lee K, Jeon Y, Tu L, Son G, Choi S, Lim Y-P, McGregor C, Lee G-J (2019) Whole genome resequencing of watermelons to identify single nucleotide polymorphisms related to flesh color and lycopene content. PLoS ONE 14:e0223441–e0223441

Sun Y, Fan M, He Y (2019) Transcriptome analysis of watermelon leaves reveals candidate genes responsive to cucumber green mottle mosaic virus infection. Int J Mol Sci https://doi.org/10.3390/ijms20030610

U.S. National Plant Germplasm System GRIN-Global. Available online: https://npgsweb.ars-grin.gov/gringlobal/search.aspx. Accessed on 8 March 2021

Wu S, Wang X, Reddy U, Sun H, Bao K, Gao L, Mao L, Patel T, Ortiz C, Abburi VL, Nimmakayala P, Branham S, Wechter P, Massey L, Ling K-S, Kousik C, Hammar SA, Tadmor Y, Portnoy V, Gur A, Katzir N, Guner N, Davis A, Hernandez AG, Wright CL, McGregor C, Jarret R, Zhang X, Xu Y, Wehner TC, Grumet R, Levi A, Fei Z (2019) Genome of 'Charleston Gray', the principal American watermelon cultivar, and genetic characterization of 1,365 accessions in the U.S. National Plant Germplasm System watermelon collection. Plant Biotechnol J 17:2246–2258

Yadav V, Wang Z, Lu G, Sikdar A, Yang X, Zhang X (2021) Evaluation of watermelon germplasm and advance breeding lines against powdery mildew race'2F'. Pakistan J Agric Sci 58

Yan L, Zhai Q, Wei J, Li S, Wang B, Huang T, Du M, Sun J, Kang L, Li C-B, Li C (2013) Role of Tomato Lipoxygenase D in wound-induced Jasmonate biosynthesis and plant immunity to insect herbivores. PLoS Genet 9:e1003964

Yong X, Guo S (2017) The Watermelon Genome. In: Grumet R, Katzir N, Garcia-Mas J (eds) Genetics and genomics of cucurbitaceae. Springer International Publishing, Cham, pp 199–210

Zakeri A, Mousanejad S, Olfati J-A, Khodaparast A (2022) Evaluation of some cucurbit genotypes for resistance to downy mildew based on AUDPC. mdrsjrns 11:29–37

Zhang Y, Zhang H, Guo S, Ren Y, Zhang J, Geng L, Liang Z, Yong X (2013) Developments of molecular markers tightly linked to Fon-1 for resistance to fusarium oxysporum f. sp. niveum race 1 in watermelon. Scientia Agric Sinica 46:2085–2093

Zhang Z, Mao L, Chen H, Bu F, Li G, Sun J, Li S, Sun H, Jiao C, Blakely R, Pan J, Cai R, Luo R, Van de Peer Y, Jacobsen E, Fei Z, Huang S (2015) Genome-wide mapping of structural variations reveals a copy number variant that determines reproductive morphology in cucumber. Plant Cell 27:1595–1604

Zhang M, Liu Q, Yang X, Xu J, Liu G, Yao X, Ren R, Xu J, Lou L (2020) CRISPR/Cas9-mediated mutagenesis of Clpsk1 in watermelon to confer resistance to Fusarium oxysporum f.sp. niveum. Plant Cell Rep 39:589–595

Zhang M, Xu J, Ren R, Liu G, Yao X, Lou L, Xu J, Yang X (2021) Proteomic analysis of fusarium oxysporum-induced mechanism in grafted watermelon seedlings. Frontiers Plant Sci 12

Zheng Y, Wu S, Bai Y, Sun H, Jiao C, Guo S, Zhao K, Blanca J, Zhang Z, Huang S, Xu Y, Weng Y, Mazourek M, K. Reddy U, Ando K, McCreight JD, Schaffer AA, Burger J, Tadmor Y, Katzir N, Tang X, Liu Y, Giovannoni JJ, Ling K-S, Wechter W P, Levi A, Garcia-Mas J, Grumet R, Fei Z. (2018) Cucurbit genomics database (CuGenDB): a central portal for comparative and functional genomics of cucurbit crops. Nucleic Acids Res 47:D1128–D1136

Zhu H, Song P, Koo D-H, Guo L, Li Y, Sun S, Weng Y, Yang L (2016) Genome wide characterization of simple sequence repeats in watermelon genome and their application in comparative mapping and genetic diversity analysis. BMC Genomics 17:557

MIX
Papier aus verantwortungsvollen Quellen
Paper from responsible sources
FSC® C105338

If you have any concerns about our products,
you can contact us on
ProductSafety@springernature.com

In case Publisher is established outside the EU,
the EU authorized representative is:
Springer Nature Customer Service Center GmbH
Europaplatz 3, 69115 Heidelberg, Germany

Printed by Libri Plureos GmbH
in Hamburg, Germany